The
Ultimate SIX
SIGMA

The Ultimate SIX SIGMA

Beyond

Quality

Excellence

to Total Business Excellence

Keki R. Bhote

AMACOM

American Management Association

New York • Atlanta • Brussels • Buenos Aires • Chicago • London • Mexico City • San Francisco
Shanghai • Tokyo • Toronto • Washington, D.C.

This publication is designed to provide accurate and authoritative information in regard to the subject matter covered. It is sold with the understanding that the publisher is not engaged in rendering legal, accounting, or other professional service. If legal advice or other expert assistance is required, the services of a competent professional person should be sought.

Library of Congress Cataloging-in-Publication Data

Bhote, Keki R.
    The ultimate six sigma : beyond quality excellence to total business excellence / Keki R. Bhote.
        p.   cm.
    Includes bibliographical references and index.
    ISBN 0-8144-0677-7
    1. Process control. I. Title.
    TS156.8 .B49   2001
    658.5'62—dc21

                                    2001037306

Printing number

10   9   8   7   6   5   4   3   2   1

To my revered and beloved mother and father—
whose lives were a perfect Six Sigma of excellence
and service to humanity

Contents

❖ ❖ ❖

LIST OF ILLUSTRATIONS

❖ ❖ ❖

FOREWORD BY ROBERT W. GALVIN, CHAIRMAN OF THE BOARD EMERITUS, MOTOROLA

❖ ❖ ❖

Motorola's launch of its world-renowned Six Sigma process was another milestone in its long march to quality excellence. The company has been generous sharing its success factors with thousands of executives from all over the globe. Other companies and their consultants, however, chose different and controversial approaches to Six Sigma. The result has been a considerable amount of confusion about the true intent and spirit of Six Sigma.

This has urged Keki Bhote in this landmark book—*The Ultimate Six Sigma*—not only to set the record straight, but also to take Six Sigma a big step further—to go beyond quality excellence to the pursuit of total business excellence.

Keki is eminently suited to the task. His forty-two–year career at Motorola included twenty-three innovations in the field of quality. He developed a series of powerful tools in the crucible of Motorola's drive for a 10:1, 100:1, and 1,000:1 improvement. In 1995, he was chosen as one of the new quality gurus of America by *Quality Digest* magazine. After retiring from Motorola, he formed his own consulting firm, with more than 400 companies as his worldwide clients.

The thrust of *The Ultimate Six Sigma* is to take excellence beyond product quality: to excellence in customer loyalty—not just in customer satisfaction; to excellence in leadership—not just in management; to excellence in a nurturing organizational infrastructure; and to excellence in releasing the inherent creativity of employees to reach their full potential. The book also takes the disciplines of design, manufacturing, suppliers, services, and the field to new heights with breakthrough innovations. A highlight is a chapter on the ten tools for the 228 metrics. Each chapter is illustrated with a case study of a benchmark company that excels in one of the twelve areas of a company's business. Finally, there is a self-assessment/audit guide by which a company can measure its overall—and area by area—corporate health.

—Robert W. Galvin
Chairman of the Board Emeritus, Motorola

PREFACE:
BLOSSOMING OF THE
ULTIMATE SIX SIGMA

❖ ❖ ❖

This book is the culmination of twenty years of hands-on experience: First, it encompasses my eleven years launching and nurturing Six Sigma at Motorola, Inc.; second, after retirement from Motorola nine years ago, my consultation business led me to work with more than 400 companies on four continents; and third, with my research expanding the frontiers of Motorola's Six Sigma into the development of the Ultimate Six Sigma and its implementation in some of the world's leading companies.

Motorola's Six Sigma

The compelling necessity to go beyond the dribble of conventional, marginal, and incremental quality advances of the past was Motorola Chairman Bob Galvin's clarion call in 1981 to improve quality across the board at Motorola by a factor of 10:1 in five years. Nobody anywhere—not in the United States, not in Japan, not in Europe—had set such reach-out goals in such record time, much less achieved them. For the first year, we floundered using traditional quality techniques and got nowhere. We needed a breakthrough. The breakthrough was our Six Sigma process.

The details of that Six Sigma process we developed at Motorola are explained in Chapter 3. In brief, they included:

❖ Faith that the lofty 10:1 quality improvement target could be achieved

* Total customer satisfaction
* Powerful new tools, especially design of experiments
* Cycle-time reduction
* Designs for ease of manufacturability
* Manufacturing innovations
* True partnerships with key suppliers
* Training for all employees

The result was that most Motorola divisions were able to achieve the 10:1 quality improvement goal in five years. The momentum continued. We went on to register an average of 800:1 improvement in ten years. Six Sigma was now in full swing.

Our success positioned us to apply for and win the prestigious Malcolm Baldrige National Quality Award in 1988—the first company to win that award. Even today, Motorola remains the only company in the large company category to win the award for a corporation as a whole, not just for one or two of its divisions. As the Baldrige Award examiners told us, the most impressive aspect of our Six Sigma process was its pervasiveness throughout the organization. Every line employee knew the importance of quality and the role he or she played in furthering it.

Motorola's Six Sigma Renewal

In keeping with Bob Galvin's vision that a company must renew itself from time to time, we started to renew our Six Sigma process after winning the Baldrige Award. These innovations are also summarized in Chapter 3. In brief, they included:

* Extending quality improvement to business processes and white-collar work
* Capturing the "voice of the customer" at the concept stage of design
* Extending cycle-time disciplines from the production floor to all business processes and to the design function
* Reducing the number of management layers between the CEO and the line worker
* Extending training to a minimum of forty hours per year per employee

After my retirement from Motorola in 1992, I formed my own consulting company for quality and productivity improvement. In my numerous contacts with other companies in the United States and internationally, I began to see the need for a revamping of Motorola's Six Sigma process.

The impetus for an overhaul came from an unexpected direction. One of the requirements for a Malcolm Baldrige winner is to share your success factors with the outside world. At Motorola, we duly organized several hundred tutorial sessions for thousands of companies to study our Six Sigma methodologies.

Six Sigma Consulting Companies: Hype and Hyperbole

As often happens, exposure is not expertise. Companies with little background waded into their own Six Sigma pools. The worst distortions have been perpetrated by opportunistic consulting companies that, with limited vision but gargantuan pecuniary appetites, foisted their limited grasp of Six Sigma processes onto their gullible clients. Of course, some improvements were made for some clients, yet many questions about the processes remain. Is a 10:1 quality improvement enough? Is a 1:1 or even a 2:1 return on huge, up-front investment costs enough? Is a break-even point of two, three, or four years for cash flow recovery timely enough? Has the bottom line really improved? The jury of actual users, not the top brass of the clients, has rendered a realistic judgment on these consulting companies. They are guilty of hype and hyperbole.

The Need for the Ultimate Six Sigma

There were other observations that came to light during my consulting work among even the leading companies that necessitated my restructuring of Motorola's Six Sigma process and led to my Ultimate Six Sigma blueprint and, eventually, the writing of this book. Among them:

❖ The emphasis on quality for the sake of quality alone was driving companies away from the two fundamental obligations of a corporation—customers and profit. Just a few years ago Motorola learned that lesson the hard way when its divisions became arrogant and did not heed the customer. They paid the price in the marketplace.

❖ Many companies were not even aware that customer satisfaction was inadequate, that customer loyalty—and retention—was the name of the game.

❖ Six Sigma could no longer be confined to just a quality initiative. It had to become a business initiative—indeed, a total business focus.

❖ Many companies were saturated with too many managers and too few leaders. Fantastic corporate power could be liberated if this ratio were reversed.

❖ Company after company talked about the importance of their employees yet allowed employee creativity to atrophy. Why couldn't industrial democracy be more than a pipe dream?

❖ The tools for quality, cost, and cycle time were antiquated. The powerful tools we fashioned at Motorola as a starting point for the Ultimate Six Sigma were desperately needed in industry at large.

❖ The functional areas of design, supply chain management, manufacturing, field operations, and services needed to go from timid, evolutionary steps to a veritable revolution.

The Ultimate Six Sigma—A Chapter by Chapter Synopsis

Based on these observations and my work with several client companies, the Ultimate Six Sigma was launched as an alternative, both to the venerable Motorola Six Sigma and the machinations of certain Six Sigma consultants. This book captures the essence of the Ultimate Six Sigma in seventeen chapters organized in two parts.

Part 1—the introduction to Ultimate Six Sigma concepts—has five chapters:

Chapter 1 deals with the definition of Six Sigma and the five heresies of that definition, as perpetrated by Six Sigma consultants.

Chapter 2 lists the needs, objectives, and benefits of the Ultimate Six Sigma. The benefits, particularly, are captured in a set of comprehensive customer and business metrics.

Chapter 3 details the origin and development of Motorola's Six Sigma.

Chapter 4 is devoted to an exposé of the hyped Six Sigma perpetrated by consulting companies. It lists a few of their strong points and many of their weak points—in point-counterpoint fashion—as well as their specific weaknesses in each of twelve key areas of business performance.

Chapter 5 is a summary of the scope, structure, and methodologies of the Ultimate Six Sigma. It also includes a unique feature of this book—a self-assessment guide (which is fleshed out further at the end of each of the remaining twelve chapters) and scoring system—by which a company can measure its business health.

Part 2 contains twelve chapters. It is specifically the product of my own research on leading companies' expansion of Six Sigma from the "small Q" of quality to the "Big Q" of business excellence in twelve vital areas: customers, leadership, organization, employees, measurement, tools, design, suppliers, manufacturing, field operations, service, and results. Each chapter in Part 2 also contains a case study of a truly benchmark company to illustrate that area of business excellence.

Chapter 6 deals with growing from mere customer satisfaction to customer loyalty. Traditional Six Sigma practice is confined to customer satisfaction. It does not take into account that 15–40 percent of "satisfied" customers defect from the average company. My 1996 book *Beyond Customer Satisfaction to Customer Loyalty—The Key to Greater Profitability* (also published by AMACOM) was enthusiastically received by the public. It showed how the two fundamental responsibilities of a company—customers and profit—are inextricably linked. Chapter 6 shows how just a 5 percent increase in customer loyalty (defined as customer retention and customer longevity) can improve a company's profits by 30–85 percent. It goes on to define various types of customers, the principles of customer loyalty, and the infrastructure that must be constructed for a company to capture customer loyalty. Lexus is the case study benchmarked in this area.

Chapter 7 examines leadership—from stifling to inspiring. Traditional Six Sigma gives token recognition to leadership. Often, it is nothing but warmed-over "bossy" management. Companies today are often saturated with too many managers and starved for true leaders. This chapter highlights twelve personal philosophies and unchanging values that are the attributes of a true leader. It goes on to develop six principles of leadership in a corporation. It ends with seven key roles by which a leader can unlock the creative energies of employees and dismantle the oppressive bureaucracy of Taylorism. Fortunately, for me and for Motorola, Bob Galvin was the ideal role model of an outstanding leader. Companies could do no better than to try to clone him in order to fill the yawning vacuum of business leadership. Predictably, he is profiled as the benchmark in this area of leadership.

Chapter 8 is devoted to organizational infrastructure and examines ways to move away from Taylorism and free the human spirit. Traditional Six Sigma leaves an organization's infrastructure alone—so it is simply a continuation of the old bureaucratic model. To introduce a viable Six Sigma process, a corporate culture must be radically changed. To change a corporate culture, the values and beliefs of its people must be changed. To change peoples' values and beliefs, the way in which people are hired, trained, evaluated, compensated, and promoted must be changed. This chapter outlines a ten-step process by which a culture can be changed and an empowerment infrastructure constructed. This involves moving from a vertical to a horizontal organizational structure with cross-functional teams, from recentralization to decentralization, from the tall organizational pyramid to the flat pyramid, and even further, to the upside-down pyramid. The company highlighted as a benchmark in the area of organization is Chaparral Steel.

Chapter 9 looks at employees. Traditional Six Sigma does nothing significant for motivating the employee who remains passive at best and alienated at worst. The Ultimate Six Sigma model changes employee motivation from extrinsic to intrinsic, from "how to" motivate to "want to" be motivated. The chapter's objective is to release the genie of creativity resident in every employee, currently locked up in a bottle of bureaucracy, and allow the employee to reach a potential of growth that has no finish line. It details three innovations launched in progressive companies: Open Book Management, Self-Directed Work Teams, and the Mini Company. All three innovations can lead to the highest stage of empowerment—stage 10 (industrial democracy). There are two case studies highlighted as benchmarks: Semco of Brazil and Springfield ReManufacturing Corporation.

Chapter 10 examines the ways to measure a company's performance. Traditional Six Sigma confines the important subject of measurement to just quality. The only concession to cost is the gathering and analysis of the cost of quality. Cycle time is off its radar screen. This chapter begins by listing important measurement axioms and principles. It distinguishes between lagging and leading indicators and between weak and robust indicators. It then develops generic measurements (covering service effectiveness, innovation effectiveness, and empowerment effectiveness) and specific measurements for various areas/departments. FedEx Corp. is the company highlighted as a benchmark in the area of effective measurement.

Chapter 11 describes quality tools. One of the fundamental limitations of traditional Six Sigma is its use of old, obsolete, and ineffective quality tools such as the seven tools of quality control (QC), PDCA (plan, do, check, act), Ford 8-D, and so on. Even those few companies that venture into design of experiments (DOE) only use classical or Taguchi DOE, both of which are costly, time-consuming, and statistically weak. The Ultimate Six Sigma uses ten powerful tools that my colleagues and I researched and introduced at Motorola. The implementation of these tools was one of the major causes of Motorola's success in achieving almost 1,000:1 improvement in quality, along with accompanying improvements in cost and cycle time. Chapter 11 briefly summarizes ten powerful tools introduced at Motorola:

1. Design of experiments (DOE)—for quality breakthrough
2. Multiple Environment Over Stress Testing (MEOST)—for reliability breakthrough
3. Mass customization and Quality Function Deployment (QFD)—for capturing the voice of the customer

4. Benchmarking—for a short-cut implementation of practices in best-in-class companies

5. Total Productive Maintenance (TPM)—for process/equipment productivity breakthrough

6. Poka-Yoke—for prevention of operator-controllable errors

7. Next Operation as Customer (NOAC) and Business Process Reengineering—for quality cost and cycle-time breakthrough in white-collar work

8. Total Value Engineering—for enhancing customer "wow" at lower cost

9. Supply Cain Optimization—for quality, cost, and cycle-time breakthrough in upstream and downstream supply chains

10. Lean Manufacturing/Cycle-Time Reduction—for revolutionizing manufacturing productivity

Each tool is briefly summarized, it being beyond the scope of this book to detail them. That would require a book for each one of the tools.

The case study of a benchmark company in the area of tools is TELCO.

Chapter 12 deals with design productivity. Traditional Six Sigma skirts, to a large extent, methods by which design productivity can be vastly improved. For quality improvement in design, for example, it only uses classical design of experiments—a modest approach at best. For reliability improvement in design, it uses failure mode effects analysis (FMEA), a kindergarten technique. For cycle-time improvement in design, it concentrates on efficiency rather than effectiveness. The Ultimate Six Sigma puts design at front and center stage in an organization and is in the best tradition of prevention (rather than correction) of defects, cost overruns, and launch delays. The chapter is divided into six sections:

1. Organization (highlighted by concurrent engineering)

2. Management guidelines for design (highlighted by six "do's")

3. Customers (highlighted by capturing the voice of the customer and developing new ways to establish realistic specifications and tolerances)

4. Design quality (highlighted by Design of Experiments [DOE] and Multiple Environment over Stress Tests [MEOST])

5. Design cost (highlighted by total value engineering, group technology, and other techniques)

6. Design cycle time (highlighted by parallel—versus serial—development, outsourcing, and early supplier involvement)

This chapter on design looks at the Japanese car industry in general, and Toyota in particular as a benchmark.

Chapter 13 examines the role of the supplier revolution in customer-supplier relationships and practices. Traditional Six Sigma gives only token recognition to the role of suppliers. At most, it extends training and quality auditing for its suppliers. It does not address partnership with suppliers and does not extend real help to them. The Ultimate Six Sigma represents a radical departure from conventional approaches to both upstream suppliers and downstream suppliers (e.g., distributors and dealers). It does so by:

❖ Advocating and strengthening outsourcing—not just at the piece-part level but at the black-box level, including design

❖ Strengthening true partnership with suppliers, not just in name, but in fact

❖ Creating a viable infrastructure for supply chain management

❖ Rendering active, concrete help to suppliers in return for continual quality, cost, and cycle-time improvement

❖ Using the principle of cost as a ceiling for a customer company and profits as a floor for a supplier company

The case study of a benchmark company in the area of supply chain management is Toyota.

Chapter 14 is about manufacturing. In fairness, manufacturing is the principal area for traditional Six Sigma concentration. Yet it has not succeeded in restoring manufacturing to its rightful place in the corporate sun. It continues to use worn-out methods, such as mass production, automation, material requirements planning (MRP II), and weak manufacturing disciplines. The Ultimate Six Sigma milks all the support it can gain for manufacturing from leadership, organization, employees, measurement, tools, design, and supply chain management. Further, it concentrates on the twin engines of quality and cycle time to propel manufacturing forward at high speed. In addition, this chapter lists at least eight disciplines to accelerate quality improvement and at least twelve disciplines to accelerate cycle-time improvement. Here, Hewlett-Packard is the case study for a benchmark company displaying business excellence in manufacturing.

Chapter 15 concentrates on field service, one of the most neglected and underrated areas of a company. Traditional Six Sigma practically ignores the existence of field service. It does not address field reliability, other than FMEA, which is an inconsequential tool. It does little for field constituencies, such as frontline troops, distributors, dealers, servicers, installers, and—above all—users. The Ultimate Six Sigma develops a process that revitalizes field operations by:

- ❖ Concentrating on twelve product characteristics and their associated techniques to dramatically improve field reliability and other associated elements to produce customer "wow"
- ❖ Describing six parameters by which predelivery services are assured
- ❖ Listing thirteen parameters needed to strengthen the downstream supply chain
- ❖ Focusing on four techniques that are vital for the consumer.
- ❖ The case study of a benchmark company in the area of Field Services is Caterpillar.

Chapter 16 deals with services. Although traditional Six Sigma extends to services—that is, the white-collar operations within a company—it uses a "six steps for Six Sigma" format, which is about as useful as the Ford 8-D eight-step format for production problem solving. It provides a framework—an empty shell—but little substance. The Ultimate Six Sigma uses the Next Operation as Customer (NOAC) discipline to significantly improve the quality, cost, and delivery of a business process in services. It does so by:

- ❖ Enumerating six NOAC principles that govern business processes
- ❖ Providing a necessary organization infrastructure for NOAC's success
- ❖ Providing a ten-step road map for improvement of any business process including, most of all, "out-of-box thinking" to creatively bypass and completely revamp traditional business process flows
- ❖ Employing three truly creative tools to implement out-of-box thinking
- ❖ Discussing (if briefly) business process reengineering as an alternative to NOAC.
- ❖ The case study of a benchmark company in the area of services is Solectron.

Chapter 17, as the book's last chapter, fittingly addresses results. Traditional Six Sigma consultants and their clients are divided into two groups in terms of results: One group concentrates solely on quality advances, generally in the modest range of 10:1. They also are mired in counting defects per opportunity, a semi-fraudulent way to artificially boost a sigma score! The second group extends results to financials, such as profit and return on investment, but it goes no further in examining the whole spectrum of results. The Ultimate Six Sigma approaches results in far more significant and meaningful ways. It provides a large and comprehensive list of parameters (i.e., metrics); companies can then choose those parameters that are in line with their priorities and their culture. In addition, Chapter 17 lists four primary characteristics (with a total of fifteen success factors) for assessing business results; embellishing and circumscribing these primary factors are thirteen secondary factors, broken out by area and discipline:

Four Primary Characteristics
1. Customers
2. Leadership
3. Employees
4. Financials

Secondary Factors, by Area
1. Leadership: 25 metrics
2. Organization: 11 metrics
3. Employees: 14 metrics
4. Tools: 28 metrics
5. Design: 29 metrics
6. Supply Chain Management: 28 metrics
7. Manufacturing: 24 metrics
8. Field: 14 metrics
9. Services: 13 metrics

Secondary Areas, by Disciplines
1. Quality: 11 metrics
2. Cost: 14 metrics
3. Cycle Time: 5 metrics
4. Innovation: 12 metrics

The result is a total of fifteen primary metrics and 228 secondary metrics covering the entire results, each with a quantified rating (albeit, mainly for large companies) by which a company can assess its results. It represents the grand finale of the Ultimate Six Sigma treatment. The case study of a benchmark company in the area of results is General Electric.

From Liabilities to Opportunities for Industry in the Twenty-First Century

This first decade of the twenty-first century, to paraphrase Charles Dickens, "is the worst of times and the best of times" for industry. It is the worst of times because of:

- ❖ The dizzying technological revolution, which can obsolete a company's product overnight. (The U.S. Air Force has a saying about its research: "If it works, it is obsolete.")
- ❖ Globalization, where, apart from the hot breath of international competition and a consumer penchant for never-ending price reductions, there is a rising cacophony from environmentalists, labor unions, and the Third World decrying corporate practices that pursue private profit at the expense of Mother Earth and the downtrodden.
- ❖ The human dislocation caused by layoffs and restructuring, which is occurring simultaneously with unseemly bonuses and stock options for top management who have failed at the marketplace.

Yet, this can also be the best of times, with unlimited opportunities for outstanding business success to be achieved by:

- ❖ Converting micromanagement into inspiring leadership
- ❖ Abandoning Taylorism and freeing the creative spirit of the worker
- ❖ Escalating customer satisfaction into customer loyalty and retention
- ❖ Enhancing shareholder value with stakeholder (e.g., customer, supplier, employee, investor) value
- ❖ And, most important, transforming the unfilled needs of the less fortunate into socially redeeming projects that result in corporate profit

This book can change the worst of times into the best of times. It offers a "how to" approach to make these opportunities come alive in twelve vital areas of the business world. No company that ignores the guidelines of this book can survive long. No company that assiduously follows its guidelines can ever lose its business or its soul.

Keki R. Bhote

ACKNOWLEDGMENTS

❖ ❖ ❖

As with my fourteen other books, my first and perennial gratitude goes to my revered mother—a famous author and journalist in India—who sparked the love of writing in me. She will always remain my inspiration, along with my noble father who has been a role model and whose life was the "Ultimate Six Sigma."

The second inspiration in my life has been Bob Galvin, Chairman Emeritus of Motorola, a captain of world industry par excellence. Among his many, many innovations in industry was his launching and nurturing Motorola's Six Sigma drive—a concept that has reached all corners of the globe. He has been my guiding light in my fifty years of discipleship under him.

The germ of the idea to write this particular book came from the eminent C. Jackson Grayson, Chairman of the American Productivity and Quality Center and advisor to three U.S. Presidents. His encouragement sustained me in venturing beyond Motorola's Six Sigma to new heights and broader horizons.

I owe a deep debt of gratitude to Jim Nelson, publisher of Strategic Directions, Zurich, Switzerland, who urged me to enlighten my readers to jettison quality for the sake of quality alone and pursue it for customer loyalty and profit.

In some ways, this book is a memorial to my late and esteemed "guru," Dorian Shainin, the world's foremost problem-solver in industrial quality. His Design of Experiments and Multiple Environment Over Stress techniques captured in this book are the tools to achieve breakthrough levels in quality and reliability, respectively.

Let me also take this opportunity to salute my colleagues—Janet Fiero, Mark Schleicher, John Lupienski, Reed Hope, and the late Carlton Braun—for their signal contributions in advancing the real Six Sigma at Motorola. We were encouraged by the vision of Bill Wiggenhorn, President of Motorola University, to put Six Sigma on the world map.

Following my retirement at Motorola, a number of distinguished clients helped me forge my Ultimate Six Sigma in the crucible of fierce industrial competition. Foremost among them is Bill Beer, President of Maytag Appliances, who patterned his adaptation of the Ultimate Six Sigma as the "Maytag Constitution." At Philips, the high octane think tank of Willy Hendrickx, Franz Wouters, Sid Dasgupta, and Mike Chew, worked with me to introduce their "Best" process in European and Asian plants.

At Caterpillar, the Corporate Quality and Reliability team of Prakash Babu, Ted Taber, and Carl Saunders has successfully embraced my techniques, as have President Harvey Kaylie of Mini Circuits, Vice President Mike Katzorke of Cessna Aircraft, Vice President J. C. Anderson of Whirlpool, Brian McGuire of Harley Davidson, and Vice President Christian Gianni of Fisher and Paykel in New Zealand.

I owe special thanks to Ratan Tata, Chairman of Tata Industries, India's largest corporation, for introducing the spirit and substance of my Ultimate Six Sigma in several of his divisions, especially at the Tata Engineering and Locomotive Co. (TELCO), which has the distinction of manufacturing India's popular car—the Indica—with 100 percent India fabricated parts. Under the brilliant leadership of Vice President Nath and Ramesh Parkhi, TELCO became an Ultimate Six Sigma benchmark in India.

I especially cherish the support I've had from Celerant Consulting, specifically Henk De Jong, Sal Puarr, Alois Deubert, and Gert Hut, for the firm's willingness to strengthen its own Six Sigma initiatives with mine in its widespread consultations in Europe, as well as training and coaching the firm's consulting teams. In addition, I humbly thank many of my 400 clients in thirty-three countries for testing, honing, and polishing the practice of the Ultimate Six Sigma, culminating in this book.

Without the encouragement and guidance of Vice President Hank Kennedy and Senior Acquisitions Editor Neil Levine of AMACOM Books at the American Management Association, this book may not have made it to the launching pad. I would be totally remiss if I did not acknowledge the most tangible contribution of my associate, Jean Seeley, whose professionalism, perfection, and uncanny speed produced this manuscript in record time.

Finally, my heartfelt gratitude to my family—to my daughter, Shenaya Bhote-Siegel of Bhote-Siegel, Inc., who, as a noted graphic designer, has produced the attractive jackets for all my books including this one; to my son, Adi, who coauthored my book, *World Class Quality*, with me; to my daughter Safeena, and son Xerxes, for their boundless faith in their father crossing the finish line; but above all, to my beloved wife Mehroo of 46 years, for her patience, encouragement, and boundless caring and love.

You cannot have a Six Sigma company with a three sigma manufacturing, a two sigma design and a one sigma management. . . .

—Keki R. Bhote

The Ultimate SIX SIGMA

PART 1

DEFINITIONS AND CONCEPTS

Part 1 contains five chapters that
introduce the different definitions and
concepts of Six Sigma.

❖

One

WHAT IS SIX SIGMA?

❖ ❖ ❖

We are witnessing Six Sigma Houdinis who can conjure up the magic of transforming any old quality level to an ideal of Six Sigma!

—KEKI R. BHOTE

Will the Real Six Sigma Please Stand Up?

During the presidential campaign of 2000, the flagship journal *The Economist* asked the rhetorical question, "Will the real Al Gore stand up? Is he the pragmatic politician; or is he the dreamy idealist; or is he the sour, partisan, reactionary, attack dog; or is he the panicky panderer?"

The same type of question may well be asked of the many faces of Six Sigma. Its conceptions and misconceptions; its applications and misapplications; its claims and counterclaims have created so much confusion in industry that even quality professionals are yearning for the real Six Sigma to stand up. Three widely different perceptions of Six Sigma exist in the business world: the statistical Six Sigma; the hyped Six Sigma peddled by some consulting companies; and the Ultimate Six Sigma. The Ultimate Six

3

Sigma, the subject of this book, goes beyond just quality excellence in pursuit of total business excellence.

The Statistical Six Sigma

Sigma (σ) is a statistical parameter that measures the standard deviation of a group of data, associated with a quality characteristic, from its average ($\overline{X}$). Technically, if a very large group of data (or population) is involved, its average is called μ and its standard deviation is σ. A sample drawn from such a large population has its average designated as $\overline{X}$ and its standard deviation is called s. It has become conventional to use both $\overline{X}$ as average and σ as standard deviation in both cases.

A distribution is a bell-shaped curve of that parameter or characteristic (see Figure 1-1) showing the area under a typical normal curve that falls

Figure 1-1. Relationship of area within/outside a normal curve and sigma limits.

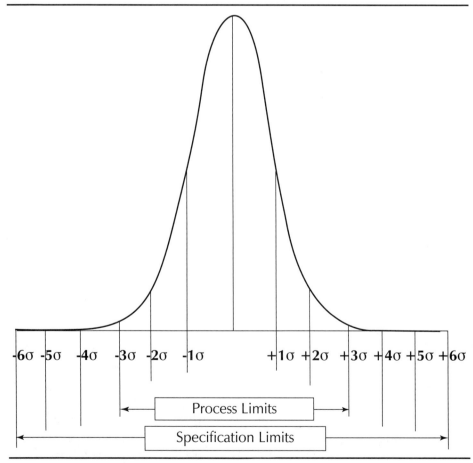

Table 1-1. Quantitative relationship between sigma, percentage (ppm/ppb) defective, and $^c p_K$ (for process limits at $\overline{X} \pm 3\sigma$).

Specification Limits	Amount Defective Outside Sigma Limit		$^c p_K$
	%	PPM/PPB*	
$\overline{X} \pm 1\sigma$	31.74	317,400 PPM	0.33
$\overline{X} \pm 2\sigma$	4.56	45,600 PPM	0.67
$\overline{X} \pm 2.5\sigma$	1.24	12,400 PPM	0.83
$\overline{X} \pm 3\sigma$	0.27	2,700 PPM	1.00
$\overline{X} \pm 3.3\sigma$	0.096	60 PPM	1.10
$\overline{X} \pm 4\sigma$	0.0063	63 PPM	1.33
$\overline{X} \pm 5\sigma$	0.00057	0.57 PPM	1.67
$\overline{X} \pm 6\sigma$	0.0000002	0.02 PPM (2 PPB)	2.00

*A 1% defective = 10,000 parts per million (ppm) defective.

within and outside the X: $\pm 1\sigma$, $\pm 2\sigma$, $\pm 3\sigma$, $\pm 4\sigma$, $\pm 5\sigma$, and $\pm 6\sigma$ limits. The area outside the curve is the percentage—that is, parts per million (ppm) or parts per billion (ppb)—of that parameter that is considered defective. This is shown quantitatively in Table 1-1.

$^c p$ measures the ratio of a specification width (highest allowable–lowest allowable value) to the process width (highest observed–lowest observed value). $^c p_K$ equals $^c p$ if the process average and the design center (or target value) are the same. If not, a slight formula correction lowers $^c p_K^*$ relative to $^c p$. Table 1-1 shows that if process limits are at $\overline{X} \pm 3\sigma$ and the specification limits are narrower, at $\overline{X} \pm 2\sigma$, a defect level of 4.56 percent can be expected for that parameter. This used to be a standard or target level of quality for design engineers in the early 1970s. A defect rate of around 4.5 percent (in $^c p_K$ terms: 0.67) was considered "good enough." If the process limits and the specifications limits are both at $\overline{X} \pm 36$—i.e. a $^c p_K$ of 1.0, the defect level is reduced to 0.27 percent or 2,700 ppm. This was considered a reach-out target level of quality up to the mid-eighties.

Later, however, with global competition driving quality levels toward zero defects, process limits at $\overline{X} \pm 3\sigma$, and specification limits at $\overline{X} \pm 4\sigma$ (i.e., a $^c p_K$ of 1.33), the defect level is further reduced to 63 ppm. This was considered a target level in the mid-1990s. In fact, Quality Systems Requirements QS-9000—the automotive quality system introduced by the Big Three automobile manufacturers for their key suppliers—requires a minimum $^c p_K$ of 1.33.

World-class companies are striving for process limits reduced to $\overline{X} \pm 3\sigma$ (relative to specification limits of $\overline{X} \pm 5\sigma$), resulting in a defect level of a miniscule 0.57 ppm.

*$^c p_K$ is also a ratio of specification width(s) to process width, but corrected for the deviation of an average $\overline{X}$ from a design center.

The ultimate is for process limits at $\overline{X} \pm 3\sigma$ to be no more than half the specification limits at $\overline{X} \pm 6\sigma$—a defect level of 2 ppb (parts per billion) or a $^{c}p_K$ of 2.0. I introduced the concept of $^{c}p_K$ at Motorola in 1983—the prelude to its Six Sigma thrust—as group director of quality and value assurance at its automotive and industrial electronics group. By 1984, we achieved $^{c}p_K$ rates of 2.0 in several important parameters in our product lines. For all practical purposes, this is zero defects, which is the statistical meaning of Six Sigma. (For a convenient shorthand symbol the $\pm$ sign in sigma levels is often dropped. Thus $\pm 6\sigma$ is shortened to 6σ.)

The challenge we faced at Motorola, as we became the first company to win the prestigious Malcolm Baldrige National Quality Award, was how to extend a sigma or defect level (a $^{c}p_K$) associated with a single parameter to an entire product line with several parameters. Even more challenging was extending the concept of sigma to include the addition of all defects across a whole production line—from start to finish. Thus was born a reach-out concept of Six Sigma where:

1. The total number of defects per unit of product (for all parameters) would be converted into a percentage, or ppm, defective, with Table 1-1 used to calculate the corresponding sigma level.

2. Going even further, the total number of defects at each of several checkpoints on an entire production line would be added and divided by the number of units on a product that were produced. This figure, called total defects per unit (TDPU) would be converted into a percentage, or ppm defective, with Table 1-1 used to calculate the corresponding sigma level.

As an example, if a cellular phone line checked 100,000 units in its outgoing test station and found 5 defects, the defect level would be 0.005 percent or 50 ppm or 4.1σ. If the same production line had 10 checkpoints (or workstations) and found 5 defects at each checkpoint, there would be 50 defects per 100,000 units produced or 0.05 percent or 500 ppm or approximately 3.48σ.

The Hyped Six Sigma

For an industry struggling to achieve even as poor a quality level as $\pm 3\sigma$ or 2,700 ppm for a single parameter (the present industry average at the dawn of the twenty-first century is still below $\pm 3\sigma$), this escalation from a single parameter to an entire product containing several parameters, and a further escalation to include the total defects found at all checkpoints of a product line, presented so difficult a quality challenge that Motorola lowered its pristine quality standards of excellence.

The Motorola ideal also began to be systematically watered down by several consultant companies with shallow Six Sigma roots. Their approach deserves the appellation "the hyped Six Sigma." In Chapter 4, there is a full exposé of the whole convoluted approach to the hyped Six Sigma and its "Black Belt" apparatus. This section will confine itself to the five progressively worse heresies perpetrated by these consulting companies to so water down the definition of Six Sigma that any quality level—high or low—could be passed off as Six Sigma.

Heresy 1: The 1.5 Sigma Shift

The hyped Six Sigma approach asserts that it is too difficult to hold the average, $\overline{X}$, at the target value (i.e., design center) of a parameter's distribution because of inherent shifts in materials or processes. So it blithely allows a 1.5 sigma shift of $\overline{X}$ from the target value. This increases the defect level from 2 ppb to 3.4 ppm—a quality adulteration of 1,700:1. Yet its adherents have the impudence to call this Six Sigma!

True practitioners of problem solving know that it is relatively easy to correct for an $\overline{X}$ shift from a target value through a minor adjustment or tweaking and eliminate the 1.5 sigma noncentering. It is much more difficult to reduce the spread (or range, or variation, or sigma) of a parameter. That requires the application of design-of-experiment (DOE) techniques, of which 80 percent of companies are not even aware. Still, the Six Sigma hypers brazenly perpetuate the fraud of sprinkling holy water on a 3.4 ppm defect level and baptizing it as Six Sigma. (A few of us at Motorola strongly opposed all of these quality dilutions, but our compromising colleagues won the tug-of-war.)

Heresy 2: Diluting the Defect Level and the Associated Sigma Level With Parts Counts

The sigma level, artificially boosted by the 1.5 sigma shift, is given a further boost by the hyped sigma people who divide the actual product defect ppm level by the number of parts in a product and equate the artificially-low defect level to a fictitiously-high sigma level. For example, if a product containing 100 parts, has a defect level of 10,000 ppm, or 1 percent or a true sigma level of 0.86, this unnecessary 1.5 sigma shift would artificially raise the sigma level from 0.86 to 3.75. Now, if the parts count of 100 were taken into consideration the artificial defect level would be reduced from 10,000 ppm to 100 ppm on this product and the sigma level raised from 0.86 to 3.75 and on to a most respectable 5.2.

The higher the parts count (complex products can have over 50,000 parts) in a product, the greater is the likelihood that an average- or poor-quality product will fly under the radar detection screen and masquerade as a perfect Six Sigma product. In fact, engineers in hyped Six Sigma companies do not want to lower the parts count in their products for fear of lowering their sigma performance. In a few cases, they even attempt to add to the parts count (and increase costs) to make their sigma score look good. The result is that the customer, upon hearing a manufacturer's Six Sigma siren song, later discovers to his chagrin that the quality of the product received is as bad as ever.

Heresy 3: Fudging With Opportunities per Part to Reduce Defect PPMs Even Further

Sadly, this is not the end of the con game. Beyond the 1.5 sigma center shift dilution and the parts count "dilution squared," there is a further "dilution cubed." This is done by dividing the defect level of heresy 2 by the number of opportunities each part can have for defects. In the previous example, if each of the 100 parts has five opportunities for defects, the true defect level of 10,000 ppm would be artificially and progressively reduced first to 100 ppm by parts count and then reduced to 20 ppm by opportunity, while the sigma level would be fictitiously increased from 0.86 to 3.75 to 5.2 and on to an incredulous 5.6.

Table 1-2 shows the progressive magnitude of the dilution. What an easy rubber yardstick for the Six Sigma hypers to juggle.

Heresy 4: Definition of an Opportunity for Defects

There is yet another fudge factor in the hyped Six Sigma sleight-of-hand. It has to do with how different people may count an opportunity for defects on a part. For example, does a resistor with two leads have one opportunity for defects or two (because it has two leads, each of which supposedly has an opportunity for a defect)? Does a microprocessor with 80 leads have one opportunity or 80 opportunities for defects? The hyped Six Sigma con artists choose the latter to make themselves look good with

Table 1-2. The anemic Six Sigma: the "hyped" approach. The preposterous reduction in defect levels and increases in σ levels. The product in this case is a cellular phone.

Configuration Based on:	True Defect Level (PPM)	True Sigma Level	"Hyped Six Sigma" (Fictitious) Defect Level (PPM)	"Hyped Six Sigma"* (Fictitious) Sigma Level (σ)
Total Product (TP)	10,000	0.86σ	10,000	3.75σ (Dilution #1)
100 Parts/Product	10,000	0.86σ	100	5.2σ (Dilution #2)
Opportunities/Parts	10,000	0.86σ	20	5.6σ (Dilution #3)

*Allowing a 1.5σ shift of $\overline{X}$ from target value.

a fictitiously high sigma level. Yet does the customer give a hoot about how many parts are in his product or how many opportunities each part has for generating defects? If his supplier claims the product achieved 3.4 ppm—a supposedly stellar Six Sigma performance—then such a customer interprets that the 3.4 ppm refers to the product as a whole and is not based on an artificial parts count, or even worse, on the number of opportunities for defects.

CASE STUDY

Heresies Extended to Business Processes: Technical Publications

Hyped Six Sigma companies make a similar hash of Six Sigma when they apply their convoluted methodology to white-collar work. The following case study will illustrate the sham:

A hyped Six Sigma company ordered its publications department to launch and measure a Six Sigma process. The department is responsible for the writing, composition, and overall production of technical manuals that accompany the product to the company's customers.

For the computation of sigma levels, the department decided that:

❖ The manual would not be a unit of product, but a page would be a viable unit. (This was rationalized on the basis that defects in manuals with differing numbers of pages would not be a fair measurement; counting the number of pages per manual would normalize the metric.)

❖ A typical text page was defined as containing 400 words, each containing an opportunity for error.

❖ There was a further subdivision of the words, based on the number of letters (or keystrokes) per word—an average of seven opportunities per word for error.

The average number of pages per manual was 160 and the average number of defects found was four. The department ignored the defect rate per manual of 4 million ppm. It ignored the defect rate per page of $4 \times 10^6/160$ or 25,000 ppm. It ignored the defect rate per word of 25,000/400 or 61 ppm and went on to calculate the defect rate per keystroke or 61/ 7 or 8.7 ppm. This was equated to a stunningly high sigma level of 5.8. Table 1-3 shows the fictitious reduction of defect levels and the simultaneous falsifying of higher and higher sigma levels.

In the end, the publication department's manipulated sigma levels did not help. Its customers were willing to overlook the few "typo" errors (four in this case), but they were dissatisfied with the meaningfulness of the manual, its clarity, and its tortured language. Finally, the opportunities count was abandoned and a comprehensive customer satisfaction index was established as a far more effective metric.

Heresy 5: A Mountain of Costs for a Molehill of Benefits

There is yet another grave weakness that results from tracking defects per opportunity—spiraling costs. A cardinal rule in any measurement is that the cost of the measurement should be at least one order of magnitude (1:10) lower than the expected tangible benefits. A companion rule is that any activity that does not benefit customers or positively impact the bottom line should be eliminated. Measuring marginal activities of dubious value with a Six Sigma per opportunity yardstick is total waste.

This is especially true in white-collar work, where hyped Six Sigma practices have infiltrated peripheral areas such as clerical work, training, employee expense statements, and stockroom picks. Here are two examples that, like the technical publication case study cited previously, illustrate hyped Six Sigma measurements carried to their illogical conclusions.

Six Sigma for Secretaries

Hyped Six Sigma crusaders measure the sigma level of secretaries by the number of errors per one million opportunities, with the latter count

Table 1-3. Technical publications department case study. Camouflaging high-defect (ppm) levels with politically desirable high sigma levels.

Configuration Based on:	True Defect Level (PPM)	True Sigma Level	"Hyped Six Sigma" (Fictitious) Defect Level	"Hyped Six Sigma" (Fictitious) Sigma Level (σ)*
Tech Manual as a Whole	4 defects = 4 million	<0.2σ	4 million PPM	1.2σ
160 Pages/Manual	4 million	<0.2σ	25,000 PPM	3.5σ
400 Words/Page	4 million	<0.2σ	61 PPM	5.25σ
7 Keystrokes/Word	4 million	<0.2σ	8.7 PPM	5.8σ

*Allowing a 1.5σ shift of $\overline{X}$ from target value.

based on the number of memos or the lines of type or the words per line. Who should detect these errors? The secretary? The boss? An external auditor? And at what cost? It would be simpler and more effective, by far, to have the boss or internal customer evaluate the secretary's work in a subjective, nonthreatening manner, especially if the work does not directly affect the customer.

Six Sigma for Training

Hyped Six Sigma companies measure the sigma level of classroom training by determining the score of 1 to 5 (with 1 the lowest and 5 the highest) given by the students to the instructor, per one million opportunities, with the latter counts based on the number of students multiplied by the number of line items per evaluation form. Here again, student feedback is, at best, an insufficient measure of training effectiveness, especially if there is no follow-up of the on-the-job implementation. Instead, training effectiveness should be assessed by using a three-month and six-month follow-up of the managers of the students and their perceptions of actual implementation and by measuring tangible results (e.g., improvements in quality, cost, cycle time, customer retention, and longevity).

Hyped Six Sigma companies have become so obsessed with the minutiae of Six Sigma measurements, especially in support services, that they can eventually bankrupt the companies they heroically try to reform. As the saying goes, figures do not lie, but liars can figure. True parts per million and sigma levels don't lie, but hyped Six Sigma liars can figure and deceive. It is worth repeating the heresies so you can be wary of them.

A Summary of Hyped Six Sigma Concepts

Heresy 1 — A 1.5 sigma shift of $\overline{X}$ from the design center—in effect deteriorating the defect level of 2 ppb to 3.4 ppm and pulling a fast one by calling this Six Sigma.

Heresy 2 — Dividing true product defect levels by the number of parts in a product and equating the lower artificial defect level to an inflated sigma level.

Heresy 3 — Dividing the already-diluted defect level of Heresy 2 by the number of opportunities for defects in each part and equating the even lower artificial defect level to an even higher sigma level.

Heresy 4 — Dividing the even-more-diluted defect level of Heresy 3 by increasing and fudging the opportunity count on each part to an absurdly high sigma level.

Heresy 5 — Instead of concentrating on those products and services most important to their customers, hyped Six Sigma advocates attempt to measure all areas, including peripheral white-collar work, with the same sigma yardstick, causing high costs without corresponding value.

Advocates of the hyped Six Sigma can conjure up any sigma level that suits their Machiavellian gamesmanship. But, in the final analysis, they get hoisted on their own petard. Customers, dissatisfied with the same tired defect levels on products that are dressed up in fancy "hyped Six Sigma" attire, vote with their feet and their pocketbooks to dump such companies and switch to their competition.

The Ultimate Six Sigma

The Ultimate Six Sigma is the real Six Sigma. As explained in depth in subsequent chapters of this book:

1. It goes way beyond conventional quality systems practiced by companies with a "follow the leader" mentality such as:

❖ ISO-9000

❖ The Ford 8-D methodology

❖ QS-9000

❖ The Malcolm Baldrige National Quality Award

❖ The European Quality Award

❖ The Deming Prize

❖ Total Quality Management (TQM)

2. It goes way beyond traditional quality and reliability to breakthrough levels, with increased customer retention, value, and "wow" and the potential to double corporate profit

3. It goes way beyond the hyped Six Sigma

4. It goes way beyond the narrow confines of product quality, and lifts its sights to the pursuit of excellence in all facets of a business

5. It goes way beyond token decreases in cycle time to 10:1 and even 50:1 decreases, with resultant increases in inventory turns of 10:1 and 20:1

6. It goes way beyond customer loyalty to include the loyalty of all stakeholders—customers, employees, suppliers, distributors, dealers, and investors

7. It goes way beyond the narrow scope of traditional Six Sigma. By contrast, Ultimate Six Sigma moves:

❖ From autocratic management to inspiring, visionary leadership

❖ From an organizational straitjacket to breaking the chains of bureaucracy

❖ From stifling Taylorism to releasing the creative spirit and innovative talent of all employees

❖ From the tedium of lagging metrics to the excitement of leading, challenging indicators

❖ From the frustration of weak tools to revolutionary tools with sledgehammer power

❖ From the poor design world of built-in defects, cost overruns, and late launches to the era of robust designs, cost caps, and streams of new products frequently introduced

❖ From the downward spiral of a confrontational customer-supplier relationship to the upward spiral of substantial improvements in customer quality, cost, and cycle time alongside higher profitability for partnership suppliers

❖ From manufacturing as a setting sun to manufacturing as a rising sun

Figure 1-2. Evolution of the purpose of a business.

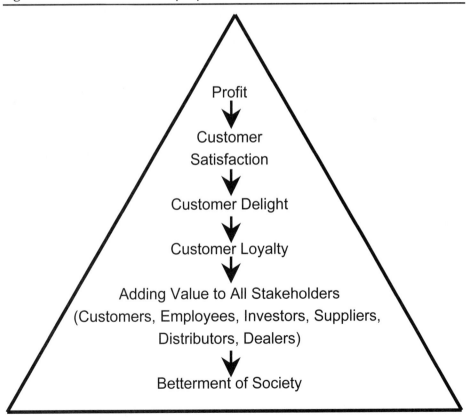

❖ From the field as out-of-sight, out-of-mind to a hallowed place next to the customer throne

❖ From white-collar blues to a happy, productive workplace

❖ From the lows of marginal, incremental results to the highs of best-in-class performance

❖ From the narrow view of quality in the past to, in this twenty-first century, a scope that caters to the betterment of society

The progression to the Ultimate Six Sigma is portrayed in Figure 1-2. Ever since Peter F. Drucker, the management guru, proclaimed fifty years ago that the purpose of a business was not profit—important as it is—but the customer, industry has witnessed an enlargement of its business horizons in each decade. I characterize this progression as follows:

In the 1970s, customer satisfaction began to replace profit as the purpose of business.

In the 1980s, customer delight—or customer "wow"—became a business mantra.

In the 1990s, customer loyalty—or customer long-term retention—became the rallying cry for progressive companies.

In the first decade of this twenty-first century, customer loyalty is metamorphosing into the loyalty of all stakeholders: customers, employees, suppliers, distributors, dealers, and investors.

In the first half of the twenty-first century, businesses will go beyond their self-imposed boundaries to address the many dimensions of a society yearning for the betterment of humankind. That is the future direction of the Ultimate Six Sigma.

TWO

THE NEED, OBJECTIVES, AND BENEFITS OF THE ULTIMATE SIX SIGMA

❖   ❖   ❖

*American management is
out of touch with its
customers, out of touch
with its employees, out of
touch with its suppliers.
Other than that, it is in
pretty good shape.*

—TOM PETERS

The Need for the Ultimate Six Sigma

In business, there exists a half-life phenomenon that is little known in the United States:

❖ The half-life of customers is only five years.

❖ The half-life of employees is only four years.

❖ The half-life of investors is only one year.

❖ The half-life of a corporation is only twenty years.

In fact, of the Fortune 500 companies listed in 1900, only one remains. Of companies that made this list in 1980, only 20 percent remain.

There are many reasons for this appalling attrition: Mergers, hostile takeovers, rapidly changing technologies, and globalization all play their part. The root cause is what Tom Peters dramatically states in the quote I chose to open this chapter—the failure of management. More precisely, it is the failure of leadership, the failure to pursue excellence in all areas of a business. Capturing this excellence is the central theme of the Ultimate Six Sigma.

Quality for the Sake of Quality Alone, Not for Customers and Long-Term Profit

In his historic acceptance speech at the 1964 Republican Convention, Barry Goldwater stated that the pursuit of liberalism, in the cause of human betterment, is no virtue and the pursuit of conservatism in the same cause is no vice. To paraphrase Goldwater, the pursuit of quality for the sake of quality alone is no virtue and the pursuit of quality for customers and long-term profit is no vice. History is replete with examples of companies nearly shipwrecked on the rocks of a narrow quality focus while paying little attention to customers and business imperatives. Consider these examples:

❖ *Florida Power and Light (FPL)*. This electrical utility spent more than $7 million in preparing for the Deming Prize—the Holy Grail of quality awards. It won this coveted award (with the help of several Japanese consultants, who also doubled as Deming Prize examiners). Shortly thereafter, FPL almost went bankrupt. Preoccupied with people preparing quality charts and tied up in meetings, customer complaints were ignored. Power blackouts in the middle of winter—a mortal sin—went unanswered. FPL had to be rescued by new management who applied the wrecking ball to its top-heavy quality structures that ignored the customer.

❖ *Wallace Company: Quality Laurels and Chapter 11*. This oil equipment company was among the first companies to win the Malcolm Baldrige National Quality Award. It became a quality role model. Two years later, as the cost of its quality obsession soared, it filed for Chapter 11 bankruptcy. (This triggered a joke in the media that winning the Baldrige Quality Award was a sure ticket to bank-ruptcy.)

❖ *United Parcel Service (UPS): Fast Delivery Results in Loss of Market Share.* UPS linked its quality efforts to fast delivery. Detailed time-and-motion studies became its definition of quality. UPS knew exactly how long it took for customers to answer their doorbells. The company even redesigned seats in its delivery vans so drivers could get out faster. Yet UPS steadily lost market share because what its customers really wanted was more interactive time with the drivers and more practical advice on best shipping methods. In a sharp departure from its fixation on fast delivery, the company now allows its caring drivers to spend extra time with customers; as a result, UPS has increased its revenues-to-cost by a factor of 10:1.

❖ *Varian Associates.* This is a classic case where quality charts headed north while business charts went south. This company went the whole hog for total quality management (TQM). It cycled 1,000 of its managers through a four-day course on quality. The language of quality circles and cycle time replaced the focus on electronics and customers. The upshot? A profit loss of 1.5 percent versus a profit gain of 3 percent before total quality management.

These four examples illustrate, in stark terms, the danger of pursuing quality when it is isolated from the customer, isolated from employees, and isolated from bottom-line results. The inoculation against this virus is the Ultimate Six Sigma.

Fads, Potions, and Nostrums of the Quality Movement in the Last Fifty Years

This disillusionment with quality is by no means new. The history of the quality movement in the past fifty years has had a program-of-the-decade flavor as organizations seek quality salvation with each new fad replacing a discarded one. For example:

1950s	*Sampling Plans*—A whole host of sampling plans were developed, each claiming to be the alpha and omega of quality control. Luckily, they soon faded in popularity.
1960s	*Zero Defects*—Launched by the U.S. Department of Defense, this psychobabble stated that if workers could only pledge themselves to zero defects, the quality genie would rise. The zero defects movement turned out to be all show and no substance.
1970s	*Quality Circles*—Based on teams tackling daily line prob-

lems, quality circles achieved some success in Japan (yet the quality tools given to their workers were elementary at best). Transplanted, however, from the group solidarity "rice culture" of Japan to the rugged John Wayne soil of American individualism, quality circles became a hothouse plant and withered on the vine.

1980s   *Statistical Process Control (SPC)*—In the mistaken notion that Japan's quality success was because of statistical process control and control charts, American companies started papering their factory walls with control charts. Automotive companies such as Ford Motor spent billions of dollars on control charts and registered millions in savings. That is a 0.1 percent return on investment. They would have fared better investing in a failed savings and loan bank.

The Age of Standards: 1980s–1990s

Starting with their granddaddy, the defense department–sponsored Mil-Q-9858A, a whole plethora of quality standards has been spawned all over the world from the 1980s to the 1990s. They include:

❖ *ISO-9000.* This quality standard was launched in 1987 as a worldwide standard by a consortium of forty-five countries within the International Standards Organization (ISO). It is an elementary quality system—the least common denominator of mediocrity for these squabbling nations. Even though revisions and improvements have been made every few years, it has set back the quality movement by twenty years. As its critics say, if you want to freeze unacceptable defect levels, ISO-9000 will enable you to do so consistently!

❖ *QS-9000.* This is the quality standard developed in the 1990s by the Big Three automotive companies—General Motors, Ford Motor Company, and Chrysler Corp. —for their first-tier suppliers. It is an improvement over ISO-9000, with a minimum $^{c}p_K$ of 1.33 required of these suppliers. (Never mind, that the Big Three cannot meet its requirements in their own plants!) This standard is still couched in bureaucratic speak and contains little guidance on how to help its hapless suppliers achieve even modest levels of quality.

❖ *ISO-14000.* Another international standard patterned after ISO-9000, its worthy objective is to improve industry's contribution to the environment. The jury is still out on the case of Mother Earth versus industrial greed, as the failure of the United Nations environmental conference in Tokyo in the late 1990s so amply proved.

The true impact of these standards is yet to be determined. They have generated baby ISOs in niche industries, such as the aerospace industry, but their authoritarian language has put companies into straitjackets—the very antithesis of creativity.

The Birth of Quality Systems: 1960s–1990s

Formal quality systems are the outgrowth of the haphazard quality manuals that individual companies had created in the early days of modern quality control (the 1960s to the 1990s). Examples are:

❖ *The Deming Prize.* Named after Dr. W. Edwards Deming for his seminal quality work in Japan, it is still the Holy Grail pursued by Japanese companies for the past forty years. It is now open to companies and individuals in all countries. It emphasizes statistics, but is not focused on leadership or on the principal stakeholders—customers, employees, and suppliers.

❖ *The Malcolm Baldrige National Quality Award.* The award was established by the United States Congress in 1987 as America's answer to Japan's Deming Prize. Although the Baldrige Award guidelines are considerably superior to the requirements of ISO-9000 or QS-9000, they are far from a world-class quality system because they touch only superficially on the customer and on leadership. They fall short in the all-important "how to" of tools and techniques to achieve quality breakthroughs. Furthermore, the ambiguities of language would make a lawyer look like a novice! In the past several years, the bloom is off the Baldrige Award because mandatory conformance to ISO-9000 has taken precedence over the voluntary incentive to try for the Award. The number of applicants for the award—never higher than a paltry 110—has dropped to under a depressing fifty today.

❖ *The European Quality Award.* Created within the European Union in the early 1990s, it is patterned after the Baldrige Award. With a focus on business results, it is an improvement over Baldrige, but is nowhere near world-class quality requirements.

❖ *Total Quality Management (TQM).* A late-blooming quality system in the 1980s and 1990s has been total quality management. Heralded with much fanfare as the decisive solution to the quality challenge, it has proved to be just the latest in a long line of fads, potions, and nostrums. Table 2-1 is a summary of various surveys, conducted on several hundred leading U.S. companies at the height of the movement, by prestigious organizations on TQM ineffectiveness.

Table 2-2 lists a subjective—but relevant—score of the effectiveness of the various quality standards and systems. It is based on my consultations

Table 2-1. Surveys of TQM ineffectiveness.

Surveyor	Results
American Electronics Association[1]	73% of companies had TQM under way. Of these, 63% had lowered quality defects by less than 10% (world-class: 50% to 80%).
Arthur Little[2]	Only one-third of the companies stated that TQM had improved their competitive position.
McKinsey & Co.	67% of the companies had stalled or fallen short of real improvement.
American Quality Foundation[3]	77% of the companies failed to achieve results and business objectives.
Quality Magazine[4]	Companies have trained thousands of employees and launched numerous quality processes, and have been disappointed in the results. Despite their best intentions, these companies have floundered. They need annual improvements at a revolutionary rate, but their achievements are stunted.
Rath & Strong[5]	Rated companies on TQM efforts to improve market share, lower costs, and make customers happy. On a scale of 1 to 5—with 1 as worst and 5 as best—most companies rated only a 2.

with more than 400 companies all over the world. *My conclusion is that there is a compelling need for a world-class system that only the Ultimate Six Sigma can best provide.*

Tools—Driving a Nail in the Wall Without a Hammer

Bill Conway, former chairman of the Nashua Corporation, once received a delegation of vice presidents from Ford Motor Company, who wanted to

Table 2-2. Relative effectiveness of quality standards/systems.

Quality Standard/System	Effectiveness (Scale: 1 = least effective; 100 = most effective)
ISO-9000	5
QS-9000	15
ISO-14000	Too new to be assessed
Deming Prize	30
Malcolm Baldrige National Quality Award	35
European Quality Award	40
Motorola Six Sigma	50
Hyped Six Sigma	30
Ultimate Six Sigma	90

explore the reasons for Nashua's outstanding quality success. He told his listeners, "I'm going to select two of you for a contest. The winner will get a free vacation in Hawaii. The loser will get nothing. Since you're all successful vice presidents, motivation will not be a factor in this contest which involves driving a nail into the wall behind me. Each of you will get a nail. One will get a hammer, the other—nothing. Who will win?" Tragically, industry tries to solve its chronic quality problems with weak tools—hammering with a wet noodle instead of a sledgehammer. A catalog of these wet-noodle tools follows:

❖ *The Seven Tools of QC (of Kindergarten Effectiveness).* The Japanese have packaged a set of quality techniques, labeled the seven tools of quality control (QC), to solve quality problems in production. While a case can be made for the seven tools of QC to solve the simplest of problems, they are totally ineffective in solving chronic quality problems. The seven tools are:

1. PDCA (plan, do, check, act)
2. Data collection and analysis
3. Graphs/charts
4. Check sheets, tally sheets, and histograms
5. Pareto charts
6. Cause-and-effect diagrams
7. Control charts

❖ *The Seven Quality Management Tools (Another Wet-Noodle Hammer).* The Japanese have an amazing penchant for making simple things complicated. In the 1980s, they collected planning tools, which they borrowed from organizational development disciplines, and called them the seven quality management tools, the next generation in problem solving. These tools are useful in planning and correlation studies, but they are even less useful than the seven tools of QC in problem solving. They are:

1. The affinity diagram
2. The interrelationship diagram
3. The tree diagram
4. The matrix diagram
5. The matrix data analysis plot
6. The process-decision program
7. The arrow diagram

❖ *The 8-D Methodology: A Ford Boondoggle.* In the early 1990s, Ford Motor Company introduced its 8-D (eight disciplines) problem-solving technique for its plants and suppliers. Many companies, attracted by the Ford name and/or pressured by its rigid, bureaucratic approach to quality, adopted 8-D as a problem-solving mantra, spending days and weeks in costly seminars. At best, Ford 8-D provides a framework or outline for problem solving. It provides no tools—or weak tools—for finding the root cause of a problem, verifying it, or validating the effectiveness and permanency of the corrective action. The results are frustration and disillusionment. The eight steps of this methodology are:

1. Use a team approach.
2. Describe the problem.
3. Contain the symptom.
4. Find and verify the root cause.
5. Choose corrective action and verify.
6. Implement permanent corrective action.
7. Prevent recurrence.
8. Congratulate the team.

❖ *The Engineering Approach (Guesses, Hunches, Opinions, and Theories).* Sometimes called the "scientific approach," the engineering technique spans an "observe, think, try, and explain" cycle:

1. The "observe" step is okay, provided the engineers "talk to the parts" (as will be explained later in Chapters 11 and 14). The problem is, they don't know how to talk to the parts.
2. The "think" step involves guesses, hunches, opinions, fads, and theories that are often seized on to curry favor with the ideas of a boss, who is several organizational layers removed from the problem.
3. The "try" step usually varies one factor (or variable) at a time, keeping all other factors constant. Besides taking an inordinate amount of time, this method can miss significant interaction effects between factors and render the experiment weak or downright wrong.
4. The "explain" step often attempts to rationalize the results to fit a preconceived theory. There is no attempt to verify the improvement by turning the problem on and off.

❖ *Worker Involvement.* In effective problem-solving, the author has a proven rule: The first priority is to "talk to the parts." If that does not succeed, talk to the workers on the job. Only as a last, desperate resort, talk to the engineers and managers who are too remote from the action.

Behavioral scientists have indicated that the team approach is an excellent organizational building block. The Japanese, in particular, use teams under various labels: quality circles, *kaizen,* minicompanies, glass-wall management, and small group improvement activities (SGIA). But even workers and teams, good as they are, need powerful tools. This is illustrated by Henley's law (named after an industrial philosopher, Wes Henley). It states that the dominant characteristics of a team tend to grow exponentially. If the dominant characteristics are ignorance and misinformation, there will be an exponential growth of ignorance and misinformation. If the dominant characteristics are knowledge and sound information, there will be an exponential growth of team knowledge and sound information. A company needs and deserves good leadership. It needs and deserves dedicated, enthusiastic workers. However, giving excellent workers weak-noodle tools will only produce confusion, frustration, and disenchantment.

❖ *Computer Simulation.* The computer age is upon us in earnest. We seem to have developed a blind faith in the computer's ability to do anything, to solve any problem, instead of relying on the God-given computer between our ears—the brain. For problem solving, a computer must be programmed with the mathematical equation that governs the relationships between the dependent variable and the independent variables. Without that formula, the computer is reduced to a guessing game, at best. This is especially true when interaction effects are present between variables. Unfortunately, in complex products and processes, not even an Einstein can develop a formula that fits. Finally, even if a computer can provide an approximation to the formula, its results must be verified, using design-of-experiment (DOE) techniques.

❖ *Control Charts.* In 1980, the NBC television documentary *If Japan Can, Why Can't We?* provided a historical continental divide between traditional and statistical quality control. It attributed Japan's success to its mastering control charts. Hundreds of thousands of U.S. companies were blackmailed into using control charts at the point of a gun by blinkered customer companies. Yet control charts, in their seventy-five-year history, have never been effective in solving chronic problems. At best, they are only somewhat useful in monitoring quality, once solved with techniques such as design of experiment.

❖ *Kepner-Tragoe.* Some companies have dabbled in problem solving by using the Kepner-Tragoe methodology that specializes in making observations and detection. Similar to the methodology used in solving crimes, Kepner-Tragoe is wonderful for writing fictional novels, but not for industrial problem solving.

Table 2-3 lists a subjective but realistic score of the effectiveness of various problem-solving tools—again, based on my problem solving for companies on four continents. As with systems, there is a need to jettison the tired tools of the twentieth century and embrace the powerful tools of the twenty-first century, such as the design-of-experiment methods discussed in Chapters 11, 12, and 14.

Objectives of the Ultimate Six Sigma

The Ultimate Six Sigma process has several reach-out objectives in industry:

❖ Develop a comprehensive infrastructure that goes well beyond the narrow confines of quality (the small Q) to encompass all areas of business excellence (the Big Q).

❖ Maximize all stakeholder loyalty—customer loyalty, employee loyalty, supplier loyalty, distributor/dealer loyalty, and investor loyalty.

Table 2-3. Relative effectiveness of problem-solving tools.

Problem-Solving Tools	Effectiveness (Scale: 1 = least effective; 100 = most effective)
Seven Tools of QC	3
Seven Management Tools of QC	2
8-D	2
Engineering Approach	5
Worker Involvement	10
Computer Simulation	15
SPC/Control Charts	1
Kepner-Tragoe	10
Design of Experiments (DOE)	
1. Classical	30
2. Taguchi	20
3. Shainin/Bhote*	100

*This technique is detailed in Chapter 11, Chapter 12, and Chapter 14.

❖ Maximize business results: profits, return on investment, asset turns, inventory turns, sales/value-added per employee.

❖ Minimize people turnover and bring joy to the workplace, especially to the line worker.

❖ Go beyond modest and mediocre quality standards/systems to devise an ideal yet practical quality system.

❖ Go beyond the tired problem-solving tools of the twentieth century to forge powerful new tools for the twenty-first century.

❖ Go beyond the propaganda and results-with-mirrors of the hyped Six Sigma consulting companies to usher in an Ultimate Six Sigma—low in implementation costs and high in business results.

❖ Provide keys to critical success factors in each of twelve areas:

1. Customer loyalty and long-term retention
2. Quality of leadership (to provide vision and inspiration, which facilitates employees' reaching their full potential)
3. Quality of organization (to revolutionize the ways people are hired, trained, evaluated, compensated, and promoted)
4. Quality of employees (to provide empowerment on the road to industrial democracy)
5. Quality of metrics (to assess business excellence)
6. Quality of tools (to achieve quality, cost, and cycle-time breakthroughs)
7. Quality of design (to maximize customer value and "wow")
8. Quality of supplier partnerships (to improve customer quality, cost, and cycle time while enhancing supplier profits)
9. Quality of manufacturing (i.e., overall) effectiveness
10. Quality of field reliability (toward zero field failures)
11. Quality of support service (i.e., business/white-collar) effectiveness
12. Quality of results (to develop and rate world-class metrics)

❖ Conduct periodic audits and self-assessments to achieve continuous, never-ending improvement.

Benefits of the Ultimate Six Sigma

Concentrating on the principles, methodologies, and actions described in this text will enable a company to create metrics for business, customer loyalty, and quality. The benefits of each of these metrics are as follows:

Business Metrics

❖ Enhance the business's long-term profits by factors of 2:1 to 5:1 and from 4 percent to 20 percent of sales (after tax)

❖ Enhance return on investment by 3:1 to 8:1 and from 10–15 percent to more than 50 percent

❖ Enhance asset turns from four to over fifteen

❖ Increase inventory turns from six to ten to more than 100

❖ Reduce people turnover from 20 percent to 10 percent, and eventually down to less than 0.5 percent per year

❖ Increase productivity (i.e., value-added) per employee per year from $100,000 by 30 percent, eventually to more than $500,000

Customer Loyalty Metrics

❖ Improve customer loyalty and retention levels from below 75 percent to 99 percent

❖ Increase customer retention longevity from less than five years to over fifteen years

❖ Increase the satisfaction rating of all stakeholders by 2:1 to over 90 percent

❖ Increase market share position to number one or two in each business line

Quality/Reliability/Cycle Time Metrics

❖ Reduce outgoing defect rates from the 1–10 percent range down to 10 parts per million (ppm) and lower.

❖ Reduce total defects per unit (TDPU) on an entire product line from the one to five range down to 0.1

❖ Increase c_{p_k} of critical parameters from the 0.5 to 1.0 range up to 2.0 to 5.0

❖ Reduce field failures from the 2–20 percent range per year down to 100 ppm per year

❖ Reduce the cost of poor quality (as a percent of sales dollars) from the 8–20 percent range down to less than one percent

❖ Reduce cycle times in production and business processes (in multiples of theoretical cycle time—that is, direct labor time) from the 10–100 range down to 1.5–2.0

These benefits are not just pie-in-the-sky numbers. They can and are being attained by world-class companies.

Three

THE ORIGIN, DEVELOPMENT, AND RENEWAL OF MOTOROLA'S SIX SIGMA

❖ ❖ ❖

*The mind to imagine, the
will to do.*

—MOTOROLA CREDO

The story of Motorola, as the pioneer of the Six Sigma process, provides the perfect backdrop to the development of the Ultimate Six Sigma methodology. The genesis of Motorola's Six Sigma was the intense challenge of Japanese competition. They were eating our lunch in several businesses in the 1970s and getting ready to eat our dinner as well. "Meet the Japanese Challenge" became our call to arms. Bob Galvin, then chairman of the board at Motorola, decided that he would fight the Japanese with the same tool with which they had captured market share in the United States—namely quality.

10:1, 100:1, 1,000:1 Quality Improvement

In 1981, Galvin established a quality improvement goal of 10:1 in five years. Motorola's previous quality record had been considered respectable, improving at 10 percent yearly. But Galvin's goal was to improve it—not by 50 or 100 percent—but by 1,000 percent in five years. Many skeptics declared it an impossible goal. Yet, by 1986, most of Motorola's divisions had met that goal. This author, then the group director of Motorola's Automotive and Industrial Electronics Group, achieved the 10:1 improvement in three years. Yet, when we benchmarked ourselves vis-à-vis the Japanese, we found that they were still ahead in quality. Consequently, in 1987, Galvin raised the height of the bar with a further 10:1 improvement, but this time in two years. And in 1989 he again furthered our sights for a third 10:1 improvement by 1991. Thus, starting in 1981, Motorola had to improve quality by an incredible 1,000:1 in ten years. The goal was not completely achieved throughout Motorola's worldwide operations, but the average improvement was 800:1, starting from an already-respectable quality base in 1981. The entire progression to these lofty quality heights was given the appellation Six Sigma—less as a statistical term, more as a rallying cry for perfection.

The Pot of Gold

Many in the media were critical of Motorola's "obsession" with quality. "Wouldn't there be a severe cost penalty for such a magnitude of quality improvement?" they smirked.

Motorola stood firm and never lost sight of its customers or the bottom line in this quality pursuit. Since 1979, Motorola had been tracking the cost of poor quality, which includes warranty, scrap, repair, inspection, and test costs along with the cost of carrying inventory. In ten years, it saved more than $9 billion by reducing this cost of poor quality. That was the pot of gold that enabled Motorola to recapture customers previously lost to the Japanese through lower prices. That was the pot of gold that gave our employees higher incentives and wages. That was the pot of gold that allowed us to keep our loyal investors—with Motorola stock having appreciated 24:1 in thirteen years. Several years later, a leading news journal asked: "Mr. Galvin, you have led the company to many quality peaks and have won many honors, including the Malcolm Baldrige [National Quality] Award. What is it that you regret the most in your quality drive?" Galvin's amazing answer was, "I did not set high enough goals!"

Success Factors in Motorola's Six Sigma Drive

There are several reasons for Motorola's spectacular quality achievement. Success has many fathers; failure is an orphan. Nevertheless, the highlights are detailed here.

Success Factor 1: Inspiring Leadership

Foremost was Bob Galvin's dynamic leadership and complete dedication to quality as the central thrust of a business engine. Without his vision, inspiration, and charisma, we would have been mired in mediocre improvements. He led the quality charge rather than delegating it to lower levels. He was on the frontline. His ethics, abiding trust in people, and guidance in getting people to reach the maximum of their potential were legendary. He freely admitted that he did not know how to reach the lofty goals he had set. Yet, he had a robust faith in his people and believed that by leading into pathways no one had ventured before, he would inspire them to develop a road map to success.

Success Factor 2: Twenty-first Century Tools Forged in the Twentieth-Century Crucible

A second success factor was our conviction that the old tools for quality, cost, and cycle time, developed in the earlier second half of the twentieth century, were obsolete. I took responsibility and leadership in researching and implementing powerful new tools to get ready for the twenty-first century. (A description of these tools, along with their objectives, benefits, and a brief methodology, is given in Chapter 11.) The road to implementation of these tools proceeded as follows:

Timeline to Quality

1982   My associates and I institutionalized design-of-experiment (DOE) techniques to solve chronic quality problems in Motorola's U.S. plants. In the next five years, we extended DOE to most of our fifty worldwide plants.

1984   We introduced Multiple Environment Over Stress Testing (MEOST) to vastly improve field reliability.

That same year we introduced the Next Operation as Customer (NOAC) in support services to improve quality, cost, and cycle time in all white-collar areas.

1985   We introduced quality function deployment (QFD) to capture the "voice of the customer."

1986   We introduced benchmarking to model success factors found in the "best in class" companies that we visited.

1987   We introduced poka-yoke to prevent operator-controllable errors.

> 1988   We introduced cycle-time reduction as the other side of the quality "gold coin."

As a result, Motorola benefited by orders-of-magnitude improvements in quality, more than $30 billion in cost reduction, and a greater than 10:1 reduction in manufacturing cycle time.

Success Factor 3: Total Customer Satisfaction

In the 1980s, Motorola revolutionized its main objective from profit to total customer satisfaction. It was not that profit wasn't important. But the realization grew that if we truly took care of our customers, our profits would follow. And they did. High-level executives, next in importance only to the sector general managers, were appointed as chief customer officers. The quality function was designated as the customer's advocate. Galvin would personally visit ten of our most important customers each year, talking not to his corresponding CEOs but to the people who "felt, smelled, and directly dealt" in our products. He would return with pages of notes and sensitize us to address what our customer wanted, not what Motorola thought they needed. All executives followed suit with their most important customers.

Success Factor 4: Empowerment of People

Motorola had always enjoyed a worldwide reputation as a "people" company. Its open-door policy to the chairman, its freedom from unions (with employees rejecting unions time and again), and its financial team incentives were hallmarks of a caring management. But in its Six Sigma era, the caring became even more explicit. Individual dignity entitlement assured a mutual evaluation of supervisor and employee, with a growth path charted for the latter. Baldrige examiners were amazed that the entire direct labor population was singing from the same sheet of Motorola quality music with an enthusiastic dedication to quality at all levels.

The jewel of the crown was Motorola's total customer satisfaction (TCS) competition. Small teams would choose a worthy improvement theme and voluntarily pursue it to successful conclusion. Each plant would field fifty to sixty such teams, with winning teams working their way up from plant to regional to sector and eventually to corporate levels. Each year 6,000 teams would participate and the excitement and enthusiasm they generated was awesome to behold. Motorola benefited enormously, too, with $2 billion to $3 billion per year in savings.

Success Factor 5: Robust Designs

The design function generally controls 70 percent of a product's total cost and causes 80 percent of its quality problems. Therefore, its infrastructure was changed to a concurrent engineering team approach while powerful tools such as DOE, MEOST, design for manufacturing (DFM), and other techniques were used to minimize problems in production and the field. Early supplier involvement (ESI) was introduced to allow the full participation of partner-suppliers very early in the design cycle.

Success Factor 6: Win-Win Partnerships With Suppliers

Jettisoning the old win-lose contest with suppliers, a new era of win-win partnerships with key suppliers was ushered in to pursue mutual objectives of quality, cost, cycle time, and profit improvements.

Success Factor 7: Standardized Metrics

With vastly different product complexities and parts counts in Motorola's fifty manufacturing plants, sigma levels had to be "normalized" so that each plant would be entitled to a level playing field in its march to Six Sigma. In addition, a standard reporting format was established worldwide. Called the 5-up charts, this reporting format covered:

1. Outgoing quality
2. Total defects per unit (TDPU)
3. Field failure rates
4. Cost of poor quality
5. Cycle-time reduction

The 5-up charts were watched for trends with the same scrutiny applied by financial analysts.

Six Sigma Renewal at Motorola

Motorola's history, led by founder Paul Galvin and institutionalized by his son Bob Galvin, was to renew itself every few years—renew its vision, its goals, its vigor. Our Six Sigma process was renewed in several imaginative ways following our success in winning the prestigious Malcolm Baldrige National Quality Award in 1988.

❖ *Extending Six Sigma to All Business Processes.* With our spectacular success in improving quality in manufacturing, attention turned to a woeful area of neglect in corporate America—business processes. The same Six Sigma methodology for improvement (see Chapter 16) was used to improve quality and productivity in all white-collar operations—from accounting to management information systems, from legal and patent departments to human resources. Metrics similar to those used in the 5-up charts in production were created to measure progress in each white-collar area.

❖ *Capturing the Voice of the Customer.* Companies have been notorious in failing to consult customers at the start of design. As a result, 80 percent of products introduced into the marketplace fail. I was a delegate, representing Motorola, in two international study missions headed by Ford Motor Company to evaluate Quality Functions Deployment (QFD) as a viable method to capture the voice of the customer in design. Ford introduced QFD to America and I introduced it at Motorola. (Today, however, mass customization is gaining ascendancy over QFD whenever a large base of customers with common requirements cannot be found.)

❖ *Pursuing a 10:1 Improvement in Cycle Time.* One of the by-products of our 10:1, 100:1, and 1,000:1 quality drive was the discovery, almost by serendipity, that cycle time could also be simultaneously reduced. Cycle time is the total elapsed clock time for a process, from start to finish, and is several multiples (50:1 to 100:1) of the actual direct labor time, known as theoretical cycle time. We became convinced that quality and cycle time were inseparably linked—they were two sides of the same coin. Motorola's CEO went on to establish another reach-out goal—a 10:1 improvement in cycle time in five years, not only in manufacturing, but also in business processes and even in design.

❖ *Flattening the Organizational Pyramid.* Recognizing that having fifteen to twenty management layers between a CEO and the line worker promoted a suffocating bureaucracy, Bob Galvin again led the charge to reduce these layers and promote freedom for employees to achieve corporate goals their own way (see Chapter 8).

❖ *Training, Training, Training.* Motorola has always believed in training, but Bob Galvin institutionalized it. Over the objections of Motorola's board of directors, he had authorized the creation of a $12 million learning center on our campus for training Motorolans in a number of subjects, with quality in the lead. The cost then was 1.5 percent of payroll. Today, this seedling has flourished into eleven learning centers around the globe, costing 5 percent of payroll, with forty to 120 hours per year per employee allotted for training. The focus is Motorola University, ably captained by Bill Wiggenhorn, with a syllabus that would put an academic university to shame.

❖ *Tackling the Quality of Education in the Nation's Schools.* Stepping outside the confines of industry, Motorola undertook to inject quality into the curricula and teaching methods of high schools and univer-

Table 3-1. Quality-effectiveness scores, by type of organization.

Institution	Quality Effectiveness (Scale: 1 = least effective; 100 = most effective)
Manufacturing	70
White Collar	45
Service	25
Government	20
Schools	15
Hospitals	10

sities. The importance of quality had only been vaguely understood in these institutions and the word *customer* was off their radar screen. I was personally involved in introducing Six Sigma to a prestigious university in the Chicago area and was amazed by the magnitude of the challenge. Table 3-1 lists quality-effectiveness scores, subjective though they are, that I have found, in my subsequent consultations, generally apply to different institutions. To schools, as they say in show business: "You have a long way to go, baby!"

❖ *Sharing Motorola's Success With the Outside World.* One of the obligations of our winning the Malcolm Baldrige Award was sharing our methodology with other companies. Motorola was very open and generous in its revelations and has presented hundreds of seminars and tutorials, with companies from all over the world coming to us to learn about our Six Sigma practices. Our Six Sigma was now on the world map.

Four

The Hyped Six Sigma: From the Pure Six Sigma to the Sick Sigma

You can fool some of the people all the time; and you can fool all the people some of the time; but you can't fool all the people all the time.

—Old proverb

Hyped Six Sigma Strengths: A Mixed Bag

Because of the enormous publicity and self-promotion attached to the hyped Six Sigma, it becomes necessary to set forth, in some detail, the few strengths and the many weaknesses of this approach. The format is a point-counterpoint demarcation.

1. *Point 1: Supposed Fantastic Savings.* Hyped Six Sigma consultants claim savings in the millions and even billions of dollars for their clients in the space of three or four years.

Counterpoint 1. Inside information from personnel doing the work in their client companies reveals that the savings are achieved "with mirrors"—illusion rather than reality. As an example, all savings, even though there is no relation to the hyped Six Sigma, are included as Six Sigma savings. Furthermore, the Six Sigma savings are not audited, especially by independent auditors. In the final analysis, a profit improvement of 3 percent—from, say, 10–13 percent—in three years is, at best, a modest achievement.

2. *Point 2: Return on Investment.* Hyped Six Sigma consultants claim a small loss in savings versus funds invested by their client companies in the first year, a break-even point in the second year, and a 2:1 to 3:1 return in subsequent years.

Counterpoint 2. Returns of 2:1 and 3:1 are not uncommon in the implementation of any worthwhile technique. As a matter of fact, savings of $9 billion in ten years in just quality cost reductions and returns of up to 10:1 were registered by us at Motorola, and I've seen similar results in my subsequent consultations with other companies. These results were achieved without the expenditure of millions of dollars in fees charged by hyped Six Sigma consultants.

3. *Point 3: Commitment From Top Management.* The commitment on the part of top management to the implementation of hyped Six Sigma has been remarkable. This is a necessary condition for any technique to succeed.

Counterpoint 3. The main reason that some client companies embrace the hyped Six Sigma consultants has been the pressure applied by the financial analysts. Time and again these financial analysts—lured by the reported financial gains—virtually command these companies to adopt hyped Six Sigma if they expect to get a good report card from these tyrannical gnomes of Wall Street. After a few years, as reality sinks in, some of the companies become disillusioned with their hyped Six Sigma consultants and either continue the program on their own or scrap the entire Six Sigma movement. Several discerning companies see through the consultants' con game and tear up the million-dollar-plus contract checks. There is real danger that the pure Six Sigma baby can be thrown out with the hyped Six Sigma bathwater.

4. *Point 4: The Black Belt Organization.* One of the strengths of the hyped Six Sigma consultants is a comprehensive infrastructure to sustain the program. Labeled the black belt approach, it employs terms such as deployment champions, project champions, master black belts, black belts, and green belts to create an aura of invincibility. Such an organizational framework is necessary to justify the cost of the intensive training imparted and to assure the program's continuity.

Counterpoint 4. Establishing such a formal organization promotes a bureaucracy that is the antithesis of flexibility and teamwork. And while teams are formed for problem solving, the burden falls on the black belt employees, with the rest of the team more or less passive. Even more important, the entire labor force at the line operator level is left out of the picture instead of being trained and encouraged to contribute to solutions. What a tragedy it is when all of direct labor cannot be turned loose for problem solving, as happens in our Ultimate Six Sigma practice. Why not make all employees black belts? That is what we tried at Motorola and what I have done in some of my client companies.

5. *Point 5: Hyped Publicity.* No quality fad of the past fifty years has been publicized with as much fanfare as the hyped Six Sigma, nor have the media and quality professionals been as uncritical in their evaluations of the program. A virtual chorus of fawning voices—including the American Society for Quality—accompany the hyped Six Sigma siren song.

Counterpoint 5. Fortunately, a negative reaction is slowly building against the hype, led by seasoned and veteran professionals. Articles challenging both the methodology and the results are beginning to appear in the trade press. Especially important are the responses of the hyped Six Sigma's real customers—the implementers and the doers who are perceiving that their hyped Six Sigma emperor wears no clothes!

6. *Point 6: Publicizing the Main Tool: Design of Experiment.* The hyped Six Sigma consultants use design of experiment (DOE) as their main—some say only—tool. Invented by Sir Ronald Fisher more than seventy years ago, classical DOE, as it has been called, has languished in university lectures for the most part, while Taguchi DOE has received warm receptions among several U.S. companies. The hyped Six Sigma consultants have brought back the luster of classical DOE vis-à-vis the statistically weaker Taguchi DOE as a problem-solving tool.

Counterpoint 6. However, the difference between classical DOE and Taguchi DOE is the difference between Tweedledum and Tweedledummer. Both are complicated, costly, and time-consuming. Both have statistical weaknesses. In the presence of strong interactions you get confounding results that lead to weak or wrong outcomes. Neither approach talks to the parts—the real clue generators. As will be pointed out in Chapter 11, the Shainin/Bhote DOE practiced in the Ultimate Six Sigma is much simpler, more cost-effective, and far more statistically powerful.

7. *Point 7: Robustness of Design.* The hyped Six Sigma approach rightly points out that 80 percent of quality problems are attributable to poor design and that the challenge is to make a design "robust."

Counterpoint 7. The very definition of robustness in the hyped Six Sigma approach is muddled. A robust design must be made impervious (or relatively so) to uncontrollable factors such as temperature, humidity, operator errors, and so on that collectively constitute "noise." In electronic terms, the signal-to-noise ratio (i.e., the robustness of the result compared to the uncontrollable factors) must be as high as possible. Classical DOE does not even know how to list these noise factors, much less assess their interacting impact. As a result, a robust design is rendered emaciated.

Hyped Six Sigma Weaknesses: In Salient Features

Having disposed of the strengths (limited as each one is), we can now turn to the many weaknesses of the hyped Six Sigma, first in terms of salient features and second, in each of the twelve areas of a company.

Weaknesses in Salient Features

There are ten salient features in a company that, when analyzed, reveal deficiencies in the hyped Six Sigma approach.

1. *Objective*. The hyped Six Sigma objective is to improve a company's bottom line while increasing customer satisfaction. On the surface these seem worthy objectives, however, there are two failings. First, customer satisfaction is inadequate as compared to customer loyalty (as explained in Chapter 6). Second, you cannot have even limited customer satisfaction without the satisfaction of other primary stakeholders—employees, suppliers, distributors, and dealers. These important stakeholders are glossed over in the hyped Six Sigma calculus. By contrast, the Ultimate Six Sigma elevates customer loyalty to the jewel in the company crown. At the same time, it aggressively seeks the loyalty of employees, suppliers, distributors, and dealers—in short, the entire upstream and downstream supply chain.

2. *Scope*. The hyped Six Sigma scope is confined mainly to the product, making only feeble overtures to business areas. On the other hand, the scope of the Ultimate Six Sigma extends to leadership, people, organization, tools, measurement, supply chains, field operations, support services (i.e., all white-collar areas), and in-depth results. In fact, each of Chapter 6 through 17 is devoted to one of twelve key areas of a company in this book. The overriding belief of Ultimate Six Sigma is that companies are people, not just inanimate products.

3. *Defect Goal*. The hyped Six Sigma defect goal is a lofty 3.4 parts per million (ppm). But this is seldom for a product as a whole. It is a defect *per part per opportunity* for defects—a meaningless number that converts defect levels that are 100 to 5,000 times worse to 3.4 ppm, Houdini fashion! In our Ultimate Six Sigma, outgoing defect levels of zero to 10 ppm—as seen by the customer—*are the goal for an entire product*.

4. *Field Failure Goal*. The hyped Six Sigma conveniently bypasses even a mention of a field failure level goal. Worse, it has no methodology for

achieving field reliability other than elementary techniques, such as failure mode effects analysis—a preliminary paper study and an engineering guessing game. The Ultimate Six Sigma leaves the hyped Six Sigma miles behind with an ambitious but attainable reliability goal of no more than 10 ppm to 100 ppm per year. It employs powerful tools, such as Multiple Environment Over Stress Testing (MEOST), to achieve breakthroughs in reliability hitherto unheard of in industry.

5. *Implementation Cost.* One of the criminal sins of the hyped Six Sigma is a Himalayan mountain of fees charged by the consultant companies to the clients they have dazzled. These fees can extend over $10 million. Black belt training costs $40,000 to $50,000 per student, and books from hyped Six Sigma consultants can be priced at more than $3,000. Their contents are not much above entry-level quality 101 courses taught in colleges for the past fifty years and that are used less than 10 percent of the time even by quality practitioners. Compare this with the Ultimate Six Sigma implementation cost of less than $2,000 per student (including hands-on workshops); that training can be supplemented with textbooks and lecture notes costing no more than $60.

6. *Inability to Solve Chronic Quality Problems.* Time and time again, I've been called on to bale out companies with critical problems that the master black belts have been unable to solve. It isn't that they lack brainpower. The real problem is that they have been saddled with weak, wet-noodle tools. The Ultimate Six Sigma practitioners use the powerful sledgehammer tools described in Chapter 11. Their success leads to more success, along with the thrill of self-accomplishment.

7. *Torturously Long Training Time.* In hyped Six Sigma companies, training extends to six whole months—an unconscionable waste. The output is a black belt graduate who is able to solve one problem. As opposed to this Rube Goldberg training contraption, the Ultimate Six Sigma training is done in two sessions of two days each, with twice the number of participants. Each training session includes nontechnical direct labor who sometimes outperform the engineers in workshop exercises—including workshops that directly tackle the client's chronic problems.

8. *Poor Training Curriculum and Methodology.* The hyped Six Sigma curriculum is a rigid set piece. The first six weeks are spent in management orientation and championship training. For the black belts, there are four rounds of one-week instruction and three weeks of practice for each round. For example:

❖ In round one (i.e., week one) and the three weeks of on-the-job practice, the focus is solely on measurement. While measurement accuracy is important, the methodology of gage R & R (reproducibility and repeatability) is cumbersome and incomplete. In our Ultimate Six Sigma curriculum, measurement accuracy methods employ instrument multivari scatter plots that are simpler, easier, and more powerful than gage R & R. They can be taught with one hour of instruction and three hours of practice.

❖ In round two, the instruction is on finding the causes for a problem. Tedious statistics, along with process mapping and control charts, are of little help in problem solving. In the Ultimate Six Sigma approach, the motto is: "Don't let the engineers do the guessing; let the parts do the talking." The parts give you solid clues to the cause of the problem, far more effectively than engineering guesses, hunches, and opinions.

❖ In round three, improvement is sought. The main tool is classical DOE that, as stated previously, is time-consuming, costly, and statistically weak. Furthermore, there is no mechanism to verify the effectiveness or permanency of the improvement, as is done in the Ultimate Six Sigma approach using the Shainin/Bhote DOE (specifically, the B versus C technique).

❖ In round four, the attempt is to control the gains achieved by documenting procedures and using the obsolete tool of control charts. (By contrast, the Shainin/Bhote DOE controls improvements by freezing the gains with positrol, increasing the signal-to-noise ratio with process certification, and monitoring improvements with precontrol.)

9. *System Scoring and Audits.* Almost all respectable quality systems/standards such as Malcolm Baldrige, the European Quality Award, and so on, have a scoring system by which a self-assessment or an external, independent assessment can be made on a company's quality health. The hyped Six Sigma has none—a fundamental flaw. A cardinal feature of the Ultimate Six Sigma is its comprehensive audit and self-assessment. Another feature of respectable systems/standards is the periodic audit to see how well a company is progressing, year by year, on the road to quality perfection. The hyped Six Sigma has no provision for any longitudinal audit. Some of my clients are in the third and fourth years of such annual audits.

10. *Sustainability of Culture.* In order to change a company's culture, the values and beliefs of its employees must be changed. For these values and beliefs to be changed, the company's approach to hiring, evaluating, compensating, and promoting its employees must undergo a radical change. The hyped Six Sigma completely bypasses these fundamentals. As a result, it has no mechanism to sustain a new and elevated corporate culture. This lack of depth or lack of roots is manifested by the attenuation—even abandonment—of the hyped Six Sigma effort soon after the consulting companies leave. Their exit is matched by the exit of the black

belts who become organizational misfits, get disillusioned, and leave the company.

Hyped Six Sigma Weaknesses: Area by Area

Having examined the weaknesses of the hyped Six Sigma methodology in terms of salient features, we can now turn the searchlight on to its weaknesses in each of twelve important areas of a company: customers, leadership, organization, employees, measurement, tools, design, suppliers/distributors/dealers, manufacturing, field operations, service (i.e., service industries and support services in manufacturing industries), and results.

1. *Customer Satisfaction Versus Customer Loyalty*. The hyped Six Sigma companies go one step further than companies that pursue quality for the sake of quality alone. They affirm that quality must be linked to the customer. Yet they have no knowledge of customer loyalty and its profit multiplier, no concept of the necessity to reduce the customer base, no idea of the infrastructure required for customer retention, and no clue about SWAT teams to recapture defecting customers. The Ultimate Six Sigma fully explores how to maximize customer retention and profitability in Chapter 6.

2. *Management Versus Leadership*. The hyped Six Sigma client companies are, for the most part, headed by CEOs who are autocratic, headstrong managers. They are not leaders. They are still glued to the ills of Taylorism. As a result, their people are micromanaged and cowed into meek compliance with the wishes of their bosses. In the Ultimate Six Sigma, the focus shifts from managers—a dirty word—to leaders, an ennobling concept. Leaders empower their people, unleash their human spirit, and encourage them to reach their highest potential, as expanded in Chapter 7.

3. *Organization: Vertical Versus Horizontal Management*. The hyped Six Sigma companies, guided by their consultants, do nothing to change the vertical, bureaucratic, departmental, cubbyhole structure of organizations into the horizontal, freedom-filled, synergistic team organization that is essential to the Ultimate Six Sigma approach, discussed in Chapter 8 on organization.

4. *Employees: Passive Versus Empowered Employees*. The hyped Six Sigma companies do little to promote a culture change in their employees. Their consultant companies do not address how to revolutionize a culture. As a result, the entire workforce—outside of the very thin black belt population—remains passive, with employees stuck with the same dull, boring jobs they always had. In the Ultimate Six Sigma process, described in Chapter 9 on employees, a road map leading to true empowerment and injecting joy into the workplace is charted.

5. *Measurements: Rubber Yardsticks Versus Firm Yardsticks*. The rubber

yardstick of the hyped Six Sigma crowd—which blesses 3.4 ppm defects/ part/defect opportunity as Six Sigma—has already been explored in depth in Chapter 1. Additional sins of omission are the complete absence of metrics for reliability, customer loyalty/retention/longevity, design, supply chain management, cycle time, innovation, and service. The only worthwhile metric given some prominence is the cost of poor quality (COPQ). By contrast, the Ultimate Six Sigma covers each of these metrics in detail in Chapter 10 on measurements.

6. *Tools: Rubber Hammer Versus Sledgehammer.* The only major tool taught by the hyped Six Sigma consultants is the classical design of experiments. This is like equipping a whole problem-solving toolbox with just one tool—a screwdriver. The limitations of classical DOE have been briefly mentioned in previous sections of this chapter. There is no mention of reliability (as distinguished from quality), mass customization, total productive maintenance, poka-yoke, Next Operation as Customer (NOAC), supply chain management, total value engineering, and cycle-time reduction. By contrast, a whole set of tools—each fitted for a particular problem—constitutes a major advantage of the Ultimate Six Sigma approach. My Chapter 11 highlights these powerful tools for the twenty-first century.

7. *Design: A One-Dimensional Versus Multidimensional Design.* As in the case of tools, the only approach to design by the hyped Six Sigma consulting companies and their clients is the weak classical DOE ahead of production. There is no simultaneous engineering team structure, no emphasis on reliability, no focus on design for manufacturability, no mention of diagnostics, no attention to product characterization/optimization, and no emphasis on design cost reduction and design cycle-time reduction. All of these disciplines are an integral part of the Ultimate Six Sigma process, and are highlights in Chapter 12 on design.

8. *Suppliers: Win-Lose Contests Versus Win-Win Supplier Partnerships.* The hyped Six Sigma consultants are still in a Rip van Winkle slumber when it comes to supply chain management and the advances made in the last twenty years. Other than black belt training and pro forma audits extended to suppliers, there are no partnerships forged with key suppliers, no management infrastructure, no commodity teams, no early supplier involvement (ESI), no idea incentives, and no hands-on, active, concrete help to suppliers in improving quality, cost, and cycle time. These disciplines are vital in Ultimate Six Sigma and are emphasized in Chapter 13 on suppliers.

9. *Manufacturing: Breakthrough Manufacturing Strategy: Mirage versus Reality.* The hyped Six Sigma consultants introduce a so-called breakthrough strategy to their client companies with great fanfare. Underneath the glitter, it is the same warmed-over traditional problem-solving approach of recognize, define, measure, analyze, improve, and control. It is just two

small steps better than the PDCA (plan, do, check, act) cycle or one step better than the Ford 8-D problem-solving framework. They also use tired and discarded techniques such as gage R & R, control charts, and burn-in. There is no guidance on clues generated from talking to the parts, and there is no use of positrol, process certification, precontrol, poka-yoke, truncated MEOST, proof of improvement, field escape prevention, and cycle-time reduction disciplines. (These techniques are clearly described in Chapter 14 on manufacturing, as part of our Ultimate Six Sigma methodology.)

10. *Field a Black Hole Versus a Shining Star of Reliability*. Reliability is more important than quality. It has two additional dimensions—time and stress. The hyped Six Sigma consultants have no reliability targets or means of achieving them. In fact, reliability seems to be a black hole for them. By contrast, in the Ultimate Six Sigma, reliability is a shining star that burns bright in the customer's firmament. In terms of field support, the hyped Six Sigma crowd pays no attention to predelivery audits, customer/dealer/installer/servicer training, service effectiveness and speed, diagnostics (especially built-in diagnostics), and parts support. The Ultimate Six Sigma covers all these disciplines in Chapter 15 on field service.

11. *Support Services as an Appendage Versus Maximizing Service to All Stakeholders*. It is well known that quality, cost, and cycle time are in the dark ages in all white-collar areas as compared to manufacturing. The hyped Six Sigma crowd attempts marginal improvements in support services, such as determining customer needs and process mapping. Here, too, its philosophy is weak and—as expected—its tools are even weaker. For example:

❖ Hyped Six Sigma has not grasped the importance of the internal customer as the means to satisfy the external customer; consequently, it fails to strengthen the needed cooperation in the internal customer-internal supplier link.

❖ It does not go beyond kindergarten tools to solve white-collar problems and limits itself to elementary Pareto charts and cause-and-effect diagrams.

❖ It does not use out-of-box thinking or creative tools to achieve breakthrough improvements. One suspects that it is even ignorant of techniques to get these results, such as total value engineering, force field analysis, process redesign, and job redesign.

The Ultimate Six Sigma covers the whole service waterfront in Chapter 16.

12. *Results: From Mediocrity to World-Class Results*. If the game of bowling were played using industry norms, the bowling pins would be removed

so that there was no target to shoot at, team interaction would be discouraged, and the score would be kept by the supervisor who would demand that each person do better. The hyped Six Sigma results are couched in somewhat similar fashion. There are no firm targets for quality (except the diluted 3.4 ppm), reliability, cycle time, customer retention/longevity, c_{p_K}, cost/productivity, innovation, or even for some of the more traditional business parameters. On the contrary, the Ultimate Six Sigma not only measures each of these parameters, it expresses five levels of achievement for each—ranging from a primitive company's results to world-class levels. There is a list of four primary factors and fifteen primary metrics; thirteen secondary factors with 228 secondary metrics that constitute the most detailed measurements/results yardsticks—a highlight of Chapter 17.

Five

THE SCOPE, STRUCTURE, AND METHODOLOGY OF THE ULTIMATE SIX SIGMA

❖ ❖ ❖

*Human progress is seldom
linear. Often it is reversible.
But it comes through bright
and shining at the final
goal post. . . .*

OLD PROVERB

The Unfolding Tableau of the Ultimate Six Sigma

Having won the coveted Malcolm Baldrige National Quality Award and earned the plaudits of a grateful America for launching its Six Sigma, a question inevitably arose at Motorola: "What do we do for an encore?" There was general agreement, from the chairman down, that we could not

44

rest on our laurels. Success is often the father of failure. Winning companies tend to get complacent and flabby. Some colleagues suggested we try for the Deming Prize, which was being opened up to nonJapanese companies. Others wanted to try for another Baldrige Award after the mandated five-year waiting period. (The only company to win the Baldrige award twice is Solectron Corp., a contract manufacturer. It outperformed the Standard & Poor's 500 in terms of stock price appreciation by almost 10:1—a 425 percent increase in three years. Motorola was second, outperforming the Standard & Poor's 500 by 8:1—a 373 percent increase in five years.) Carried away by Six Sigma fever, other enthusiasts wanted to push the envelope to Seven Sigma, even Twelve Sigma. Bob Galvin, Motorola's then chairman, suggested that a few of us research extending the frontiers of product quality to all areas of a company—including the quality of leadership, the quality of employees, the quality of organization, and the quality of business processes.

As a result, I started working with Dr. C. Jackson Grayson—a great American who had served in the cabinet of two U.S. presidents and who persuaded the Congress of the United States, along with the American Society for Quality, to create the Malcolm Baldrige National Quality Award. Our goal was to go beyond the Baldrige Award, which was becoming somewhat ossified. Grayson, who is now the chairman of the American Productivity & Quality Center (APQC) in Houston, was most gracious with his suggestions and encouragement. My research with Grayson was the foundation of the Ultimate Six Sigma.

After forty-two years spent in an exciting and fulfilling career at Motorola, I decided to retire and formed my own consulting company, Keki R. Bhote Associates. My goal is to help companies worldwide solve and prevent chronic quality problems using the enormously successful Shainin/Bhote design of experiments, and other tools we had launched at Motorola in the 1980s. While this goal was achieved at many of the more than 400 companies where I've consulted, I came to the conclusion that companies needed to operate on a broader canvas than just quality perfection. They should, instead, reach for business perfection.

Genesis of the Ultimate Six Sigma

Motorola's Six Sigma was good. Hundreds of companies had come to learn about it and many copied it in various stages of implementation. Yet, we needed to take Six Sigma to graduate school. This became even more urgent because the hyped Six Sigma consultants were retrogressing their clients from college to high school! Six Sigma was degenerating into another quality fad. As a result, I escalated my research into going beyond Six Sigma. My clients liked the expansion and urged me to formalize it.

My work came to the attention of Jim Nelson, the far-seeing publisher of Strategic Directions of Switzerland. He encouraged me to write two booklets as part of a huge opus, *The Total Quality Portfolio*. These booklets were:

> *Plan for Maximum Profit: The 12 Critical Success Factors That Guarantee Increased Profits From Total Quality*[1]
>
> *Quality Project Alert: The Early Warning Signals for Any Quality Initiative in Danger of Costing More Than It Earns*[2]

The favorable reception accorded to these two volumes and the enthusiastic encouragement from one of my clients provided the launching pad for this book.

Scope of the Ultimate Six Sigma

Conventional Six Sigma differs from the Ultimate Six Sigma in the following ways:

❖ *Conventional Six Sigma stops at quality—the little Q.* The Ultimate Six Sigma expands its horizons to optimizing all facets of a business—the improvement of quality, cost, cycle time, and all business parameters—the Big Q.

❖ *Conventional Six Sigma is limited to customer satisfaction.* The Ultimate Six Sigma takes in the full sweep of not only customer satisfaction but also customer loyalty, a much more important objective, as well as employee loyalty, supplier loyalty, distributor and dealer loyalty, and investor loyalty.

❖ *Conventional Six Sigma concentrates primarily on product, with token overtures to management, employees, and business processes.* The Ultimate Six Sigma turns the searchlight on all twelve areas of a company's endeavor: customers, leadership, organization, employees, measurement, tools, design, suppliers, manufacturing, field operations (e.g., distributors/dealers/servicers), service (i.e., service industries and support services in manufacturing industries), and results. These twelve areas are intricately linked, as shown in Figure 5-1. An analogy can be drawn with a 12-cylinder engine. If just one cylinder misfires, the engine is likely to sputter. The Ultimate Six Sigma engine requires all twelve areas to be firing at peak performance.

Structure and Methodology of the Ultimate Six Sigma

The succeeding chapters of Part 2 (Chapters 6 through 17) are devoted to each of the twelve areas. Each chapter sharply contrasts conventional

Figure 5-1. The Big Q: the pursuit of business excellence and the twelve focus areas in the Ultimate Six Sigma.

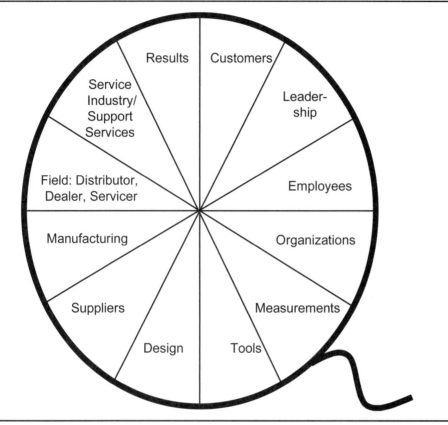

practices in industry today with the uplifting vision and the revolutionary content of the Ultimate Six Sigma.

Self-Assessment/Audit

A distinct feature of the Ultimate Six Sigma is its ability to measure the health of a company, not only in narrow terms of quality health, but in broader terms of business health. This is done by a self-assessment performed by a company's own management or by an external audit performed by outside examiners. The assessment is similar to the Malcolm Baldrige National Quality Award guidelines, but it is an order magnitude more comprehensive in scope and in content. It can also be used, longitudinally, to measure yearly progress of the company's march to business excellence and its dedication to creating an enveloping culture. Tables 5-1 through 5-3 can be used as worksheets for doing a self-assessment.

(text continues on page 51)

Table 5-1. The Ultimate Six Sigma self-assessment chart and scoring system.

Area		Key Characteristics	Importance (Points)
			125
1. Customer	1.1	Importance of Customer Loyalty	20
	1.2	Inviolate Principles of Customer Loyalty	25
	1.3	Customer Differentiation	30
	1.4	Customer Requirements	15
	1.5	Company Structure for Customer Loyalty	20
	1.6	Defection Management Control	5
	1.7	Public as Customer	10
			125
2. Leadership	2.1	Personal Philosophies and Values	60
	2.2	Corporate Principles	30
	2.3	The Corporate Role of Leadership	35
			75
3. Organization	3.1	Dismantling Taylorism	10
	3.2	Revolutionizing the Organizational Culture	65
			75
4. Employees	4.1	Motivation: From "How to" to "Want to"	20
	4.2	Job Security	5
	4.3	Empowerment-Readiness	5
	4.4	Team Competition	5
	4.5	Empowerment Systems	35
	4.6	Empowerment Stage	5
			75
5. Measurement	5.1	Measurement Axioms	5
	5.2	Measurement Principles	30
	5.3	Financial Statements	5
	5.4	Core Customers	10
	5.5	Generic Measurements	20
	5.6	Team/Department Measurements	5

6. Tools for the 21st Century	6.1	Design of Experiments	15
	6.2	Multiple Environment Over Stress Test	10
	6.3	Mass Customization/Quality Function Deployment	10
	6.4	Total Productive Maintenance	5
	6.5	Benchmarking	5
	6.6	Poka-Yoke	5
	6.7	Business Process Reengineering and Next Operation as Customer	10
	6.8	Total Value Engineering	5
	6.9	Supply Chain Management	5
	6.10	Lean Manufacturing/Inventory/Cycle Time Reduction	5
			75
7. Design	7.1	Organization for New Product Introduction	5
	7.2	Management Guidelines	5
	7.3	Voice of the Customer	10
	7.4	Design Quality/Reliability	25
	7.5	Design for Cost Reduction	15
	7.6	Design for Cycle Time Reduction	15
			75
8. Supply Chain Management	8.1	Importance of Supply Chain Management	15
	8.2	Supply Partnership Principles	10
	8.3	Types of Mutual Help	5
	8.4	Selection of Partnership Suppliers	15
	8.5	Infrastructure	15
	8.6	Supplier Development	15
			75
9. Manufacturing	9.1	Manufacturing Resurgence	10
	9.2	Quality Improvement in Manufacturing	40
	9.3	Cycle Time Reduction in Manufacturing	25

(continues)

Table 5-1. (Continued).

Area	Key Characteristics	Importance (Points)
		75
10. Field Operations	10.1 Product Reliability	15
	10.2 Predelivery Services	15
	10.3 Services to Downstream Supply Chain	20
	10.4 Services to User	25
		75
11. Service Industries/ Support Services in Manufacturing	11.1 Basic Principles of Next Operations as (NOAC)	30
	11.2 NOAC Structure	30
	11.3 NOAC Implementation	15
		75
12. Results—Primary	12.1 Customers	20
	12.2 Leadership	20
	12.3 Employees	20
	12.4 Financials	15

Table 5-2. Success factor rating.

Rating	Criteria
1	No knowledge of the success factor.
2	Only a conceptual awareness of the success factor.
3	Success factor started, with less than 50% implementation.
4	Success factor 50% to 80% implemented.
5	Success factor implementation more than 80% complete, along with reflected business results.

Table 5-3. Total rating: a corresponding business health and equivalent sigma level.

Total Company Rating	Equivalent Business Health	Equivalent Sigma Level
800–1,000	Robust health	6 Sigma
600–799	Good health, but periodic physical checkups (audits) urged	5 Sigma
400–599	Poor health; continual monitoring needed	4 Sigma
200–399	Major surgery needed	3 Sigma
Below 200	Terminally ill	2 Sigma

Case Studies of Benchmark Companies

Another distinctive feature of our Ultimate Six Sigma is a short case study of a benchmark company, in each of the twelve areas of an industry. These case studies, which are included in Chapters 6 through 17, are based partly on reputation and partly on my own experiences in over 400 consultations, but always on performance excellence.

ULTIMATE SIX SIGMA SELF-ASSESSMENT FORMAT

Each of the twelve areas depicted in Figure 5-1 is divided into key characteristics that are the salient features of that area. In all, there are sixty key characteristics. Each characteristic, in turn, is circumscribed by several success factor statements that elevate that characteristic to world-class standards. Table 5-1 is a capsule summary of the self-assessment, divided into the same twelve areas and further subdivided into key characteristics. In subsequent chapters, each key characteristic of a particular area is framed by three to twelve success factor statements that are necessary to propel that characteristic to a world-class level.

Scoring System and Importance Scales

The maximum score in the self-assessment is 1,000 points, distributed between the twelve areas, according to the importance of each area. Even though all areas contribute to business success, two areas—customers and leadership—are accorded the highest importance or weight of 125 points each. This is because without customers no company can exist, and without leadership no company can long survive. All the other areas—employees, organization, measurements, tools, design, suppliers, manufacturing, field operations, services, and results—are given a somewhat lower but equal importance, or weight, of seventy-five points each. This is to avoid endless arguments as to which of these remaining areas are more important or less important relative to one another.

(Note: It is not necessary that these importance scales be cast in concrete for all companies to slavishly follow. Several of my clients have modified these importance scales to better fit their unique corporate culture. But it is recommended that all divisions/businesses within a company adopt a uniform importance scale for purposes of comparing performances with a uniform yardstick.)

Each key characteristic in each area is assigned an importance weight (i.e., points). The characteristic points add up to the importance points of each area. This is also shown in Table 5-1. The success factors in each key characteristic are each given a maximum importance of five points. These success factors in each key characteristic are explained in detail in subsequent chapters devoted to each of the twelve key areas, so that there is no ambiguity of meaning to misguide a potential assessor.

Rating

Each success factor should be given a rating of one to five by assessors, with one being the worst and five being the best, using the criteria shown in Table 5-2.

The ratings of all the success factors in a key characteristic are added to give a rating subtotal for that characteristic. All the subtotal key characteristics are added to arrive at an area rating. Finally, all the area ratings are added to compute the total assessment score, for the company as a whole or for each of its plants, businesses, divisions, groups, or sectors. The twelve areas contain a total of sixty key characteristics (listed in Table 5-1) and a total of 200 success factors. With a maximum rating of five points per success factor, the maximum total rating possible is 1,000 points.

Table 5-3 shows how a company's total rating or assessment score can be interpreted in terms of a business health chart and as a recognition of sigma metrics. An interpretation of sigma levels can also be (nonmathematically) extrapolated.

PART 2

THE ULTIMATE SIX SIGMA—TWELVE AREAS OF BUSINESS EXCELLENCE

The twelve chapters in Part 2 cover the panorama of the Ultimate Six Sigma, not just for quality excellence, but for total business excellence in areas vital to industry.

❖

Eight: From Taylorism to Empowerment Creation in the Organization

Nine: From Passivity and Boredom Among Employees to Industrial Democracy

Ten: From Traditional Indicators to Robust Metrics

Eleven: From Obsolete Tools of the Twentieth Century to the Powerful Tools of the Twenty-First Century

Twelve: From Historic Levels to Designs in Half the Time With Half the Defects, Half the Costs, and Half the Manpower

Thirteen: From a Customer-Supplier Win-Lose Contest to a Win-Win Partnership for the Entire Supplier Chain

Fourteen: From Second-Class Citizen to Manufacturing as a Major Contributor to Business Excellence

Fifteen: Field Operations: From an Appendage to a Maximum Service to Downstream Stakeholders

Sixteen: From the Black Hole of Little Accountability to Service as a Productivity Contributor

Seventeen: From Mediocrity to World-Class Results

Six

From Mere Customer Satisfaction to Customer Loyalty

❖ ❖ ❖

*You may think that you
make products, but you
really make loyal customers.
You may think that you
make sales, but you really
make loyal customers.*

—Mack Hanan and
Peter Karp

Corporate Myopia—Failing to Perceive the Inadequacy of Customer Satisfaction

It is nothing short of amazing, in this age of so-called enlightenment when communications travel at the speed of light, to observe how many CEOs:

❖ Fail to see the woeful shortcomings of mere customer satisfaction

❖ Fail to witness the hemorrhaging defection rate of their customers

❖ Fail to recognize, much less fathom, the titanic leak of their profits caused by customer defections

❖ Fail to identify customers in the process of leaving the company

❖ Fail to take action in luring them back

❖ Fail to appreciate the synergy between customer loyalty, employee loyalty, supplier loyalty, distributor/dealer loyalty, and investor loyalty

Here are the facts on the depressing negatives of mere customer satisfaction:

1. According to a Juran Institute survey, more than 90 percent of the top Fortune 200 companies are convinced that maximizing customer satisfaction maximizes profitability and market share. Yet fewer than 2 percent are able to measure bottom-line improvements from documented increases in levels of customer satisfaction.

2. Of customers who say they are satisfied, 15 to 40 percent of them defect from a company each year.

3. In the U.S. auto industry, the average repurchase rate of satisfied customers from the same car company is less than 30 percent. The corresponding figure for the appliance industry is below 45 percent.

4. The defection rate of people over the age of sixty-five is 40 percent; for those older than thirty-five it is 60 percent; for those between ages twenty to thirty-five it is more than 85 percent.

5. There is little correlation between customer satisfaction and customer loyalty, as shown by the fat parallelogram in Figure 6-1.[1]

By contrast, here are the facts on the enormous positives of customer loyalty:

1. There is a very close correlation between customer loyalty and profitability, as shown by the very thin parallelogram in Figure 6-2.[2]

2. A 5 percent reduction in customer defection can result in profit increases from 30–85 percent, as shown in Figure 6-3.[3]

3. If customers increase their customer retention (the opposite of customer defection) by 2 percent, it is the equivalent of cutting their operating costs by 10 percent.

4. Loyal customers provide higher profits, more repeat business, higher market share, and more referrals than do just satisfied customers.

Figure 6-1. Customer satisfaction is not a predictor of customer loyalty.

Source: Keki R. Bhote, *Beyond Customer Satisfaction to Customer Loyalty—The Key to Greater Profitability* (New York: AMACOM, 1996).

5. It costs five to seven times more to find new customers than to retain customers.

6. One lifetime customer is worth more than $850,000 to a car company (based on 10 purchases of $25,000 each, multiplied by three referrals, plus $100,000 in profits from finance charges, parts, and service).

The sine qua non of corporate concern, therefore, should be a paradigm shift:

❖ From customer satisfaction to customer loyalty, customer retention, and customer longevity

❖ From zero defects to zero customer defections

Examples of the Highest Loyalty Leading to Meteoric Financial Success

Table 6-1 shows a sampling of some of the companies that have put customer loyalty on a pedestal and, as a result, reaped amazing financial suc-

Figure 6-2. Close correlation between customer loyalty and profitability.

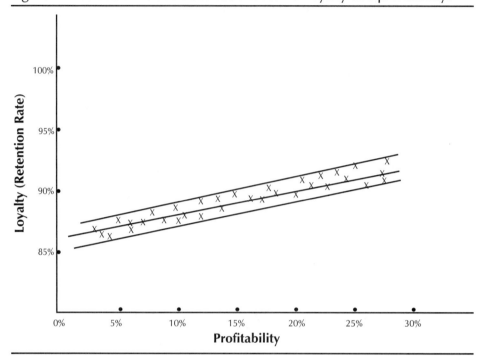

Source: Keki R. Bhote, *Beyond Customer Satisfaction to Customer Loyalty—The Key to Greater Profitability* (New York: AMACOM, 1996).

cess. Is there any longer a doubt that customer loyalty is the centerpiece of corporate priorities and the centerpiece of our Ultimate Six Sigma tableau?

Who Is a Customer?

The very visualization of who is, or should be, a customer has undergone a metamorphosis. Customer differentiation, mass customization, former customers, noncustomers, other stakeholders, and even the public carry today the appellation of "customer."

Not All Customers Are Worth Keeping: Customer Differentiation

Companies have been convinced for the past twenty years that not all suppliers are worth keeping. Yet it comes as a shock to most companies to realize that the same applies to customers as well. Let me illustrate this point. I was consulting with a company that was losing money. I asked its CFO how many customers were on the company's roster. Eight hundred was his reply. "How many of these 800 customers are profitable?" I queried. "About 200," he said. "Then, why do you keep most of the rest?" I parried. He answered with a straight face, "To keep the plant busy!" Un-

Figure 6-3. Profit increases with a 5 percent reduction in customer defections.

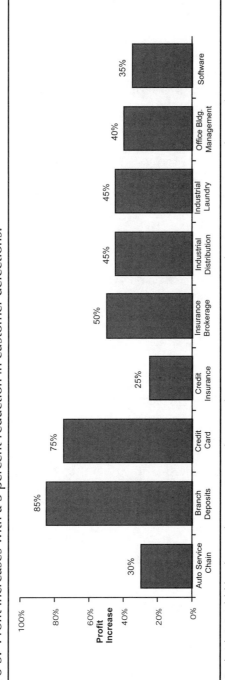

Source: Frederick F. Reichfeld and W. Earl Sasser, ''Zero Defects: Quality Comes to Services,'' *Harvard Business Review*, September–October 1990.

Table 6-1. Companies with the highest customer loyalty and reflected financial highlights.

Company	Highlights	Customer Retention Rate
Leo Burnett *Advertising*	❖ Highest Cash Flow—No bank debt ❖ Highest paid employees in industry ❖ Senior Managers spend majority of time with a single client	98%
MBNA *Credit Cards*	❖ From 38th place to Number 2 in industry ❖ Profit up 20 times in 9 years ❖ Earnings/Share Growth: 18%/year ❖ Investors get 30% return of equity ❖ Concentration on affinity organizations ❖ Bad loans: 3% vs. 6% for best competition	Highest in its industry
John Deere *Tractors*	❖ Focus on right dealers ❖ Seeks not just lifetime customers, but also generations of families ❖ Low mark-up on spare-parts customers more important than parts profit	77% (in 11 years 98%/year)
State Farm *Insurance*	❖ $20 billion capital-internally generated ❖ Covers 20% of nation's households ❖ Policies never canceled after national disasters ❖ Homes rebuilt for higher than policy ❖ Agents as partners—highest in industry	95%
Chick-Fil-A *Fast Food*	❖ From one restaurant to >600 ❖ Employee turnover: 4–6% vs. industry average of 40–50% ❖ Employees earn 50% over others in industry ❖ Financed by internal cash flow only ❖ Each store puts up on $5,000 in earnest money	???
USAA Insurance *Investment in Military*	❖ From $207 million in assets to $34 billion ❖ Yet employee base up on 5 times ❖ Employee defect rate from 43% to 5%	99%

Source: Keki R. Bhote Associates.

less a company is in the charity business for its customers, the rationale makes no business sense.

A. T. Kearney[4] uses a metal analogy to differentiate and segment a customer base. These distinctions into four categories of customers are illustrated in Figure 6-4 and outlined here:

❖ *Platinum customers* constitute 10 percent of the total customer volume of a company but contribute 30 percent of its profits. These are the crown jewels—the most loyal and the most difficult customers for competition to dislodge.

❖ *Gold customers* span 15 percent of the company's customer volume and provide 20 percent of its profits. They are almost as important as the platinum customers, but a company may not forge strategic alliances with them as it does with its platinum customers. To-

Figure 6-4. Platinum, gold, silver, bronze, and tin customers: customer differentiation and contribution to profitability.

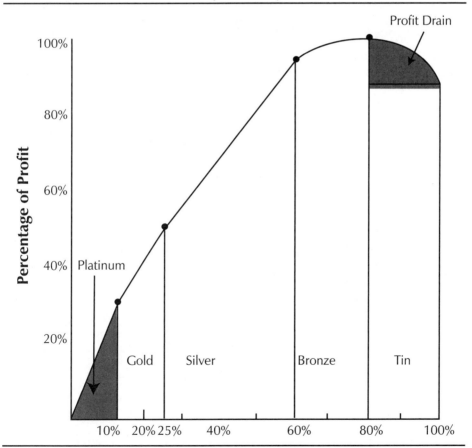

Source: A. T. Kearney, *The Company Satisfaction Audit* (Zurich: Strategic Direction Publishers, 1994).

gether, the platinum and gold customers comprise the core customers, and, according to Konosuke Matsushita, the founder of the
Matsushita Company empire, a company must "take [this] customer's skin temperature every day!"

❖ *Silver customers* account for 35 percent of the customer base and
about 40 percent of the company's profits. Relationships with these
customers are, at least, maintained if not assiduously cultivated.

❖ *Bronze customers* make up 20 percent by customer volume but barely
5 percent of the company's profits. For these customers, the company is on a slippery slope from black ink to red ink.

❖ *Tin customers* total 20 percent by customer volume and actually drain
profits by 15 percent. Often sentimental ties or the future potential
of converting them into profitable customers may hold a company
back from terminating them. At the least, an attempt should be
made to increase prices to facilitate their exit or enhance profitability.

Another classification of customers with varying degrees of loyalty has
been developed by Keki R. Bhote Associates as part of our Ultimate Six
Sigma approach. Table 6-2 delineates the distinction between various categories of customers—loyalists, ship jumpers, complainers, defectors, mer-
cenaries, whiners, gorgers, hostages, and terrorists—each with their
special characteristics, along with a correspondingly appropriate action required by the company.

From Mass Marketing to Mass Customization

Henry Ford transformed manufacturing in the twentieth century with
mass production. Mass customization may effect a similar revolution for
customers and manufacturing in the twenty-first century.

Over twenty years ago, the Toyota production system replaced mass
production with lean manufacturing. Instead of the old economy of scale
and large economic build quantities (EBQ), lean manufacturing engi-
neered innovations by which an EBQ of one could be as cost-effective as
large EBQ runs of 1,000 or 100,000.

Another movement, started thirty years ago, is quality function deployment (QFD). Its purpose is to capture "the voice of the customer"
rather than "the voice of the engineer," but QFD is still based on a foundation of mass marketing. There may be segmented customer bases, but
each base requires a large number of customers with the same options to
go through the maze known as "the house of quality."

Mass Customization

This movement started more than ten years ago. It recognizes that each
customer is unique and has highly individualized requirements—that is,

Table 6-2. A company's best friends and worst enemies among its customers.

Customer Category	Customer Satisfaction Score*	Likely Loyalty	Customer Characteristics	Required Company Action
Loyalists	5	100%	❖ There is a complete fit between their needs and a company's offerings. ❖ Their expectations are so exceeded that they become the company's missionaries and apostles.	❖ A company's crown jewels. ❖ Core customers whose needs, expectations, and ideas must be individualized and continually tapped.
Ship Jumpers	3 or 4	25% to 50%	❖ Their loyalty is only skin-deep. ❖ They jump ship the moment competition offers a lower price or other perceived benefits.	❖ 50% of them can be wooed back and converted to loyalists by showering attention on them.
Complainers	2 or 3	10% to 25%	❖ These are usually conscientious people who feel an obligation to help a company correct its problems.	❖ Acting promptly on their complaints and informing them of the specific changes made can win them back as loyal customers.
Defectors	1 to 2	<5%	❖ Turned off by negative experiences. ❖ Their complaints have been ignored.	❖ Even assessing the underlying reasons for their dissatisfaction, it may be too late for these customers, but the lessons learned are valuable for the future.
Mercenaries	3 or 4	<5%	❖ They chase low prices, buy on impulse, pursue fashion trends, or seek change for the sake of change.	❖ Little or no attention need be paid to these customers.
Whiners & Gougers	1	1%	❖ They are attention seeking; some have ulterior motives for wringing unwarranted concessions.	❖ Ignore or get rid of them.

(continues)

Table 6-2. (Continued).

Customer Category	Customer Satisfaction Score*	Likely Loyalty	Customer Characteristics	Required Company Action
Hostages	1	100% (Temporary)	❖ They suffer the worst experiences but have no alternative supplier; they are company-captive, but escape at the first loosening of this monopolistic stranglehold.	❖ Company must be sensitive to its monopolistic position and accommodate these customers.
Terrorists	1	0%	❖ They mount a vendetta or crusade against the company in media, community, and courts.	❖ Measured response to legitimate problems is needed. ❖ Do not cave in to blackmail.

*1 = Very dissatisfied, 2 = dissatisfied, 3 = Neutral, 4 = Satisfied, 5 = Extremely satisfied
Source: Keki R. Bhote Associates. Copyright © 1999.

each customer today wants exactly what he wants, where he wants it, when he wants it, at prices he wants. Mass customization turns topsy-turvy Henry Ford's famous comment of promising customers any color car they want as long as it is black!

Mass customization conducts a dialogue with each core customer, binding producer and consumer together. "It is a switch from measuring market share to measuring a customer's lifetime value."[5] This is made even easier by the Internet, which allows the "democratization of goods and services."[6] The challenge for a company, then, is to produce a quantity of one as economically as a much larger volume with the help of information technology (IT), computer-integrated manufacturing (CIM), flexible manufacturing systems (FMS), and just-in-time (JIT) practices. Companies embracing mass customization now number in the hundreds. Among the prominent ones are Dell Computer, Levi Strauss & Co., Mattel, Paris Miki, Express Custom Tailors, Hewlett-Packard, Motorola, and The Ritz-Carlton Hotel Company. Yet, in the rush to mass customize, a few principles must be kept in mind if the proliferation of customers, options, and costs is to be avoided:

❖ Reduce the number of customers (as discussed in the section on customer differentiation).

❖ Reduce the number of options, using an ABC cost analysis. Otherwise, the number of permutations on a product family can soar to hundreds of thousands.

❖ Postpone product differentiation until the latest possible stage in the supply chain. The best such stage is product differentiation by the customer, the second is at the distributor/dealer level, the third is in final assembly, and the worst is at the start of production.

❖ Modularize the design, LEGO-style, for easy, rapid, and reliable assembly.

❖ Maximize standardization—or at least commonality—up to the point of differentiation.

❖ Produce modules in parallel (simultaneously) instead of in a series (sequentially).

❖ Carry JIT/lean production to your suppliers to reduce inventories and prevent stock-outs at the same time.

❖ Automate order entry and programs for instant changes in options (for example, using PROM and EPROM chips).

❖ Reduce cycle time to customers by 10:1 and 100:1.

❖ Increase inventory turns with "pull" systems and Kanban, up to fifty and more.

Former and Defecting Customers: A Leaky Bucket With a Rusty Bottom

Customer defection is a cancer on a corporate body that goes undetected until it is too late. I have heard company executives confidently state, "We have customers waiting in line for every unit we can produce. So why do we need to bother with some defecting customers?" The analogy of a leaky bucket is useful to illustrate this myopia. Water is being pumped in at the top. At the bottom is a leak. At a given moment in time, inflow may exceed outflow, but the flow can also reverse itself at any moment, resulting in a situation sans water, sans profit, and sans customers! A survey conducted by the REL Consulting Group on customer defections at Fortune 500 companies revealed these shockers:[7]

❖ Sixty-one percent of companies surveyed felt that customer defections would have an insignificant impact on sales.

❖ More than 33 percent of companies surveyed made no attempt to identify defecting customers.

❖ Twenty-five percent of companies surveyed did not ask defecting customers their reasons for leaving.

❖ Thirty-three percent of the surveyed companies took no action on defecting customers.

The bottom is falling out of the leaky bucket!

It is well known that more than 90 percent of unhappy customers do not complain. They vote with their feet and go to the competition. Even worse, a truly unhappy customer is likely to tell anywhere from fifteen to twenty other potential customers of his unhappy experience. At least three or four of these potential customers will listen to this word-of-mouth bashing and remain noncustomers. On the other hand, "a customer who experiences a problem that a company corrects usually ends up being a loyal customer, more than one who didn't have a problem. Moreover, if customers feel that a company has reacted rapidly to their concerns, they will be even more loyal."[8]

Later in this chapter, a whole section is devoted to the organizational infrastructure and the "SWAT team" actions needed to focus on combating customer defections with laser-like intensity.

Noncustomers: A Corporate Black Hole

Paradoxically, a fourth category of customers is the noncustomer. For most companies, this category is off their radar screen—the corporate equivalent of a cosmic black hole. The following is a list of contributing causes that result in noncustomers:

* ❖ Lack of knowledge of the product or service offered by the corporation
* ❖ Lack of advertising of products or services
* ❖ Perceived poor value relative to the price paid for products or services
* ❖ Poor corporate public image (the result of unfavorable media publicity—indictments, recalls, lawsuits, environmental infringements, safety lapses, corporate arrogance, and so on)
* ❖ Dissatisfaction with one element in the supply chain (such as an unfavorable experience with a distributor, dealer, or servicer on a similar product, with the anger transferred to the manufacturer)
* ❖ Bad experience with one product transferred to the brand as a whole
* ❖ Bad-mouthing of friends and relatives

Traditional market research bypasses this key group. Management is too preoccupied with quarter-by-quarter profits and the tyranny of the financial analysts to pay much attention to attracting the noncustomer. But it is well known that people learn more from failure than from success. Companies can learn much from their failure to capture noncustomers instead of basking in the warm glow of short-term success.

The Internal Customer—If Not a King, At Least a Prince

Companies respect—if not worship—their external customers. But their internal customers—the next operations that receive the work of the previous processes in manufacturing, business, or design—are often treated like dirt. As an example, manufacturing is logically the customer of engineering, but engineering often looks upon manufacturing as a second-class citizen.

In the Ultimate Six Sigma, the internal customer is elevated—if not to the status of a king, as is the external customer, at least to that of a prince. The internal customer specifies his requirements to the internal supplier and, once agreement is reached between them, the internal supplier must meet these specifications. The internal customer now becomes the scorekeeper and the evaluator of the internal supplier's performance. In progressive companies, internal customer appraisals are replacing older boss appraisals in determining merit increases and promotions. Sometimes, the internal customer can terminate the internal supplier's services for continued substandard performance—and, in extreme cases, can even go outside the company for service.

The internal customer, along with its metric (cycle time) is fully explored in Chapter 16, which focuses on white-collar work in support services.

Other Stakeholders

Quite often, the other stakeholders of a corporation—employees, suppliers, distributors, dealers, servicers, installers, as well as stockholders—are not looked upon as customers, either. Yet, in a larger sense, they too are customers who must be served. The interactions between these groups must be recognized, their symbiotic relationships encouraged, and their synergy nurtured. It is well known that you cannot have loyal customers without having loyal employees and loyal suppliers.

The Public

The last category—the public—is seldom perceived by a company as its customer-at-large. Yet a company derives its legitimacy from the community and country in which it operates. In an August 2000 BusinessWeek/Harris Poll, 95 percent of those polled indicated that U.S. corporations should have more than one purpose—profit. They should sacrifice some profit to make things better for their workers and their communities. Public perception of a company as a corporate citizen is a collage of the following factors:

❖ Ethics/integrity (e.g., the corporation's dealings with stakeholders, government, and the community)

❖ News media coverage (positive or negative)

❖ Employment (e.g., issues such as job stability, minimal downsizing and layoffs, fair wages, fringe benefits, health insurance, pension policies, working conditions, safety, and employee morale)

❖ Nondiscrimination in hiring, retraining, and promoting minorities

❖ Environmentalism (e.g., stance on nonpollution and waste management)

❖ Adherence to requirements of regulatory agencies

❖ Public perceptions of value (i.e., benefit to cost ratio) vis-à-vis competition

❖ Legal standing (e.g., the company is free from lawsuits and recalls)

❖ Discharging social responsibilities (e.g., through community service, charity, urban development, education, and health policies)

❖ Image as a multinational (e.g., respect earned in countries where the corporation's plants/offices are located)

Inviolate Principles of Customer Loyalty

Having shed light on the many faces of a customer and the vital importance of customer loyalty, we can now enumerate several inviolate principles of customer loyalty. Glossing over them or ignoring them will lead to customers switching to competition. (In this context, only core customers are considered.)

Five Inviolate Principles

1. There is a firm commitment to partnership between the company and its core customers—a win-win bond, not a win-lose contest.

2. The partnership is based on ethics, mutual trust, and mutual help. (The hallmarks of a true leader are covered in greater depth in Chapter 7.)

3. The company concentrates on all elements of customer "wow," with a special focus on any element missing from the company's product line or services and on unanticipated features that generate customer delight. (My students in graduate school have dubbed this Bhote's Law. It states, "It is that element of customer 'wow' missing from a company's product or service to which top management must pay maximum attention.") Figure 6-5 illustrates the elements of customer "wow" associated with products; Figure 6-6 deals with services.

4. The company's top management pays regular visits to the actual users of its product and services. Its executives spend time with customers and develop close personal relationships with them. Table 6-3 is a good test to determine what percentage of time top management spends per week on customer care versus internal preoccupations. If the former ex-

Figure 6-5. A network of elements of customer "wow"—for products.

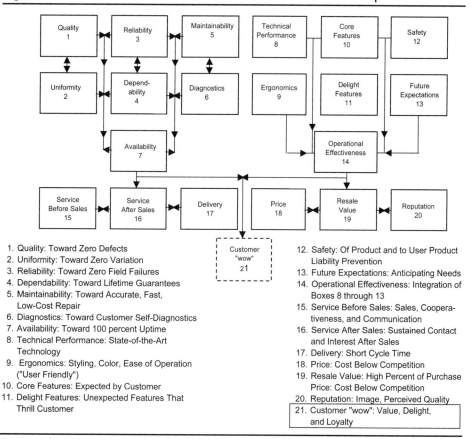

1. Quality: Toward Zero Defects
2. Uniformity: Toward Zero Variation
3. Reliability: Toward Zero Field Failures
4. Dependability: Toward Lifetime Guarantees
5. Maintainability: Toward Accurate, Fast, Low-Cost Repair
6. Diagnostics: Toward Customer Self-Diagnostics
7. Availability: Toward 100 percent Uptime
8. Technical Performance: State-of-the-Art Technology
9. Ergonomics: Styling, Color, Ease of Operation ("User Friendly")
10. Core Features: Expected by Customer
11. Delight Features: Unexpected Features That Thrill Customer
12. Safety: Of Product and to User Product Liability Prevention
13. Future Expectations: Anticipating Needs
14. Operational Effectiveness: Integration of Boxes 8 through 13
15. Service Before Sales: Sales, Cooperativeness, and Communication
16. Service After Sales: Sustained Contact and Interest After Sales
17. Delivery: Short Cycle Time
18. Price: Cost Below Competition
19. Resale Value: High Percent of Purchase Price: Cost Below Competition
20. Reputation: Image, Perceived Quality
21. Customer "wow": Value, Delight, and Loyalty

Source: Keki R. Bhote Associates.

ceeds the latter in time, the company is well positioned on the road to achieving customer loyalty and long-term profit.

5. The company displays genuine interest in the customer long after the sale is consummated.

Loyalty Metrics

When first introduced to customer loyalty and its importance, companies commonly ask, "But how do you measure loyalty? We can measure customer satisfaction readily with questionnaires and surveys." The answer is that it is actually easier to measure loyalty with two overall figures: customer retention and the longevity of that retention. Customer surveys, on the other hand, are most unreliable in predicting customer behavior and often subjective and capable of manipulation. Table 6-4 is a detailed list of customer loyalty metrics—both quantitative and qualitative.

Figure 6-6. A network of elements of customer "wow"—for services.

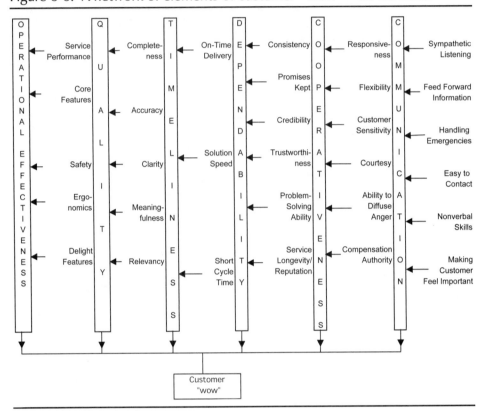

Source: Keki R. Bhote Associates.

Infrastructure to Capture Maximum Customer Loyalty

Following the metrics by which customer loyalty/retention can be assessed, the next step is to build the infrastructure within the company to develop and sustain a customer loyalty culture. Elements critical to this infrastructure are:

1. *Vision and Mission Statements*. In some companies, such statements may be platitudinous. In others, they may truly reflect the purpose and focus of a company. One acid test would be to see if the words *customer* and *customer loyalty* are part of such statements. If not, the company is handicapped at the starting line. In my book *Beyond Customer Satisfaction to Customer Loyalty—The Key to Greater Profitability,* I give an example of a typical customer loyalty statement. It reads:

> To create and nurture loyal customers who have received—and perceived that they have received—added value from us; to retain their loyalty over many years; and to do all this in a manner

Table 6-3. Profile of hours/week spent by top management on customer care versus other activities.

Activities	Hours/Week*
1. Personal contact with customers	_____
2. Feedback from customers on company's performance	_____
3. Determining what customers, former customers, and noncustomers want	_____
4. Feedback from customers on company's performance	_____
5. Recognizing outstanding customer service employees	_____
6. Cutting costs	_____
7. Meetings with other executives and managers	_____
8. Talking to stockholders and analysts	_____
9. Influencing technical improvements	_____
10. Policy and strategic planning	_____

*If the total hours in the first five items is greater than the total hours in the second five items, top management is voting for customer care with that most precious resource—time.

Table 6-4. Customer loyalty metrics.

Quantitative Metrics

❖ Maintenance ratio (number of customers retained to number defected)
❖ Defection rate (defecting customers as percent of total number of customers)
❖ Amount and continuity of core customers (by number, by dollars, or by time)
❖ Longevity of core customers (total sales generated over several years)
❖ Value to core customers (dollars saved for core customers by the company)
❖ Satisfaction (in satisfaction surveys with a 1–5 rating, use only "5" = Very satisfied)
❖ Repeat purchases by core customers
❖ Referrals by customers
❖ Percent of total company product range (brand loyalty)
❖ Customer Loyalty Index

Qualitative Measures

❖ Customer sharing her financial objectives/concerns with the company
❖ Customer offering constructive ideas/suggestions
❖ Customer loyalty despite less advantageous conditions
❖ Frequency of customer contacts with company
❖ Frequency of company contacts with core customers, before and after sale
❖ Amount of time spent by company with core customers
❖ Information on competition shared with company

that will lead them to share their positive experiences with others.

2. *A Customer Loyalty Steering Committee.* Generally, a company has a few steering committees to guide it, including business process reengineering and supply management. The most important steering committee, however, is the one charged with advancing customer loyalty. This committee should have no less a representation than the CEO and his direct reports. Its tasks should include:

- ❖ Establishing targets for maximum defection of core customers, minimum retention rates, desirable longevity (in years), and extension to brand loyalty
- ❖ Quantifying the lifetime value of a customer and its reciprocal—the lifetime loss of a defecting customer
- ❖ Appointing a "customer czar"
- ❖ Elevating the scope, importance, and effectiveness of customer-contact employees
- ❖ Appointing a customer defection SWAT team
- ❖ Reviewing progress against customer targets
- ❖ Tying incentives to increasing retention rates for senior management
- ❖ Promoting synergy between customer, employee, supplier, and distributor/dealer

3. *Chief Customer Officer (CCO).* This person is the embodiment of the mind, heart, and soul of the customer within the company. The CCO should be next in rank only to the CEO—ahead of the chief operations officer (COO) and chief financial officer (CFO). Companies without such a focal point may find themselves rudderless in navigating rough customer seas. It is also worthwhile to make the quality function the customer's advocate, as was done at Motorola.

4. *Customer Defection Management SWAT Team.* In basketball and football games, fans usually start a chorus: "Defense." It is not only in sports but also in companies that defense is important—defense against customer defections. Organizationally, a company should create a customer defection SWAT team made up of the company's best marketers and problem solvers. This should not be an ad hoc group, but a permanent entity because customer loyalty—like quality—has no finish line. Its most important task is gathering, analyzing, and reducing complaints primarily associated with core customers. Table 6-5 lists the most typical problem areas reported by customers, their causes, and the most appropriate tools

Table 6-5. Typical customer-reported problems, their causes, and corrective tools.

Area	Causes	Appropriate Corrective Tools
Poor product quality	❖ Poor design ❖ Poor manufacturing ❖ Poor material from suppliers ❖ Poor workmanship	❖ Design of experiments (DOE) ❖ Design of experiments; total productive maintenance (TPM) ❖ Design of experiments ❖ Poka-yoke
Poor product reliability	Designs not robust with time and field stresses	Multiple Environment Over Stress Testing (MEOST), DOE, and derating
Product liability potential	❖ Poor design for human and product safety ❖ Customer misapplication	❖ Product liability prevention (PLP) ❖ "Misapplication proof" design; warning label
Poor field repair service	❖ Unmotivated repair staff ❖ Poor training and tools ❖ Parts nonavailability	❖ Job redesign; reengineering the repair process ❖ Built-in diagnostics ❖ Reliability improvement, reducing necessity for spare parts
Customer misuse	Instruction not read or followed	❖ Robust designs, to reduce impact of noise factors ❖ Warranty labels; fail-safe designs
Customer non-comprehension	Nonuser-friendly features	Ergonomics and training
Late delivery	Poor forecasts; master schedules; MRP II; equipment breakdowns; supplier delinquency	Pull systems, just-in-time (JIT); TPM; small lots; reduced setup times; supply management
Disconnected distributors/ dealers	Limited loyalty to company	Distribution/dealer/customer councils; supply chain management
Order processing, billing, accounts receivable errors	Order inaccuracies, backorders, poor order tracking, routing errors, and returns processing	Flowcharting; Next Operation as Customer (NOAC); business process reengineering

(continues)

Table 6-5. (Continued).

Area	Causes	Appropriate Corrective Tools
Customer-contact employees not empowered	❖ Untrained, unmotivated, and underpaid employees	❖ Customer sensitivity training, "moments of truth" principles reinforced; more management attention (Hawthorne effect); pay incentives
	❖ Company rules and regulations as a straitjacket	❖ De-emphasis of all rules that do not or adversely affect the customer
	❖ Financial adjustments to customers unheard of	❖ Authorizing employees (up to certain limits) to compensate angry, dissatisfied customers
Dictatorial management	Management by fear; micromanagement	Management must change to leadership. Only the board of directors can effect the transition.

and techniques to correct them. Strange as it may seem, many of these techniques are not even known to most companies, much less used by them.

Another essential task of the SWAT team is to keep a constant finger on the pulse of core customers. That means knowing their changing needs and priorities; their experiences with the company's products, services, and sales/service/dealer/installation personnel; their future expectations; their referrals to other potential customers; and their involvement—in the case of original equipment manufacturer (OEM customers)—in new designs.

CASE STUDY

Lexus—A Benchmark Company in the Area of Customer Loyalty

Table 6-1 listed a few companies that excelled in customer loyalty, with retention rates of more than 90 percent per year. In the automotive market, one product—the Lexus car—stands head and shoulders above all other car manufacturers' products when it comes to customer loyalty.

Results

❖ While the retention rate for the average U.S. car company is below 30 percent, the retention rate for Lexus—Toyota Motor Corp.'s luxury car—is 75 percent (that is, 94 percent per year in a typical four-year life with a single owner). The corresponding retention rate for BMW is 43 percent and 42 percent for Mercedes.

❖ Lexus accounts for only 2 percent of Toyota's sales, but 33 percent of its profits.

❖ Its trade-ins are the highest in the industry. In fact, customers have difficulty buying a used Lexus.

Strategy

❖ Lexus studied the total car-owner cycle—from shopping, buying, driving, and servicing to trade-in—to optimize value to the customer at each step.

❖ It targets Cadillac and Mercedes customers, who have high brand loyalty.

❖ It believes that the key to customer loyalty is a loyal partnership with its dealers.

❖ It focuses on less than 200 dealerships, compared with a market saturation of up to 600 dealerships among other car companies. The greater volume per dealer affords lower markups and higher profits.

Profitable Service

❖ Lexus service charges are lower than its competition because of high service volume.

❖ Its post-warranty service captures 80 percent of its customers versus industry figures of 30–40 percent.

❖ It provides intensive training of service personnel—more personnel are trained in one month than in eighteen months with Cadillac.

❖ Its fewer dealer locations are compensated with free loaners and free pickup and delivery for servicing customers from their homes.

❖ It proves that a dealer with lower retention rates on customer-paid service loses $500,000 per year versus a comparable dealer with higher service loyalty.

❖ It has teams of service consultants sent to dealers who underperform on service profitability.

❖ It teaches dealers how to failure-analyze defection rates.

❖ Lexus is the only car in the luxury car market that tracks service retention rates for every dealer, every day.

❖ Lexus has the best parts inventory management, with higher parts availability and lower inventory costs—an average of $100,000 compared with the industry average of $200,000.

SELF-ASSESSMENT/AUDIT ON CUSTOMER LOYALTY

Table 6-6 is a self-assessment /audit that a company (or its external Ultimate Six Sigma auditors) can conduct to gauge how well it measures up in the key area of customer loyalty. It lists seven key characteristics and twenty-five success factors (each worth five points) for a maximum total score of 125 points. The term *customer* in Table 6-6 refers mainly to core customers.

Conclusion

Of the twelve areas devoted to the Ultimate Six Sigma, this area of customers and winning their loyalty is the most important. The words of Mahatma Gandhi, a giant among the outstanding leaders of the twentieth century, ring as true today as when he wrote them seventy years ago. His quotation hangs on the walls of most CEOs and division managers all over India:

> A customer is the most important visitor on our premises.
> He is not dependent on us. We are dependent on him.
> He is not an interruption of our work. He is the purpose of it.
> He is not an outsider in our business. He is part of it.
> We are not doing him a favor by serving him.
> He is doing us a favor by giving us an opportunity to do so.

Table 6-6. Customers: Key characteristics and success factors (125 points).

Key Characteristic	Success Factors	Rating				
		1	2	3	4	5
1.1 Vital Importance of Customer Loyalty	1. The company fully recognizes the importance of customer loyalty/retention; and highlights it in its vision/mission statements and its key objectives. 2. The company has quantified its customer/defection rate and customer longevity and computed the correlation of customer loyalty and profit. 3. The company has made customer loyalty so pervasive that every employee understands its importance and pursues it in his domain. 4. The company diligently emphasizes only those activities that benefit the customer and jettisons, as far as possible, those only needed for internal use.					
1.2 Inviolate Principles of Customer Loyalty	1. There is a firm commitment to partnership between the company and its customers—a win-win bond, not a win-lose contest. 2. The partnership is built on a foundation of ethics and uncompromising integrity, mutual trust, and mutual help. 3. The company adds measurable value to customers—as perceived and evaluated by them. 4. The company concentrates on all elements of customer "wow" with a special focus on missing elements important to customers and unanticipated features that generate customer delight. 5. The company's management develops close personal relations with the customer by: ❖ Paying regular visits to the actual users of its products/services ❖ Spending 50 percent of its time on customer care rather than internal matters ❖ Keeping a finger on the pulse of the customer ❖ Displaying genuine interest in the customer long after the sale is consummated					

(continues)

Table 6-6. (Continued).

Key Characteristic	Success Factors	Rating				
		1	2	3	4	5
1.3 Customer Differentiation	1. The company has recognized that not all customers are worth keeping and makes a concerted attempt to terminate customers with no potential of becoming core customers. 2. The company segments its customer base to concentrate on its core customers, especially its platinum and gold customers. 3. The company considers former and defecting customers a top priority to recapture their loyalty. 4. The company analyzes the reasons it has failed to attract noncustomers and mounts a drive to capture them and even convert them into loyal customers. 5. The company considers customer-contact employees most important, second only to external customers; it carefully selects, trains, compensates, and empowers these employees to serve the external customer. 6. The company recognizes that you cannot have customer loyalty without employee loyalty; it expands its customer reach to earn the loyalty of employees, distributors/dealers, servicers, and installers.					
1.4 Customer Requirements	1. The company pursues mass customization to fulfill the individualized needs of its core customers, thus minimizing their defection to competition. 2. The company makes mass customization effective through information technology, computer-integrated manufacturing, and lean production. 3. For its nonindividualized customers, the company uses the discipline of quality function deployment (QFD) to capture the voice of the customer and translate them into objective, quantified engineering specifications.					

1.5 Company Structure for Customer Loyalty	1. The company establishes a top-level steering committee to guide and steer it toward maximum customer retention. 2. The company appoints a customer czar or chief customer officer (CCO) to be the customer's focal point in the organization. 3. The company creates customer / distributor / dealer councils as sounding boards to strengthen its customer loyalty. 4. The company creates incentives for its management that are tied in with increases in customer retention rather than just sales.						
1.6 Defection Management Control	1. The company establishes an interdisciplinary high-level SWAT team to analyze the causes of potential customer defections and institute corrective action, using the powerful new tools to convert defecting customers into loyal customers.						
1.7 Public as Customer	1. The company has a reputation as a good employer in terms of jobs, minimal layoffs/hiring fluctuations, fair wages, health insurance, pension policies, nondiscrimination, and uncompromising integrity in its dealings with parties internal and external to the company. 2. The company discharges its social responsibility with community service, charity, educational assistance, inner-city help, and leadership in environmental improvement.						

Seven

From Stifling Micromanagement to Inspirational Leadership

We have far too many managers in industry, but far too few leaders.

—John Kotter

Leadership—The Fulcrum of the Ultimate Six Sigma

Almost equal to the importance of customers is the role of leadership in guiding a company's destiny. This is especially true today, when the captain of the corporate ship needs to raise anchor from the safe harbor of business-as-usual and guide it through the turbulent and uncharted high seas of a volatile global economy.

It is more than coincidence that both the Malcolm Baldrige National Quality Award and the European Quality Award guidelines address the subject of leadership—rather than management—as a focus and award it one of the highest scores in their assessments.

While many books and hundreds of articles have been concentrating on management, this chapter, instead, concentrates on a few specific characteristics of leadership such as:

- ❖ The personal philosophies and values of a true leader
- ❖ Leadership principles
- ❖ The enlightened role of leadership in piloting a corporation

It must be stated, at the outset, that while leadership is generally associated with the top echelons of a company, it also extends to every level of corporate rank. It even extends to an individual contributor, who, though he may have no one reporting to him, is often looked upon as a true leader by colleagues with an influence far greater than a boss with statutory authority.

Furthermore, it is not an axiom that leaders are born, not made. Leadership does not come naturally to more than a few gifted people. It can be acquired, however, through discipline, conscious persistent effort, and practice.

The Demise of Management, the Birth of Leadership: Personal Philosophies and Unchanging Values

Ask any group of people within a company for their perception of the percentage who are managers versus the percentage who are leaders. The answer ranges from 90 percent to 99 percent for managers, but 10 percent to one percent for leaders! Why is it that *manager* is a dirty word and *leader* an ennobling entity? The answer may lie in the personal philosophies and unchanging values that distinguish managers from leaders. The differences are awesome and are highlighted in Table 7-1. Some of these philosophies and values need elaboration.

Ethics: A Bedrock of Relationships, an Anchor of Stability

Every major religion of the world has an ethical code that is amazingly common to all of them. Ethics are the compass that points mankind morally to true north. It has governed relationships among people, societies, religions, and countries for more than 8,000 years of civilization. It represents an anchor of faith and stability in a changing and uncertain world.

Yet the term *corporate ethics* appears to be an oxymoron. Stories of company after company cited for financial skullduggery, bribes, kickbacks, and environmental pollution fill the news media. These firms may get by, even win out in the short run, but they lose their corporate souls and their very existence in the long run. By contrast, a corporation with sterling ethics and uncompromising integrity is not only successful over time, but is held up by the public as a role model.

Table 7-1. Personal philosophies/values: manager versus leader.

Personal Philosophies/ Values	Manager	Leader
1. Ethics	Corporate ethics—an oxymoron	Vital; uncompromising integrity
2. Trust in employees	Minimal and not reciprocated	Abiding trust that encourages people to live up to that trust
3. Help/guidance	Limited and begrudging; results in stunted growth of people	Abundant; enabling people to reach their full potential
4. Freedom	Restricted; micromanagement	Permissive; encourages people to reach goals their own way—including making mistakes
5. Superordinate values	Finessed situationally	Constant, unchanging
6. Vision	Tunnel vision; blinkered	Leading on a path never traveled; true belief and missionary passion
7. Inspiration	Little or no followership	Charisma engenders committed, loyal followership
8. Authority/power	Derived by organizational position	Gives up power—and, paradoxically, gains real power
9. Warmth	Not considered essential	Radiant; genuine interest in people
10. Humility	In short supply	Apparent in thought, word, and deed
11. Listening skills	Hears, but does not listen	Focused, with absorptive power
12. Governing style	Boss	Coach, consultant, teacher, guru

"It is a fairly new realization for corporations," says crisis communications consultant Karl Fleming, president of Prime Time Communications, "but the right *ethical decision* is also the right *business decision*." In his article, Flemings quotes Bennet Davis: "A confluence of events has rewritten the rules of the marketplace, regardless of whether companies understand this or not."[1] Every corporation is now being channeled into ethical probity by four factors:

❖ *First, the public* is beginning to judge companies by their social per-formance—their impact on the environment and their role in aggra-vating or relieving social problems—as much as by their financial performance.

❖ *Second, consumers* have become shell-shocked by a continuous bar-rage of reported shenanigans, both governmental and corporate—from Watergate to Big Tobacco—indirectly killing millions of people the world over right up to the present day.

❖ *Third, crusading special interest groups* are exerting enormous pres-sure—from challenging even venerable institutions like the Interna-tional Monetary Fund and the World Bank to judgments against cartels—on firms to toe the ethical line on human, social, and envi-ronmental problems.

❖ *Fourth, the competitive, unsparing, and technologically sophisticated media* are motivating companies to be more honest. A company's ethical lapse can now be flashed to news outlets and brokerage houses globally, before a CEO can hurry back from lunch!

In this pivotal area of ethics, managers are not necessarily unethical, but they succumb to the siren song of the company's bottom line, rather than ethical probity, when they come to a fork in the road. On the other hand, a true leader will never compromise an all-consuming integrity to stoop to the lucre of an unethical decision. Like a sturdy oak, the leader stands tall in the eyes of employees.

Trust: The Self-Fulfilling Prophecy

When a teacher looks upon a student as stupid, the student responds by fulfilling the lowest expectations of the teacher. By contrast, when a teacher has faith in a student, trusts him, and encourages him to con-stantly improve, the student rises to the challenge. This is the principle of the self-fulfilling prophecy.

In the corporate world, unfortunately, trust is a rare attribute. Manag-ers don't trust their people. In return, the people don't trust their manag-ers. Companies don't trust the unions. The unions don't trust companies. Companies don't trust governments. Governments don't trust companies. Companies don't trust their suppliers and customers. The suppliers and customers don't trust the companies. Granted, trust is not a light switch that you can turn on instantly and fill a relationship with the warm glow of trust. Granted it is fragile, and one false move can undo it. Trust has to be nurtured patiently. It is a step-by-step interactive process. In the final analysis, it has to be earned.

If managers have little trust in their employees' ability and effort, they find themselves supervising sullen employees whose work attests to the manager's low expectations. As a result, the employees' trust in the manager is equally low. True leaders trust their people, have faith in their creativity, and encourage them to grow to their full potential. Bob Galvin, chairman emeritus of Motorola, Inc., relates that his father, Paul Galvin, who founded Motorola, "subjected me to a fierce discipline. He trusted me!"[2] Trust begets trust. It generates a strong compulsion in the one trusted to live up to that trust.

Help and Guidance for Employees to Reach Their Highest Potential

Lack of trust limits a manager's desire to help his people. Insecure that they may learn so much that his own position is threatened, he renders only token help. The result is the stunted growth of his own position.

Given trust as a prerequisite, a leader enthusiastically helps people to grow. This is especially emphasized in Japan, where the CEO looks upon his primary responsibility as "the care and feeding of the young." The type of help is not a straitjacket, master-to-apprentice regimen. Rather it is guidance with a loose rein: technical, administrative, strategic, or in the arena of human relations. It is said that the average person uses the God-given computer between his ears—the brain—to only 2 percent of its potential. A genius exercises it up to 10 percent. A leader—someone who provides trust, support, encouragement, training, and help—knows that there is no earthly barrier to human potential.

Freedom to Explore, Freedom to Make Mistakes

A poor manager, preoccupied with self-preservation and a sense of omnipotence, tends to micromanage people, breathing down their necks, directing them at every turn, and sometimes even castigating them for mistakes and taking credit for their success.

The true leader, recognizing that he is not God and does not have all the answers, gives his people freedom to explore their own pathways, create their own solutions, and even make their own mistakes. Paul Galvin, Motorola's founder, says: "Do not be afraid to make mistakes. Wisdom is often born of such mistakes. You will know failure. Determine now the confidence required to overcome it. . . . Reach out. . . ."[3] Freedom, however, does not mean anarchy. The leader has the duty to lead, set the direction, establish goals, and monitor results. Then, having done that, the leader gets out of the way.

Pivotal Values—Superordinate and Unchanging

Every corporation formulates a set of values, beliefs, and principles that serve as an internal constitution. A poor manager may pay lip service to such a set of values. He may even display those values on a plaque hanging on a wall. But in reality, the poor manager finesses these values situationally—that is, he applies those values according to whatever the situation calls for. He bends and shades these values, if needed, to maximize that sacred cow—shareholder value.

A leader looks upon pivotal values as the Rock of Gibraltar—constant and unchanging. Some examples of pivotal values include uncompromising integrity, respect for the dignity of the individual, and customer worship. The leader emphasizes a few simple values that give clear direction and purpose to a group's work. The leader knows that people can only focus on just a small number of such values. For instance, President Ronald Reagan stressed, over and over, a few crucial values: less government, lower taxes, stronger defense, a view of the Soviet Union as an evil empire. He took the people with him to a "morning in America" and "that shining city on a hill." Begrudgingly, historians are reassessing him as a great leader and a great president.

Radar Vision and Missionary Passion

A manager, tied down by hundreds of operational duties and daily firefights, has no time or little appetite for developing a vision for the future. His view is blinkered, his thinking short term, his approach pedestrian. Rodman L. Drake, managing director of a management consulting firm, was amazed when he queried a number of CEOs as to where they would like their companies to be five years from now. Most of them could not formulate or even express that vision.

A leader, in sharp contrast, has a clear sense of that vision. Bob Galvin defines a leader in succinct terms as "one who takes a company on a road that no one has traveled before. It's a manifestation of the grander changes in direction, whatever they may be—geographic, product, market, or the way of doing things." That clear, farsighted vision is accompanied by true belief and missionary passion.

Inspiration and Followership to the Ends of the Earth

In the annals of time, all the great leaders—political, military, religious—have one thing in common—an ability to inspire people, bind them into a common objective, consensus, and commitment, and earn their unquestioning loyalty.

The manager with a lack of trust, a lack of vision, and a lack of passion is incapable of arousing that fervor among the troops. A pathetic managerial quote brings to mind this style: "There go my people. I must follow them for I am their leader!"

The leader, on the other hand, marshals all his personal philosophies and values and converts them into a radiant charisma that so fires up people with crusading enthusiasm that they would follow him to the ends of the earth.

Table 7-1 differentiates five additional personal philosophies and values—authority, warmth, humility, listening skills, and governing style—without the need for further commentary. Of course, not all managers are tainted in dark, sordid colors, nor are all leaders anointed with a halo. There is many a time when a Theory Y leader steps into a Theory X manager. But it is also a proven fact that such Theory X (autocratic) managers can develop, with discipline and time, into first-rate leaders.

Principles—A Corporate Genetic Code

Principles define a corporation, just as personal philosophies and values define its leader. The following principles are among the most important, but are not articulated with sufficient emphasis or clarity in many companies.

The Betterment of Society: People, Animals, Plants, and Earth Moving Toward Perfection

In the past, companies have rarely considered their obligation to society beyond the important, but limited, role of providing jobs. But there has been a gradual realization that a company derives its legitimacy from the city, state, and country in which it operates. If it ill serves this community, it will have no long-range staying power. Chapter 6 illustrated that the last, but certainly not the least, definition of a customer is the public, because the public can bestow legitimacy on a business. If not today, perhaps within the first half of the twenty-first century, farsighted leaders will concentrate (and be evaluated) on nonbusiness parameters important to the public such as:

- ❖ Ethics (e.g., freedom from lawsuits and adverse news media coverage)
- ❖ Freedom from product liability court fines, both compensatory and punitive
- ❖ Contributions to charitable organizations and community services
- ❖ Active participation in inner-city amelioration
- ❖ Contributions to effective education, in schools and colleges
- ❖ Active participation in healthcare and genetics development
- ❖ Contributions to climate, environment, and waste control
- ❖ Protection of the animal and plant kingdom and the betterment of Mother Earth

Zarathushtra, the prophet of the world's first monotheic religion of Zoroastrianism, enunciated his vision of all God's creations moving toward perfection, with man playing a central role. Perhaps today's business world can become the catalyst.

Stakeholder Value—More All-Inclusive Than Conventional Shareholder Value

Companies traditionally have focused on shareholder value, not recognizing that there are other important customers who have a stake—perhaps an even more vital stake—than the stockholder. These stakeholders are the customer, employee, supplier, distributor, and dealer. Enlightened leaders balance all of these constituencies to maximize not shareholder, but stakeholder value.

Customer Loyalty and Retention—A Forward Look (Profit Is a Rearview Mirror Look)

Chapter 6 covers, in depth, the importance of customer loyalty and retention as one of the most effective ways to maximize corporate profits. This signals a new dawn for true leadership that must look for more important leading indicators of business success, such as customer loyalty and long-term retention, rather than focus on profit alone. Profit is a lagging indicator. Profit is a result, not a cause; it is an output, not an input.

Employee Loyalty—A Prerequisite for Customer Loyalty

It is an axiom that customer loyalty cannot be achieved without employee loyalty. Chapter 9 explores the many facets of achieving employee loyalty. People-oriented leaders pay special attention to:

- ❖ Job security, with layoffs kept to an absolute minimum
- ❖ Treating employees with respect and dignity
- ❖ Employee recognition, larger responsibility, achievement, and growth
- ❖ The challenge of the job itself
- ❖ Financial incentives for performance
- ❖ True, not token, empowerment

A Win-Win Partnership With Key Suppliers, Distributors, Dealers, and Servicers

The era of confrontation between customer companies and their key suppliers is baggage that should be left behind as a relic of the twentieth century. The game of the customer winning and the supplier losing, or vice versa, is a zero sum game. Progressive leadership in both customer and supplier companies recognizes and acts on two verities:

1. The objective of a supplier company is to create and nurture satisfied, repetitive, and loyal customers who perceive that they have received added value from the supplier.

2. The objective of a customer company is to establish a firm partnership with its key suppliers and render active, concrete help to them to reduce its own defects, cost, and cycle time while simultaneously increasing the suppliers' profits.[4]

Imaginative leadership extends the same concept of partnership to a company's distributors, dealers, servicers, and installers. Chapter 13 deals more fully with how to plant and nurture these partnerships.

Goals: Few, Reach-out, and Continuous

Another distinctive principle practiced by leaders is to select just a few goals crucial to the enterprise and then establish truly challenging targets with an ambitious timetable. If people are given mediocre goals, they will persist in the same old-fashioned ways of reaching them. With reach-out targets, they will have to shift conventional paradigms and explore entirely breakthrough approaches. (See the case study of Motorola under Bob Galvin as a benchmark in leadership later in this chapter.) Furthermore, these stretched goals should not be a one-time event, but repeated continuously and systematically.

The Enlightened Role of Leadership—Unleashing the Human Spirit

Fortified by high-minded personal philosophies, values, and principles, leaders are revolutionizing the roles of traditional top management in a number of profound ways that change the very psyche of an organization.

From Taylorism to Releasing the Locked-Up Genie

For the better part of the twentieth century, management was a prisoner to the theories of Frederick Winslow Taylor, who compartmentalized the roles of managers (who did the thinking) and workers (who executed the manager's bidding). Managers designed systems, procedures, and policies to assure that the workers became as robotic and reliable as the machines they ran. These systems, which were intended to assure control and conformity, inhibited creativity and initiative. Workers became passive.

Worse, Taylorism created antagonism among the workers, even subversion!

In recent years, management has softened the polarization of Taylorism, but it still clings to top-down planning and control with a strategy-structure-systems model that chains lower levels in the organization to the captivity of conformance and cynicism. It perpetuates employees using only 5–20 percent of their capacity at work, as contrasted with the remaining 80–95 percent with which they energize their home lives.

Enlightened leaders dismantle the strategy-structure-systems model to unleash the human spirit and make initiative, creativity, and entrepreneurship a priority for their employees. Goran Lindahl, ABB Ltd.'s executive vice president, states: "My first task is to provide the framework to help engineers and other specialists develop as managers; the next challenge is to loosen the framework to let them become leaders, where they set their own objectives and standards. When I have created the environment that allows all managers to transform themselves into leaders, we will have a self-driven, self-renewing organization."[5]

Renewal and Innovation to Infuse a Corporation With New Energy

Another essential element of a leadership role is renewal of the corporation—call it a periodic corporate blood transfusion. Renewal recharges a corporation's tired batteries and infuses it with energy. It begins with "abandoning yesterday," freeing resources assigned to things that no longer contribute to results. It is closely coupled with innovation and its windows of opportunity that assess:

- ❖ A company's unexpected successes and failures and those of its competition

- ❖ Incongruities between industry efforts and the values and expectation of customers

- ❖ Demographic shifts such as collapsing birthrates and disposable-income spending

- ❖ Economic returns for consumers twenty to thirty years from now (as opposed to short-range returns)

- ❖ Global competitiveness with labor costs playing an ever-decreasing role

- ❖ The growing incongruence between economic globalization and the political nation-state[6]

Innovation is also encouraged by industry leaders among their employees. The most famous example is 3M's "skunk works," where employees are given time and freedom to innovate with no controls other than to produce a viable, salable product within one year. As a result, over half of 3M's products are introduced in the first year of life.

Creativity—A Building Block of Innovation

Creativity is inherent in all human beings. Among the most creative are children. However, the education system, rules, and regulations of the adult world knock the starch out of this early creativity and sentence people to a lifetime of humdrum conformity. Yet creativity can be learned, just as leadership can be learned. Thus, another important role for the leader is to encourage creativity in all employees through deliberate training by removing an all-pervading fear, by stimulating the challenging of policies and practices that exist purely for internal control, as opposed to benefiting the customer, by promoting the freedom to experiment, and by celebrating and rewarding creative breakthroughs.

Training—The Key to Continuous Learning

Graduating from school or college is only a passport to learning. It is said that the half-life of an engineer's usefulness is only five years after graduation (perhaps even less as new technological breakthroughs succeed old technological breakthroughs with dizzying speed). The great W. Edwards Deming said, "If you do not learn something new every day, you might as well fold your tent on Mother Earth." Most people, however, need an external stimulus to learn. A leader's role is to provide that stimulus through training: university training, corporate classroom training, and on-the-job training. Progressive leaders are allocating 3–5 percent of a company's payroll for training. Some leaders make it mandatory for every single employee to be trained for a minimum of forty hours per year. Collectively, these leaders spend more on corporate training than the entire budgets of all the universities in the United States!

A Long-Term Focus—Ignoring the Tyranny of the Financial Analyst

Most corporations find themselves prisoners of the quarter-by-quarter stock watch. Their jailers are the almighty financial analysts. (A Bernard Baruch of an earlier era or a Warren Buffet of today couldn't care less about the scrutiny of this ilk, who have never had their hands dirtied in honest production or design but who pontificate on a company's prospects with "superior" tunnel vision!) CEOs are so mesmerized by the demands of these tyrants that they spend their valuable time catering to the financial analysts' whims. One CEO goes so far as to report the number of financial analysts that are positive, neutral, or negative in terms of the company's immediate future! The horizons of these CEOs rarely go beyond two years—the half-life of most of them. Why should they labor just to make their successors look good, they say.

True leaders, on the other hand, while mindful of the dictates of the stock market, do not hesitate to sacrifice short-term gain if sustained profits and 100 percent or 200 percent greater returns on investments can be achieved in the long term. Noteworthy examples are Cisco Systems, Sun Microsystems, BroadVision, Honeywell, and R. F. Micro Devices.

Quality of Work Life—Creating Joy in the Workplace

Recognizing that employees spend nearly 50 percent of their waking hours at work, true leaders concentrate on improving the quality of work life for their people. They do this by, first, driving out fear. In his famous fourteen points of advice to management, W. Edwards Deming states: "The economic loss from fear is appalling. The fear to speak out, the fear to be upbraided, the fear to be fired, cause the employee to withdraw into his cocoon of noninvolvement. His mind is numbed, his creative juices stop flowing, and the company is the big loser."[7]

Leaders recognize that a certain amount of drudgery and boredom is inevitable in assembly-line operations, both in manufacturing and business processes. They attempt to inject a degree of job excitement by facilitating both horizontal and vertical job enrichment, creating teamwork, and giving powerful tools to the workers (see Chapter 11) so that they experience the thrills of solving problems by themselves, and by making each employee a manager in her own area. The result is an atmosphere of joy in the workplace that even a casual visitor can sense.

From Corporate Leader to Civic Leader

In the final analysis, a true leader's role is not limited to his corporate domain. A true leader discharges social responsibility to the public: the community, state, nation, and even the world. This starts with creating the conditions by which the public perceives the company to be a model employer, with a waiting list of people seeking its employment. It continues with educating the public, and especially students, on the positive and dynamic role of free enterprise in shaping the world's future. It goes on to render service to the community in terms of charity, civic causes, education, and urban development. There is even a new frontier for the creative juices of a leader—tackling chronic social problems and converting them into business opportunities!

CASE STUDY

Motorola—A Benchmark Company in the Area of Leadership

James O'Toole, noted authority on leadership and vice president of the Aspen Institute, names the modern-day CEOs who are the corporate equivalents of the four presidents memorialized on Mount Rushmore. They are Herman Miller's Max De Pree, Corning's James Houghton, Scandinavia Airlines's Jan Carlzon, and Motorola's Bob Galvin. Writes O'Toole:

> Thanks to Galvin's understated leadership, Motorola has probably done the best job of any large U.S. corporation at institutionalizing change. It became the first large company in America to enable its workers to be leaders themselves. Power is widely shared without degenerating into anarchy. . . . The results of Motorola's system are well known: perhaps the highest quality products in American industry, regular introduction of innovation, and of course, high profits.[8]

Bob Galvin has been my mentor, inspiration, and guiding light in the fifty years of my discipleship under him. This case study is but a pale reflection of the floodlight he has shone on the trail of world leadership. Galvin's personal philosophies and values as a true leader (qualities listed in Table 7-1) are as follows:

❖ *Ethics.* Galvin's uncompromising integrity is a legend at Motorola. He will not countenance any bribe, any kickback, to any customer or government official in any country in the world, no matter how prevalent and accepted a practice it might be in that country or how serious a business loss it might entail. A high Chinese government official told me that of all the foreign companies in China, Motorola was the most respected for its honesty, integrity, and transparency.

❖ *Trust.* Galvin has always had an abiding faith in his people, which in turn has engendered a compelling challenge among his associates, as he calls them, to rise to his level of trust, to earn it. When he introduced the first 10:1 quality improvement goal, he claimed he did not know how this challenging five-year target—which no industry had achieved or even undertaken—would be met. But he had such faith in his people and their ingenuity that he was confident that they will find a way. And they did. Not once, not twice, but three times, with a 10:1, 100:1, and an almost 1,000:1 improvement in ten years!

❖ *Help and Guidance.* Galvin deeply believes that there is no limit to the human potential—not just for the brainy and privileged few, but for all, no matter how low their starting point. Galvin says, "We've been about the job of evangelizing, teaching, and causing more and more people to realize that leading—taking us elsewhere—is also part of their job." I call this cloning and cultivating leaders.

❖ *Freedom.* "My normal mode," Galvin says, "is to rarely be prescriptive, because I often don't know the prescriptive answers to questions . . . My disposition has been more characterized by 'why not' than by 'why.' More often I'd say, 'Why not do it your way?' " Galvin gives his people freedom, the space to spread their wings. He reinforces his father's famous quote, bannered all over Motorola: "Do not be afraid to make mistakes. Wisdom is often born of such mistakes." Yet he expects that this freedom should result in meeting or exceeding shared goals.

❖ *Vision and Renewal.* Galvin defines a leader as someone who takes his people into the unknown, untried, and unexplored with uncanny vision. He believes vision and renewal are intertwined— they are the twin spirits of leadership. His vision is of a Motorola that can grow ten times and more in the next twenty-five years. The key to that vision is renewal, the driving thrust of Motorola. Renewal has a special meaning, too. "It does not mean mission or strategy. It speaks to genuine energy. To me, the great energy of the institution is the energy of renewing." This means change—rapid and large change— with reach-out improvements, five to ten times in magnitude. But renewal also includes recommitting to values such as respect for the dignity of the individual and total integrity. Motorola's meteoric rise is a tribute to this vision and renewal. As vital opportunities presented themselves—from battery eliminators to car radios to walkie-talkies—the basic communication link in the fields of World War II; from television to semiconductors and microprocessors; from pagers to cellular phones and iridium; from communication links in outer space to exploration beyond the solar system— Motorola's commitment to renewal has served it well.

❖ *Inspiration and Followership.* Within Motorola and among the hundreds of organizations that he has addressed,

Bob Galvin is called a second Winston Churchill, not only for his silver-tongued oratory, but also for the uplifting inspiration of his message. More important, he genuinely attracts followership. He is a role model for American industry. Among Motorolans he is an icon. One facet of that worship is the fact that Motorola—a multinational company with plants in some of the countries with the strongest labor unions—has no unions anywhere in the world. Despite repeated attempts by unions to organize its employees, Motorola remains a nonunion tower of strength because its people know the implementation of Galvin's conviction that "the company can do more for its employees than any union ever can."

A Final Word

❖ Galvin has a final word for aspiring leaders. "Be selfless; it takes a lot of confidence to be selfless."[9]

SELF-ASSESSMENT/AUDIT ON LEADERSHIP

Table 7-2 is a self-assessment/audit—similar to the one in Chapter 6 on customers—that a company can conduct to score its leadership (as distinguished from management). It lists three key characteristics and twenty-five success factors, each worth five points, for a maximum total score of 125 points.

Table 7-2. Leadership: key characteristics and success factors (125 points).

Key Characteristic	Success Factors	Rating				
		1	2	3	4	5
2.1 Personal Philosophies and Values	1. Leaders have a code of the highest ethics and uncompromising integrity and transmit their unswerving adherence to ethics to their employees and to the public at large.					
	2. Leaders have abiding trust and faith in their employees, who reciprocate that trust and who rise to fulfill that trust.					
	3. Leaders actively help their employees, through coaching and guidance to reach their highest potential.					
	4. Leaders give their employees freedom to explore their own pathways to corporate goals, including the freedom to make mistakes along the way.					
	5. Leaders establish a small set of superordinate, unchanging values to guide the corporation as a "true north" compass in the uncharted industrial seas.					
	6. Leaders display *vision* to lead their companies on unexplored paths to explore an uncharted future with a sure sense of direction, accompanied by true belief and missionary passion.					
	7. Leaders so inspire their people with such crusading enthusiasm that they are ready to implicitly follow their lead.					
	8. Leaders relinquish formal power, and by so doing, acquire real power readily conceded to them by their people.					
	9. Leaders radiate a natural warmth and genuine interest in their employees.					
	10. Leaders, by thought, word, and deed, display humility, striking a receptive chord in their people.					
	11. Leaders develop superior listening skills with an ability to capsulize what they have heard and act on the information.					
	12. Leaders do not regard themselves as bosses, but as coaches, consultants, guides, and teachers.					

(continues)

Table 7-2. (Continued).

		Rating				
Key Characteristic	Success Factors	1	2	3	4	5
2.2 Corporate Principles	1. The ultimate corporate principle is to align its basic and prime responsibility as a good corporate citizen with the need for the betterment of society as a whole. 2. A corporation must look beyond its desire to maximize shareholder value to maximize stakeholder value—including customers, employees, suppliers, distributors, dealers, servicers, and investors. 3. A key corporate principle is to focus on customer loyalty and long-term retention as a leading indicator of success rather than profit—a lagging indicator. (See Chapter 6.) 4. It is a corporate axiom that maximum customer loyalty can best be achieved by maximizing employee loyalty. (See Chapter 9.) 5. An important principle in supplier relations is converting a win-lose contest between customer and supplier to a win-win partnership with key suppliers in order to: ❖ Achieve maximum quality, cost, and cycle-time improvements for the customer. ❖ Maximize profits for the supplier with active, concrete, and mutual help. (See Chapter 13.) 6. Corporate leadership must select a few goals vital for the company's growth and establish stretch targets, with a firm timetable.					
2.3 The Corporate Role of Leadership	1. Enlightened leaders dismantle the strategy-structure-systems model of management and make initiative, creativity, and entrepreneurship a priority for their employees and move them from employees to managers to potential leaders.					

2. Leaders renew their companies by "abandoning yesterday," capturing "windows of new opportunity," and through innovations, encouraging similar innovations among their employees.
3. Leaders encourage the learning and application of creativity among their employees by removing fear, providing training, and challenging practices solely for internal control rather than for the customer's benefit.
4. Leaders focus on training their employees, allocating a minimum of 3 percent of payroll to it and monitoring payback.
5. Leaders are less concerned with quarter-by-quarter profits demanded by financial analysts and concentrate on much higher long-term profits using leading, not lagging, indicators.
6. Leaders initiate a "quality of work life" atmosphere for their employees by designing horizontal and vertical job enrichment in their work, creating teamwork, and giving them powerful problem-solving tools.
7. Leaders look beyond fulfilling their responsibilities as good corporate citizens and actively render service to the community with civic projects, charity, education, urban development, and efforts to convert chronic social problems into business opportunities.

Eight

FROM TAYLORISM TO EMPOWERMENT CREATION IN THE ORGANIZATION

❖ ❖ ❖

*We will win and you will
lose! Your failure is an
internal disease. Your
companies are based on
Taylor's principles. Worse,
your heads are Taylorized
too. You believe that sound
management means, on one
side, executives who think;
and on the other, men who
can only work. But we
know that only the
intellects of all employees
can permit a company to*

live with the ups and downs

of global challenges. Yes, we

will win and you will lose.

For you are not able to rid

your minds of the obsolete

Taylorism that we

never had.

—KONOSUKE

MATSUSHITA,

FOUNDER OF THE

MATSUSHITA

INDUSTRIAL EMPIRE

I'm not going to have

monkeys running

the zoo. . . .

—FRANK BORMAN,

FORMER CHAIRMAN OF

EASTERN AIRLINES,

DISCUSSING WORKER

PARTICIPATION[1]

The yawning organizational divide, represented by these contrasting statements, has shrunk in recent years. Matsushita's "in your face" words belie the fact that Japan remains one of the most authoritarian industrial regimes in the world. Worker participation is never allowed to break through the glass ceiling of senior management. America (Frank Borman notwithstanding) no longer thinks that workers should check their brains at the guard door while entering because they won't need them inside.

Taylorism has declined in the last twenty to twenty-five years. Today, management genuinely wants worker involvement. For the most part, however, it does not know how to get it. Organizing teams and directing its members to participate is reminiscent of King Canute of old England ordering the waves never to touch his royal feet at the seashore! Forming quality circles and adopting *kaizen* were not successful, given the rugged John Wayne individualism of Western culture. The introduction of total quality management (TQM) in the 1980s improved quality, but not the business bottom line. Business process reengineering (BPR) improved business, but was viewed by workers as a euphemism for downsizing and layoffs—a deterrent to morale.

A Ten-Step Process in Constructing an Empowerment Infrastructure

Whereas Chapter 9 concentrates on workers' needs, hopes, and yearnings—on generating the "want to" motivations, not the "how to" motivations directed by management—this chapter first builds the organizational infrastructure needed to create a milieu in which self-motivation can be sparked and sustained. For a radical change in a corporate culture—from one of stifling drudgery to one of true empowerment—employee values and beliefs must be changed. For employee values and beliefs to be changed, management systems and practices must be changed first. Table 8-1 depicts a logical ten-step sequence in constructing an empowerment infrastructure, although it is preferable to start some of these steps simultaneously.[2]

Table 8-1. A ten-step process in constructing an empowerment infrastructure.

1. Eliminate mind-numbing, energy-sapping bureaucratic rules and regulations.
2. Tear down and rebuild the organizational infrastructure.
3. Change requirements for hiring employees.
4. Train, train, train all employees.
5. Revamp performance appraisals.
6. Change the rules of compensation.
7. Design meaningful and egalitarian gain sharing.
8. Redesign promotion criteria.
9. Promote team synergy.
10. Facilitate total involvement.

Source: Tom Peters, *Thriving on Chaos* (New York: Alfred A. Knopf, Inc.), 1987.

Eliminate Mind-Numbing, Energy-Sapping, Bureaucratic Rules and Regulations

The following is a good litmus test on rules, regulations, and policies:

❖ If they benefit a customer and capture his loyalty, keep them.

❖ If they are only for internal control, challenge them and even throw them out.

There are numerous ways to get rid of unproductive rules and regulations:

❖ Get rid of the executive parking spots, the executive lunchroom, the executive washroom, and the executive limos. They only serve to puff up the managers with self-importance and—by inference—demean the workers.

❖ Avoid endless meetings. I have a rule where a company's effectiveness is inversely proportional to the number of meetings it holds.

❖ Drastically reduce policy manuals. Nordstrom, Inc., the $2 billion retailer, has just a one-sentence policy manual: "Use your own best judgment at all times!"

❖ Throw off the chains of long, written procedures—especially the tyranny of ISO-9000 and other standards that are a decided drag on productivity and have actually set quality back two decades.

❖ Cut e-mail, which is rapidly becoming junk mail, by 75 percent.

❖ Discourage employees from staring at their computer terminals like isolated robots and encourage face-to-face interactions as a healthier way to build relationships.

❖ Shorten or eliminate most reports—written or via e-mail. The Ten Commandments take only a half page; the Declaration of Independence, three-quarters of a page; and Lincoln's Gettysburg Address is only 272 words. Most reports are created in order to protect the originator's rear end.

❖ Raise spending authority for employees from zero or miniscule levels that reek of mistrust to levels that show tangible evidence of management's confidence in employee responsibility. This is especially important for those employees that come in constant contact with customers. Immediate financial settlements by such customer-contact employees can diffuse customer anger and convert people who might otherwise be alienated from the company into loyal customers.

❖ Above all, reduce the compensation ratio between top management and direct labor. It is over 100:1 to 500:1 (and rising) in the United States compared with 30:1 in Germany and 10:1 in Japan. With one exception, nothing demotivates people as much as when they see their real contributions reduced to a pittance of a pay raise while senior management dishes out huge salary increases, bonuses, incentives, and stock options to itself. The exception is even more corrosive when top management feeds at the corporate trough with obscene increases while actually laying off workers. (When Lee Iacocca was criticized by the media in 1986 for his $20 million total compensation while he cut pay for other employees, he had a blatant response: "That's the American way. If little kids don't aspire to make money like I did, what the hell good is this country!") Yet the trend in the context of the booming U.S. economy is moving exactly the wrong way. If corporate America does not heed the lessons of history, it is doomed to repeat it and perpetuate Taylorism, and severely hurt—if not kill—the golden goose of free enterprise.

Tear Down and Rebuild the Organizational Infrastructure

A second important step on the road to empowerment is to revamp the organizational structure:

❖ From the vertical organization to the horizontal organization

❖ From recentralization back to decentralization

❖ From the tall organizational pyramid to the flat pyramid

❖ And ultimately from the traditional organizational pyramid to the inverted pyramid

From the Vertical Organization to the Horizontal Organization

Figure 8-1 is a typical example of a vertical organization, with separate departments, each protected by territorial walls and moats or white spaces between them. The engineering manager does not talk to the manufacturing manager who, in turn, does not talk to the materials manager; none of them want to communicate with the quality manager; and so on. This lack of cooperation forces vertical communication—from a department to the general manager and down again to another department. Nevertheless, problems and customers are no respecters of organizations. They cut across departments horizontally.

Dr. Kaoru Ishikawa, the Japanese quality guru, remarked—during a visit to America—that the U.S. organization was flawed. It had good vertical management, but no horizontal management. He used the analogy of a piece of cloth with vertical threads but no horizontal threads. It would not be very strong.

How is a cross-linking horizontal management to be achieved? Behavioral scientists have stressed, on an effectiveness scale from one to 1,000 (with one being the least effective and 1,000 being the most effective), that holy pronouncements on mission statements and the like have an effectiveness of one. Management's exhortations for improvement have an

Figure 8-1. Traditional organization: vertical management with white spaces.

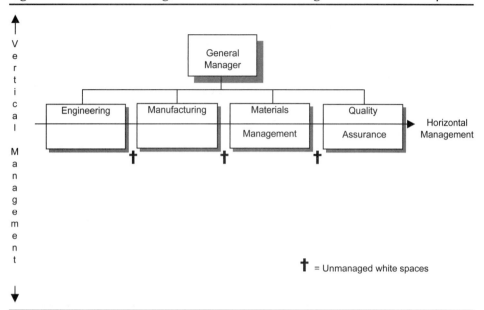

effectiveness of ten. Training has an effectiveness of 100, but the team concept has an effectiveness of 1,000. Thus, a cross-linking horizontal organization is achieved by cross-functional, interdisciplinary teams brought together to solve problems and break down departmental walls and vertical silos.

In his book *Thriving on Chaos*, Tom Peters states that "the self-managing team should become the basic organizational building block to win against other world economic powers."[3] Figure 8-2 is an example of an organization chart that has switched from departments to teams (except for a few support services that act as coaches and guides). Gone are the traditional engineering, sales, marketing, purchasing, manufacturing, and other departmental hierarchies. They are subsumed in process-oriented teams, such as the strategic development process, the product development process, the order fulfillment process, and the customer service process.

From Recentralization Back to Decentralization

Around the time of World War I, a promising organizational development, led by General Motors and DuPont, was to decentralize large companies into divisions for greater autonomy and efficiency. The virus of bureaucracy, however, attacked corporate divisions with equal vigor, causing organizational bloat. (For example, if a company does 100 units of work an hour and each of its workers does 10 units an hour, it would need eleven

Figure 8-2. A business process reengineering (BPR) organizational chart.

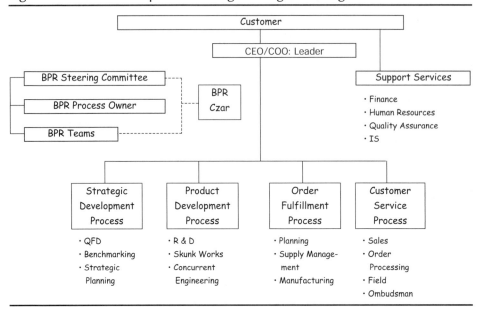

people—ten workers and one supervisor. If the demand grew to 1,000 units of work an hour, it would need not just 110 people but 194. Why? One hundred workers, ten supervisors, one manager, one assistant manager, eighteen personnel administrators, nineteen long-range planners, twenty-two people in audit and control, and twenty-three people in facilitation and expediting!)[4]

Furthermore, the end of World War II saw creeping recentralization with finance, purchasing, human resources, public relations, and quality regaining central control in the name of a new centralizing god—operations research. Even the functions of the chair, CEO, CFO, and others became the "office of the CEO" with two or three persons in each office instead of one. Adding to this organizational sprawl was the matrix organization, with each person now being subjected to the push-pull of two bosses—a functional boss and a departmental boss.

What is needed is a return to true decentralization and a simultaneous shrinkage of both corporate and divisional staff. With the information technology (IT) explosion, a direct link between senior management and the worker can be established, bypassing the ranks of middle manager, foreman, and supervisor. Using the tools of value engineering, zero base assessment, and process redesign, it is possible to cut headquarters staff by at least 75 percent as these examples illustrate:

❖ An A. T. Kearney study of forty-one large companies compared the financials of successful versus unsuccessful companies. The former had 500 fewer corporate staff specialists per $1 billion in sales.[5]

❖ Topsy Turvy, an $80 million per year company producing a women's hair product, has only three employees.[6]

❖ Mars, Inc., a $7 billion company, has only a thirty-person headquarters staff.[7]

From the Tall Organizational Pyramid to the Flat Pyramid

A third feature of organizational restructuring is flattening a total organizational entity with several layers of management levels between the CEO and the line worker (fourteen to eighteen levels for large companies; seven to eleven in smaller ones) to a maximum of five layers for large companies and three for a single plant.

For an example of what happens when a tall organization pyramid widens the gap between management layers, take the case of a young engineer who had been hired in a large company. The engineer was seven layers down in the managerial hierarchy. Twenty-five years later, he had risen to become a quality director, but he was now eight levels down from the CEO. That is bureaucracy's appetite—Sumo wrestler style! By contrast, the largest organization in the world—bar none—is the Catholic Church with 800 million members. It has only five layers of management. For 1,500 years and more, it has vetoed more layers and runs one of the tightest organizational "ships" in history.

Typically, a manager can control six to eight direct reports. The objective of the flat pyramid is to increase that span to fifty, a hundred, and more so that it becomes extremely difficult to micromanage people, direct your employees' every step, and breathe down their necks at every turn. The enlightened manager has to let go, loosen the reins, learn to trust employees, and give them the freedom and space they need—as long as corporate goals are fulfilled. The manager's role is more that of a counselor and guide. Once more, consider a few examples:

❖ Motorola's showpiece plant at Easter Inch in Scotland started a few years ago with 1,700 people and three levels, including the line worker. It now employs 4,000 people with only four levels.

❖ The A. T. Kearney study, referred to previously, found that the successful companies had three to nine fewer management layers than the unsuccessful ones—an average of 7.2 versus 11.1.[8]

❖ A Motorola organizational restriction is to discourage any plant in a single location to go beyond a 300-person headcount because of the loss of cohesion and esprit de corps that larger plant sizes could endanger.

From the Traditional Pyramid to the Upside-Down Pyramid

The futuristic organizational model takes the typical organization chart and turns it upside-down. Figure 8-3 shows this inverted organizational

Figure 8-3. The inverted organizational pyramid.

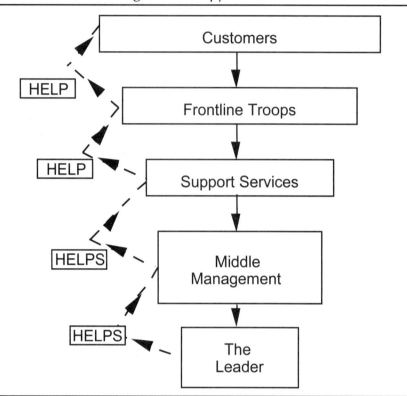

The leader (as "servant") nurtures people, helps them grow and reach out.

pyramid. The most important slot at the top is the customer, not the CEO. Next in importance are the customer-contact employees who come into frequent—often several times a day—contact with customers.[9]

Jan Carlzon, the former CEO of Scandinavian Airlines System (SAS)—and an inspiration in the managerial revolution—calls this customer-employee interface "fifty million moments of truth."[8] That is the number of contacts customers can have daily with the frontline troops of a company and which integrate in the customer's conscious and subconscious mind an indelible impression of the company as a whole—for better or for worse. Therefore, the selection, training, compensation, support, and coaching of these frontline troops are vital considerations in an organization. Instead, the current practice is to hire warm bodies off the street, skimp on their pay, throw them to the wolves of the outside world with little or no training, then berate them for not being customer-sensitive!

At the very bottom of the inverted pyramid is the CEO. The top executive is the servant of the organization, not in a menial sense, but as a servant in the best religious tradition (just as Christ served his people). The task of the CEO as leader, detailed in Chapter 7, is to trust, support, coach, and help employees reach ever-higher peaks of potential. This is a very difficult transition for the power-drunk CEOs of today to make, but—whether they like it or not—the inverted pyramid is the organizational wave of the future.

Change Requirements for Hiring Employees

With the bust-up of bureaucracy and the crafting of a viable organizational infrastructure in place, the next step is a radical departure in the way future employees are recruited. President Eisenhower once said, "War is too important to be left to the generals." As president of my Glencoe school board, I made a parallel statement: "Education is too important to be left to educators." By extension, it can be said that corporate recruitment is too important to be left to the human resources people and their psychologist testers. They can do the initial screening, but it is the line people who best know whom they need.

The following checklists outline key steps in preparing for interviews of prospective employees, techniques to use while conducting interviews, and methods to fine-tune the interview process.

Pre-Interview Preparation
- [ ] Start with the interviewer team, drawn from managers, peers, and subordinates who will be working with the interviewee.
- [ ] Make a list of the factors that work best for the proposed position, based on the characteristics of the most successful people in similar positions.
- [ ] Build a series of questions, based on such factors.
- [ ] Get coaching on effective interview techniques.

Interview "Musts"

- [ ] Deemphasize questions on professional/technical skills. Those can be taught.
- [ ] Concentrate on issues such as team player fit, customer sensitivity, alignment of personal values with corporate values, entrepreneurial "get up and go," innovation potential, and the hunger to learn and grow.
- [ ] Avoid overemphasis on PhDs and MBAs.[10] (Honda, one of the most innovative automobile companies in the world, has only three PhDs on its engineering staff. W. Edwards Deming says, "Burn all the business schools! They teach obsolete concepts, churning out MBAs with a poor grasp of the real world.")[11]

Post-Interview Audits

☐ Keep track of employee turnover to calibrate and fine-tune the hiring process in order to correlate interview questions and techniques with exit-interview employee impressions.

☐ Audit the hiring process. (Through this method, one company, NAT&D, reduced its turnover from 27 percent to 4 percent. Another reduced it from 30 percent to zero.)

Train, Train, Train All Employees

Training is the next organizational milestone on the road to empowerment. In many ways, the outcome of training, namely the elevation of worker skills, is the key differential between a company that wins and a competitor that loses. Yet, that lesson has hardly been learned in the West, with the possible exception of Germany. As examples:

❖ The U.S. government contributes—via tax incentives—3,200 times as much to companies through technology as it does through employee training.

❖ Large Japanese companies provide one year of training to new recruits before they are actually put on the job. As a result, their line workers take on the tasks of the technicians and the technicians take on the tasks of the engineers, leaving the latter free to do research and development.[10]

❖ The Japanese bring that tradition of training to their transplants. Nissan is the benchmark. It spent $63 million training 2,000 workers before its Smyrna, Tennessee plant started operations.[11]

For an effective training process, a few principles are essential:

❖ Mindware takes precedence over hardware or software (our Motorola slogan).

❖ Training should be universal within the company, with frequent retraining to prevent knowledge obsolescence (the half-life of an engineering graduate is only two and one half years, if the engineer is not recycled).

❖ Training is an investment, not a cost. If properly channeled, a 10:1 return on that investment, or greater, is possible.

❖ Training budgets should be kept to a minimum of one and one half percent of payroll, preferably 3–5 percent of payroll.

❖ Training should be universal (but focused for each employee group) and tied in with key strategic corporate goals.

❖ Effective problem-solving techniques (see Chapter 11) should be mandatory for manufacturing as well as white-collar operations. A classic example of an ineffective technique is statistical process control (SPC). The Big Three automotive companies have spent billions of dollars on control charts (the dominant feature of SPC). Their savings are in the millions. That is a 0.1 percent return on investment in thirteen years.

However, there is a giant caveat: Training without evaluation of its effectiveness is worse than no training.

Classroom training is not enough; evaluation of the instructor by students is not enough; supervisor/manager assessment is—as we say in mathematics—necessary but not sufficient. There should always be independent, quantitative audits of benefits (in terms of quality, cost, and cycle-time improvements) conducted within three months and within a year after such training.

Revamp Performance Appraisals

Dr. W. Edwards Deming railed against current systems of performance appraisal where employees are graded on a curve using, for example, a scale of one to ten, with zero being the worst and ten being the best (see Figure 8-4).[12] Deming stated that 5 percent of the people are the truly outstanding performers and are candidates for advancement. Another 5 percent are the laggard performers and should be terminated or transferred to other jobs. However, the great majority of people, 90 percent, are doing the best they can under an organizational system that is defective. Measuring these people in this manner does not really measure them; instead, this process measures the noise in the system rather than the signal (to use an electronic term, signal-to-noise ratio, where noise, in this case, is those factors beyond employees' control in the defective organiza-

Figure 8-4. Performance rating on a curve.

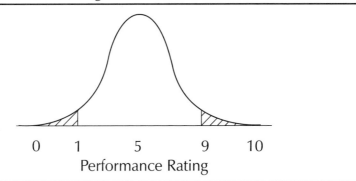

0   1   5   9   10

Performance Rating

tional system). Furthermore, these semiannual or annual performance appraisals end up doing psychological harm to the person being rated. It takes months to recover from them.

That does not mean that there should be no performance evaluation at all. One should be instituted, but the evaluation process should be:

❖ *Nonscalar*. A numeric rating scale pigeonholes employees and, as stated previously, measures the system's defects rather than those of the employees. It also pits employee against employee and destroys the fragile trust that is being built, rung by rung, on a ladder of trust.

❖ *Constant*. Deming stated that, in Japan, feedback to the employee is constant, almost daily. But it is not judgmental. The object is to provide guidance to help the employee improve. It is coaching in the best tradition of that term.

❖ *Nonthreatening*. There can and should be structure, but its aim should be to draw up an agreement including two or three vital goals, personal and team growth goals, skill enhancement (the number of different jobs that a person can handle), and interpersonal sensitivity.

❖ *360-Degree in Nature*. In the new organizational structure, where individual managers have fifty to 150 direct reports, it is impossible for the manager to cover the entire waterfront. Instead, the concept of the Next Operation as Customer (NOAC), discussed in Chapter 16, is utilized. The internal customer becomes the principal scorekeeper and the principal evaluator because the employee has greater day-to-day contact with the internal customer than with the manager. Some companies employ 360-degree evaluations, including evaluations by the manager, customer, peer, and by the people evaluating their managers.

❖ *Team Evaluations*. With the team replacing the individual as the basic building block of an organization, evaluations of teams as a whole should also be conducted by the facilitator, by the team's customers, and by the manager along lines similar to individual evaluations.

Change the Rules of Compensation

Along with radical changes in the way people are hired, trained, and evaluated, there should be corresponding changes in the way people are compensated for their work. Table 8-2 clearly shows the difference between the old, rigid practices governing pay and the new, enlightened practices.

Table 8-2. Old, rigid versus new, enlightened pay practices.

Old Practice	New Enlightened Practice
❖ Small, pro forma merit increases based largely on cost-of-living.	❖ No pro forma, "automatic" merit increases each year.
❖ No or small bonuses for employees as a whole.	❖ Large bonuses/incentives based on performance.
❖ Small differential in increases between poor and outstanding performers.	❖ Large differential in pay increases between poor and outstanding performers.
❖ Staff people command higher salaries than line people for comparable levels.	❖ Line people have equal or higher salaries than staff people for comparable levels.
❖ Salaries are based on hierarchical position and number of people managed.	❖ Salaries are based on performance/results, with individuals earning more than their manager made possible.
❖ Performance criteria are fuzzy.	❖ Performance criteria are based on results in quality, cost, cycle time, innovation, and teamwork.

In the final analysis, fair and meaningful pay is an important part of employee recognition.

Design Meaningful and Egalitarian Gain Sharing

A Yankelovich poll of American and Japanese workers on the statement "I have a need to be the best I can, regardless of pay" found, surprisingly, that the American worker outscored the Japanese.[13] However, when workers were asked, "Who would benefit from an increase in worker productivity?" the answers flip-flopped. Ninety-three percent of the Japanese workers felt they would benefit. Only 9 percent of the American workers felt they would gain. This is not surprising. The Japanese workers' bonus amounts to three to six months of extra pay per year. The workers bonus varies from zero to 10 percent per year in most American companies. (There are honorable exceptions, where a combination of incentives and employee stock ownership plans for workers results in 60–80 percent of base pay. The benchmark is Lincoln Electric Company with incentives that are 95 percent of base pay!)

By contrast, the American CEO receives a total compensation 100 to 500 times that of a factory worker (Table 8-3) while the Japanese ratio is 10:1 and the German ratio is 30:1. In the 1990s, the average CEO pay of the top 350 U.S. companies rose 535 percent while the average worker

Table 8-3. A CEO windfall: a drain on democracy.

❖ The average CEO of the 350 biggest corporations receives $12.4 million per year in salaries and stock options.

❖ The average CEO of the biggest Internet companies receives—in salaries and stock options—$15.4 million per year.

❖ The average factory worker receives $23,433 per year.

❖ The ratio of CEO to factory worker compensation in the United States is 523:1, the highest of any industrialized nation in the world.

❖ In the 1990s, the average CEO pay rose 535 percent.

❖ In the 1990s, the average worker pay rose 32 percent.

❖ In an August 2000 BusinessWeek/Harris Poll, 73 percent of those polled indicated that the compensation of top officers of large U.S. companies was "too much."

Source: Executive Excess 2000 and Bureau of Labor Statistics.

pay rose only 32 percent. The combination of bonuses and stock option plans for top management is five to ten times their salaries in leading U.S. firms. To add insult to injury, bonuses and stock option plans are often offered to executives at the same time that downsizing and laying off workers occur to shore up anemic profits. "It is a flaw in the American dream that one group should be getting such a grotesquely skewed portion of the pie while most Americans get next to nothing" says John Cavanagh, president of the Institute for Policy Studies, a Washington think tank. "It degrades our democracy."

Results as a Prerequisite

Obviously, worker incentives, bonuses, gain sharing, and stock options cannot be a socialistic giveaway. They must be earned, with significant financial gains to the company as a result of worker productivity. Table 8-4 indicates that the incentives/bonuses of companies that put people first are easily paid for by the productivity gains, measured in multiples over their industry averages.

Table 8-4. Bonuses paid versus productivity gains in leading "people first" companies.

Company	Bonus as Percent (%) of Base Pay	Productivity Gain Over Industry Average
Lincoln Electric Co.	95	3.5:1
Andersen Corp.	70	2:1
Steelcase	60	2.5:1
Worthington Industries	80	3:1
Nordstrom	80	3:1
Nucor Corp.	75	4:1

Guidelines for Maximum Effectiveness

In designing effective incentive systems—be they bonuses or stock owner-ship programs—a few guidelines need to be stressed:

❖ Base pay should be somewhat higher than comparable jobs in other companies in the vicinity.

❖ Multiple skills, creating flexibility within a team and between teams, deserve graduated incentive pay.

❖ Incentives should be based on team performance against goals.

❖ Incentives should also be tied into profit on sales and return on investments, business by business, with a floor (i.e., minimum level for the incentive gate to open) for each of these two metrics.

❖ The formula for calculating incentives should be simple and under-stood by all.

❖ A targeted incentive should be 20 percent of base pay, provided the profit/ROI gates are open.

❖ The incentives should be paid out at least every quarter, preferably monthly.

❖ Employee stock ownership plans (ESOPs) are better than cash pay-outs because they encourage employee involvement, by giving em-ployees stake in the company, as well as encourage a long-term outlook.

Redesign Promotion Criteria

Promotions, even more than salaries and gain sharing, are the most tangi-ble forms of recognition. There are two tracks for promotion. One is mana-gerial, the other professional/technical. Most people consider the managerial track as the only way up the corporate ladder, believing that position, rank, and the number of people managed are the sine qua non of advancement.

This is not necessarily true. In today's fast-changing technological world, professional contributors can be more important than managers. Progressive companies recognize this and provide a professional track for promotions that attest to its importance and prestige. Staff scientists in such companies can be worth their weight in gold and recognized as such. A second consideration is that the managerial track is not only more crowded but also much narrower today, as the ranks of middle managers get phased out at the lower rungs of the corporate ladder, and the power of information technology is available to top management.

Past Performance Is Not a Valid Reason for Promotions

In the past, there has been a tendency to reward high performers with promotions. There may be several personal attributes causing that high performance, but these attributes may not be those required for the promoted job. This is captured in the famous Peter Principle, where individuals get promoted until they reach their own level of incompetence.

Promotion Based on Potential for Leadership

Instead of past performance (which, of course, should be amply rewarded with bonuses and stock options), the criteria for promotions are very different. The litmus test is: Does the person have the potential for moving from a manager (or individual contributor) to a leader? This subject is thoroughly covered in Chapter 7. An assessment can be made by asking the following questions:

❖ Does the person have the personal philosophies and values required of leaders?

❖ Does the person, by action, advance the roles of leadership?

❖ Is the person highly people-oriented, willing to coach and help people reach higher and higher levels of potential?

❖ Is the person entrepreneurial?

❖ Is the person innovative, generating a stream of ideas?

❖ Is the person capable of creating fun and joy in the workplace?

Promote Team Synergy and Facilitate Total Involvement

These are the last two steps in the ten-step process. Team synergy is achieved when all team members feel they have gained from the team's activities and results. At this point, people recognize that interdependence, rather than independence, is sought. Total involvement becomes operative when no team member feels left out.

CASE STUDY

Chaparral Steel—A Benchmark Company in the Area of Organization[14]

The U.S. steel industry has long moaned about steel dumping from overseas companies and about how there is no level playing field because of the low labor cost abroad. The U.S. companies are constantly seeking protection and help from the Congress to bail them out.

Yet a few minimill steel companies such as Chaparral Steel Company and Nucor Corporation have captured more than 33 percent of the steel market. They have excellent profits, based on a 4:1 productivity increase over industry averages; and they pay their workers incentives averaging 75 percent over base pay. Their cost per ton of steel is lower than that of Third World countries and half the costs of a typical Japanese mill. (So much for the myth of uncompetitive American labor costs.)

Chaparral Steel has torn down its vertical silos of isolated departments and dismantled its bureaucracy. Its research operation no longer lives in an ivory tower but is on the shop floor, where ideas for improvement and technological breakthroughs are jointly evaluated by engineering, production, and maintenance and put to the test immediately using the factory as the real laboratory. In many cases, core customers are drawn into these evaluations.

Chaparral has only three management levels between the CEO and the worker—a vice president, superintendent, and foreman. The foreman has a great deal of freedom to make important decisions. The company has no quality inspectors. The line workers are responsible for their own quality—they act like owners—and the company has a reputation for the best quality in the marketplace.

Chaparral gives people freedom to perform. Its security guards, as an example, enter data into computers, double as paramedics and ambulance drivers, and reach to do some accounting functions. Chaparral's teams and their facilitators do their own hiring and training, their own safety checks, their own upgrading of old equipment and prove in (turnkey) procurement of new equipment.

A unique feature at Chaparral is its "sabbatical," where teams are given time off from their regular work to tackle special innovation projects. In the process, they visit customers, universities, suppliers, and researchers and benchmark other companies.

Over the years, I have developed my golden rule: "If you are selling in a particular country or region, it is most logical and economical to manufacture in that country or region." Chaparral is a heart-warming case study that demonstrates the validity of that rule.

SELF-ASSESSMENT/AUDIT ON ORGANIZATIONAL CAPABILITIES

Table 8-5 is a self-assessment/audit that a company can conduct to score its organizational capabilities. It has two key characteristics and fifteen success factors, each worth five points, for a maximum score of seventy-five points.

Table 8-5. Organization: key characteristics and success factors (75 points).

Key Characteristic	Success Factors	Rating				
		1	2	3	4	5
3.1 Dismantling Taylorism	1. The organization has completely eliminated Taylorism—compartmentalizing management and workers—and moved toward harnessing the power of its people. 2. The organization has moved beyond quality circles, Kaizen, total quality management (TQM), and business process reengineering (BPR) to facilitate the true empowerment of its people.					
3.2 Revolutionizing the Organizational Culture	1. The organization has eliminated bureaucracy by: ❖ Reducing executive perks that demean workers ❖ Reducing the number of meetings, memos, and e-mail ❖ Reducing policy manuals and straitjacket procedures ❖ Raising employee spending authority as an expression of trust, especially for customer-contact employees 2. The organization has reduced the compensation ratio between top management and direct labor. It has especially outlawed granting large financial incentives to top management while downsizing and laying off workers. 3. The organization has reinforced a vertical structure with cross-linking horizontal management, using cross-functional teams as a basic building block. 4. The organization has moved from a top-heavy centralized entity with large, redundant corporate staffs to a truly decentralized company that's autonomous and lean. 5. The organization has delayered its management structure, reducing the usual 8 to 16 layers of management levels between the CEO and line worker to 3 to 5 layers maximum. 6. The organization has reversed the traditional organizational pyramid and					

(continues)

Table 8-5. (Continued).

Key Characteristic	Success Factors	Rating				
		1	2	3	4	5
	created an inverted pyramid with the customer on top, the customer-contact employees next in importance, and the CEO at the bottom to serve employees as coach, guide, and counselor.					
	7. The organization has radically changed its hiring practices by deemphasizing professional skills and emphasizing team player fit, customer sensitivity, alignment of value with corporate values, entrepreneurship, and desire to learn.					
	8. The organization places great emphasis on training and continued learning for all its employees, spending a minimum of 3 percent of its payroll on training.					
	9. The organization audits its training results, assessing a benefit to cost ratio of a minimum 10:1.					
	10. The organization discards its pro forma, semiannual performance appraisal by the supervisor in favor of continual nonquantitative, nonthreatening guidance; it is moving toward customer evaluation as well as a 360-degree evaluation by customers, manager, peers, and employees.					
	11. The organization has changed the rules of compensation from small, pro forma merit raises with small differentials between employees, to no merit raises but with large bonuses/incentives that are based on performance.					
	12. The organization provides incentives to its teams in the form of bonuses and/or stock options at least every quarter, based on: ❖ Team performance and multiple skills ❖ "Floor levels" of profit and return on investment, business by business ❖ An average of 20 percent of base pay if the first two conditions are met					

❖ A maximum ratio of 50:1 between the total compensation of the CEO and the line worker

❖ No bonuses or stock options for top management during years of profit loss or layoffs

13. The organization has a promotion policy based on:

❖ A two-track system—one for managerial advancement, the other for individual professional growth

❖ Potential for leadership, not past performance

Nine

From Passivity and Boredom Among Employees to Industrial Democracy

❖ ❖ ❖

*There is no asymptotic
barrier to the human
potential, given leadership's
encouragement, trust,
support, training, and
coaching to all employees.*

—Keki R. Bhote

Our people were labeled
"ordinary," but they have
achieved extraordinary
results.

—OSCAR KUSISTO,
MOTOROLA VICE
PRESIDENT

From "How to" to "Want to" Motivation

The CEO of a large company, exposed for the first time to empowerment's power at a seminar, turned to his vice president of human resources and said, "This is good stuff. I want you to implement an empowerment culture by tomorrow!" Empowerment is not instant pudding. Nor is it a flavor-of-the-month program. It is a process that takes time, patience, and persistence. It cannot be a management-imposed process. That is where past attempts, such as TQM and business process reengineering, have failed. They did not motivate workers because motivation cannot be extrinsic. It must be intrinsic, coming from within the worker. Past managerial efforts have concentrated on how to empower, how to motivate. What is needed, however, is to go from "how to" for the worker to "want to" by the worker.

A Synopsis of Motivation Theories

Motivation theories abound on the "want to"—on needs that produce motivation. These needs range from Abraham Maslow's hierarchy of human needs to Douglas McGregor's Theory X and Theory Y. Two of the best examples of motivation theories are Frederick Herzberg's seminal work on dissatisfiers and motivators and Scott Myers's on maintenance and motivation needs.

Herzberg's Dissatisfiers and Motivators

Herzberg drew a sharp distinction between workers' needs that produce dissatisfaction when neglected or, at best, contentment when fulfilled, and the needs that truly generate motivation. Factors affecting the former are company policy, supervision, salary, and working conditions. If these factors are poor, they cause genuine dissatisfaction. Even if they are good and adequate they produce only contentment—not motivation. By contrast, Herzberg's motivation factors—achievement, recognition, the work itself, responsibility, and advancement—are the engines of motivation within workers. Figure 9-1 is a graphic portrayal of Herzberg's dissatisfiers and motivators and of the extent of the negative and positive attitudes they engender.

Scott Meyers's Maintenance and Motivation Needs

In an adaptation of Herzberg's theories, Scott Meyers developed a three-ring concept of worker needs as shown in Figure 9-2.[1] Maintenance needs in the outer ring are broken down into the following categories: physical needs—food and shelter; economic needs—a paycheck; security needs—job security; status needs—dignity of the individual, perks, and prestige; and social needs—interaction with fellow employees. These maintenance

Figure 9-1. Worker needs: dissatisfiers and motivators.

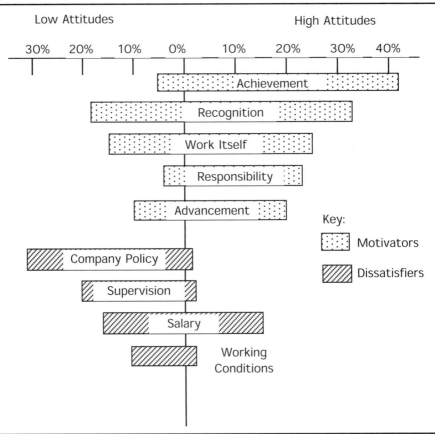

Figure 9-2. Employee needs: maintenance and motivational factors.

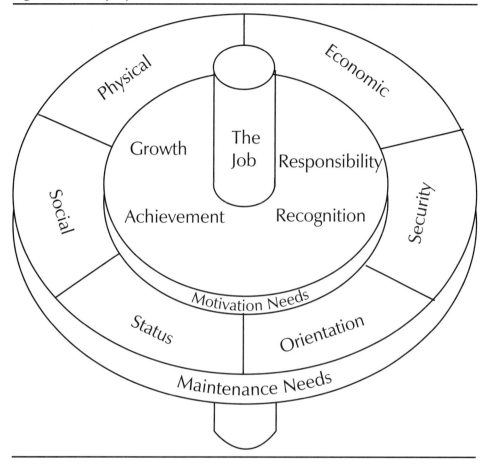

needs, similar to Herzberg's contentment factors, cause dissatisfaction when absent, but do not, on the other hand, produce motivation when present. At best, they maintain the status quo and possibly prevent employees from leaving the company. The inner ring contains the same motivation factors—growth, achievement, responsibility, and recognition—formulated by Herzberg. The innermost ring, at the very core of motivation, is the job itself—and how to make it interesting, challenging, and exciting.

Job Redesign—The Centerpiece

If, indeed, the job is the core of motivation, the challenge is how to make a dull, boring job that characterizes a production assembly line and many other routine tasks in a company, interesting and even exciting.[2]

Behavioral scientists tell us that in order to inject fun into work, you must capture three psychological states that have always made sports fun:

1. *Meaningfulness.* Workers must perceive their jobs as worthwhile and important.
2. *Responsibility.* Workers must believe that they are personally accountable for the results of their efforts.
3. *Knowledge of Results.* Workers must be able to determine, on a regular basis, whether the results of their efforts are satisfactory.

In the presence of these three psychological states, workers feel good about themselves when they perform well. They produce the all-important internal motivation. If even one of these psychological states is missing from the job, the internal motivation is reduced.

The Five "Core" Job Dimensions

Job characteristics that lead to the three psychological states necessary for internal motivation are:

1. *Skill Variety* (i.e., job calling for multiple skills with the potential for appealing to the whole person)
2. *Task Identity* (i.e., job that is identifiable as a "whole;" for example, making a whole pager unit rather than repeatedly inserting five components into a printed circuit board on a long, impersonal assembly line)
3. *Task Significance* (i.e., a job that has a greater consumer impact; for example, making or assembling parts for an aircraft as opposed to filling paper clips in a box)
4. *Autonomy* (i.e., a job where the worker has freedom and discretion in planning, executing, and improving work)
5. *Feedback* (i.e., a job that permits measurement of results, especially feedback that comes directly from the worker himself)

The Full Model of Job Redesign

The concepts of psychological states and job dimensions, essential to job interest, challenge, and fun, can be meshed to form a powerful model for redesigning jobs. Figure 9-3 is a schematic representation of how a job can be redesigned to produce true internal motivation.

At the right of Figure 9-3 are the five goals that companies and workers share: high internal motivation, high quality, high job interest and satisfaction, low absenteeism, and low people turnover. To achieve these goals,

Figure 9-3. The full model of job redesign.

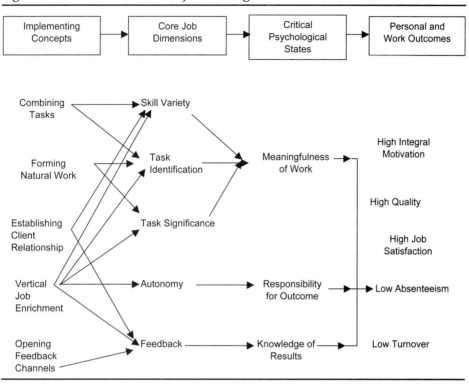

fun must be injected into the job through the three psychological states: meaningfulness of work, responsibility for outcomes, and knowledge of results. This, in turn, requires that the five core dimensions, outlined previously, must be present in the job. Finally, in order to establish these five core dimensions, five implementing concepts—combining tasks, forming natural work units, establishing client relationships, creating vertical job enrichment, and opening feedback channels—must be designed into the job for optimum results.

Injecting Fun and Excitement Into a Dull, Mind-Numbing Assembly Line Job

For most professionals, the job itself provides adequate interest, challenge, and stimulation. But what about the many routine jobs in industry—such as assembly line workers, cleaning services staff, clerks, data entry loggers, and so on—where brawn is substituted for brain. What "pride of workmanship," that management bemoans workers of today are losing, can assemblers have putting in bolts on the wheels of cars as they move past them on a car line, hour after hour, day after day, week after week, month after month?

Unfortunately, such jobs are necessary. What can be done, however, is to redesign such jobs, using the template of Figure 9-3 on a typical assembly line. Table 9-1 outlines the process step by step.

Conventional managers may argue that all of these actions take away "holy" production time. They do not realize that direct labor constitutes only 3 percent of a company's sales. A 20 percent loss of production time is only 0.6 percent of sales, whereas efforts to reduce the cost of poor quality (COPQ), improve total productive maintenance (TPM), reduce cycle time, and improve customer retention may accumulate to almost 50 percent of sales.

Job Security Versus the Damocles Sword of Layoffs

One of the roadblocks to worker motivation and to teams signing up for empowerment is their genuine fear that the resultant productivity improvements will usher in layoffs, retrenchment, and downsizing.

James Lincoln, the president and founder of the phenomenally successful Lincoln Electric Company, addressed management's responsibility in this matter:

> The greatest fear of the worker, which is the same as the greatest fear of the industrialist, is lack of income. . . . The worker's fear of no income is far more intense than the industrialist's since his daily bread and his family's depend on his job.
>
> In spite of these facts, the industrialist will fire the worker at any time he feels he can get along without him. The worker has no control over his future. Only management is responsible for the loss of the worker's job. Management, which is responsible, keeps its job. The man who has no responsibility is thrown out. Management failed in its job and had no punishment. This fact has more to do with production limitation than any other circumstances.[3]

James Lincoln's words ring as true today as they did fifty years ago. Of course, business volatility—hostile takeovers, joint ventures, and technological eruptions—cannot guarantee job security today. But progressive companies have cushioned the shock and dampened the oscillations through attrition, early voluntary retirement, generous severance pay, use of temporary workers, subcontracting as a buffer, and even shortened workweeks. More, however, can be done. Companies that are leaders in enlightened job security practices employ:

❖ An understaffing policy of 10–20 percent compensated—if need be—with overtime

Table 9-1. Redesigning a dull assembly line job to create a spark of interest, fun, and excitement.

Implementing Concept	Actions
1. Combining Tasks	❖ Obtain material. ❖ Arrange/fill bins. ❖ Check tools. ❖ Learn multiple skills, with incremental pay for each new skill. ❖ Administer self-inspection (using poka-yoke, as covered in Chapter 11). ❖ Self-test. ❖ Conduct routine preventive maintenance.
2. Natural Work Units	❖ Establish focus factories (see Chapter 14). ❖ Divide work by customer, rather than by product line. ❖ Work on all subassemblies of a product line instead of specific subassemblies of several product lines. ❖ Form teams, converting the supervisor into a facilitator.
3. Client Relationships	❖ Establish contact with external customers and external suppliers where possible, including visits. ❖ Solicit and obtain support from white-collar services. ❖ Identify and build relations with professional experts. ❖ Determine requirements of internal customers (Chapter 16) and get feedback from them on progress. ❖ Specify requirements to internal supplier (Chapter 16).
4. Vertical Job Enrichment (the most important aspect of job redesign)	❖ Help formulate work hours, including flex hours. ❖ Help determine overtime, incentives, and alternatives to layoffs. ❖ Determine line balancing/sequencing. ❖ Conduct time and motion studies (better than external industrial engineers). ❖ Form problem-solving teams. ❖ Formulate written/audio/video instructions where needed. ❖ Highlight critical quality, cost, cycle-time parameters. ❖ Work on methods improvement. ❖ Help design workstation layout. ❖ Help reduce setup/change overtime (see Chapter 14). ❖ Learn and implement the powerful tools of the twenty-first century (Chapter 11) to solve chronic problems and make presentations of results to management. ❖ Suggest and help build simple sensors (poka-yoke) to prevent operator controllable errors. ❖ Assist in Positrol, process certification, and precontrol (see Chapter 14).

(continues)

Table 9-1. (Continued).

Implementing Concept	Actions
5. Feedback	❖ Get immediate nonthreatening feedback from poka-yoke sensors. ❖ Get periodic feedback for the team from internal customers. ❖ Get guidance and coaching from the manager at least once a week. ❖ Design and maintain scoreboards for charting progress and for maximum visibility. ❖ Understand and interpret P&L statements and the balance sheet (see the section on Open Book Management).

Source: Keki R. Bhote, *Strategic Supply Management* (New York: American Management Association, 1991).

❖ A push-pull practice with subcontractors (i.e., outsourcing and in-sourcing)

❖ Sizable reductions in job classifications, obsolete work rules, and restrictive job definitions

❖ Retraining, including outside schooling

❖ Job rotation—from manufacturing to sales, from production to maintenance, even from company to community work

However, the most important assurance that management can provide is that productivity contributed by workers will never be allowed to result in a loss of their jobs. Furthermore, if layoffs are absolutely unavoidable, they should start with the ranks of management and support staff (i.e., white-collar operations) and reach the line worker last.

How Ready Is Your Company for Employee Empowerment?

Assuming that a company has good leadership and that an organizational infrastructure has been crafted, the readiness of both management and workers for employee empowerment must first be assessed.

Table 9-2, developed by Southwest Airlines Co., the most highly rated airline in America, is a test for a company's readiness for empowerment. It has nine categories (A-I), with each category containing four statements. The first statement in each category carries one point; the second statement counts for two points; the third carries three points; and the fourth merits four points. The maximum empowerment score is thirty-six points (with a rating of four in each of nine categories). Table 9-3 is the accompanying readiness score—either "not ready," "potentially ready" (with necessary conversions of management and workers), or "full ready."

Table 9-2. A test for an organization's readiness for empowerment.

Category A: Organization's Niche

1. Organization is a low cost, high-volume producer of a few commodities.
2. It produces a number of high-volume products for a few customers.
3. It produces a few products for niche markets.
4. It provides personalized products, tailored to each customer.

Category B: Employee Readiness for More Authority

1. Employees are not interested in having greater authority or responsibility.
2. A few employees want greater authority in doing their jobs.
3. A significant number of employees want greater authority in doing their jobs.
4. Most employees are champing at the bit for greater authority.

Category C: Employee Time With Customers (Internal or External)

1. Employees spend only a few minutes with each customer.
2. They spend a few hours with each customer.
3. They spend days or weeks with each customer.
4. They spend months or years with each customer.

Category D: Organizations Relationship With Customers

1. The relationship is mechanical, distant.
2. Minimum time is spent with customers.
3. There is occasional social contact with customers.
4. The company creates long-term friendships with customers.

Category E: Job Knowledge

1. A job is learned in a few short hours.
2. It takes a few days or months to learn.
3. It takes a few years to learn.
4. It takes a lifetime of learning.

Category F: Business Environment Changes

1. Business is relatively stable, unchanging.
2. Changes can be accurately predicted.
3. Only the outline of changes affecting the business is known.
4. The business environment changes continually, making employees react quickly.

Category G: Management Beliefs About Employee Motivation

1. Management manipulates people to ensure that it has power.
2. Few managers believe that employees have high needs for growth and success.
3. Some managers believe that employees have high needs for growth and success.
4. Most or all managers believe that employees have high needs for growth and success.

(continues)

Table 9-2. (Continued).

Category H: Fixed Procedures	Category I: Management Readiness to Concede Authority
1. Company procedures are fixed for every task. 2. Procedures cover most activities. 3. Procedures cover only a few tasks. 4. There are no procedures because they would not allow employees to make split-second decisions.	1. Managers not at all ready to give up power. 2. A few managers would be willing to give authority to employees. 3. Many managers would be willing to give authority to employees. 4. Most managers would readily give authority to employees.

Source: Southwest Airlines (Chaudron Associates, 1993).

Table 9-3. Empowerment readiness score.

Score (Points)	Empowerment Readiness
9 to 18	Organization is not ready for empowerment. Major changes are needed.
19 to 27	The organization is in a "twilight zone" on empowerment readiness. More work is required to convert managers into leaders and make employees want to reach for empowerment.
28 to 36	The organization is well positioned for empowerment "takeoff." Its leadership is willing to give greater authority and responsibility to workers and the latter see the tangible benefits of empowerment in their work lives.

Team Competitions—Another Gold Mine

The team concept, as the basic building block of an organization, is so pervasive in a few companies that they have proliferated them, with scores of teams in each plant, hundreds in each major division, and thousands in the corporation as a whole.

Typically, each team—generally, cross-functional—is organized on a purely volunteer basis. The desire to join a team is so compelling that few workers are holdouts. The team elects its own leader (with the supervisor graduating into a facilitator), gives itself a catchy name, and chooses its own goals (generally, improvements in quality, cost, cycle time, innovation, or some other problem) that are of concern to the team or the business in which it operates. The teams are then given time—three to six months or more—to achieve their goals. They are rewarded with bonuses, usually twice a year, with presentations to the plant management.

The Motorola TCS Competition

Motorola is, undoubtedly, a benchmark in the arena of team competition. Worldwide, it has more than 6,000 teams in friendly competition, which it calls its TCS (total customer satisfaction) competition. Each major plant has fifty to a hundred teams. The few best teams are selected to represent the plant in a regional meet. The best of these teams move on to presentations at the sector level, and the finalists from the sectors then make presentations at corporate headquarters in Schaumburg, Illinois, where the CEO evaluates the results for two full days. Six gold medals and twelve silver medals are awarded to the winners.

There has been some media criticism about this TCS competition: that it promotes competition rather than cooperation, that for every winning team there are many losing teams, that there is too much hoopla, and that the teams spend too much time with glitzy showboating. But, it is the best example of how to convert dull work into fun, generate unbelievable enthusiasm, and give workers who have never made a speech in their lives an opportunity to appear before senior management and speak with amazing aplomb. The bottom line is the savings to Motorola of $2 billion to $3 billion a year—a return on investment of 15:1.

On the Road to Industrial Democracy

In the past ten to twelve years, three positive innovations have been launched in empowerment practices—all leading to the ultimate vision of industrial democracy. They have several features in common, not the least of which is a compelling desire by the workers to play a significant role in the success of the organization and in their own success—emotional, managerial, and financial.

Open Book Management—Lightning in a Bottle

Open book management can be defined as a way of running a company that gets everyone to focus on helping the business make money "by giving every employee a voice in how the company is run and a stake in the financial outcome, good or bad."

If you could tear apart an Open-Book company and compare it with a conventional business, you'd see three essential differences, says John Case, author of *OpenBook Revolution*:

1. Every employee sees—and learns to understand—the company's financials, along with all the other numbers that are critical to tracking the business's performance. (That is why it is called open book.)

2. Employees learn that, whatever else they do, part of their job is to move those numbers in the right direction.

3. Employees have a direct stake in the company's success. If the business is profitable, they get a cut of the action. If it's not, they don't.

The underlying principle of open book management is: "The more employees know about a company, the better it will perform." Don't use information to intimidate, control, or manipulate people. Use it to teach people how to work together to achieve common goals and thereby gain control over their lives. When you share the numbers and bring them to life, you turn them into tools people can use to help themselves as they go about their business every day.[4]

Open Book Implementation

There are as many variations in how to implement open book management as there are companies practicing it. Still, a few common themes characterize most of approaches:[5]

1. *Show all employees the financials.* Put up scoreboards so that key financials can be tracked weekly. Then teach employees how to read or interpret:

❖ P&L statements and the balance sheet

❖ After-tax profits, retained earnings, debt-to-equity ratios, and cash flow

❖ Inventory turns, asset turns, and overhead chargeout rates

❖ Revenues per employee

2. *Teach the basics of business.* It is amazing how little most workers know about business. Some believe that revenues are the same as profits! When I was teaching a graduate student class at a prestigious university in Chicago, the students thought that the average profit on sales that companies made was 40 percent. When teaching the basics:

❖ Start with personal finances.

❖ Use simple case studies and business games.

3. *Empower people to make decisions based on what they know.* Plenty of companies give lip service to the concept of empowerment. They call meetings and set up project or cross-functional teams—so many teams that a wag remarked that companies could be mistaken for bowling leagues. However, since they don't share financial information, employees

do not know how their work affects the bottom line. To rectify this situation:

- ❖ Call biweekly meetings, where income and cash flow statements, and forecasts are discussed and which people take back to their own units.

- ❖ Turn the company into a collection of smaller companies, along the lines of focus factories.

- ❖ Turn departments into business centers with their customers— internal or external—as their focus, and make them responsible for their own finances, as well as making hiring and disciplinary decisions.

4. *Assure that everyone—everyone!—shares directly in the company's success and in the risk of failure.* This is not just a profit-sharing system, determined each year after the fact, at the discretion of management, and where employees don't know how the company is doing, and they don't have a clue about how their own work affects what they receive as a bonus. In the open book system:

- ❖ Annual targets are set for profits and return on assets. If employees hit both targets, employees collect payouts, ranging from 10–50 percent of their total compensation. They can track progress against these targets, month by month.

- ❖ Open book management can work without employee stock ownership, but it works better when employees own stock in the company. Equity encourages long-term thinking for staying the course and for sacrificing instant gratification in order to build for the future.

Companies Incorporating Open Book Management

The genesis of open book management began with Jack Stack, president and CEO of the Springfield ReManufacturing Corp. (SRC) in Springfield, Missouri. His missionary zeal brought several convert companies—mostly small ones at first—to SRC to study the company's miraculous turn-around. These included Cin-Made, Chesapeake Packaging, Foldcraft Company, Web Industries, Jenkins Diesel Power, and others. The impact of more than 3,000 visitors and 300 presentations by SRC has brought in the big guns: FedEx, Wal-Mart, Exxon, Frito-Lay, and Ben & Jerry's, for example. Open book management has been introduced not only in industry and unionized companies, but also in airlines, hospitals, fund-raising organizations, and government agencies—even in police and fire departments—any organization whose performance can be measured with financial statements.

It is no wonder that Chris Lee, managing editor of *Training Magazine*, after visiting several open book companies, came away dazzled. Open book management, she wrote admiringly, is "lightning in a bottle."[6]

Self-Directed Work Teams: Lifting the Oppressive Weight of Corporate Bureaucracy

If open book management has the teaching of financials to the entire workforce as its unique feature, the second innovation—the self-directed work team (SDWT)—has, as its main thrust, the dismantling of company bureaucracy and the liberation of the worker from the chains of drudgery and boredom.

Jack Welch, the CEO of General Electric Company, states that the primary cause of stagnant productivity is the deadening weight of corporate bureaucracy—"the cramping artifacts that pile up in the dusty attics of companies: reports, meetings, rituals, approvals, and forests of paper that seem necessary until they are removed."[7]

The Concept—Opposite of an Assembly Line

An SDWT is a highly trained group of employees from six to eighteen, on average, fully responsible for turning out a well-defined segment of finished work. Because every member of the team shares equal responsibility, self-directed teams are the conceptual opposite of the assembly line where each worker has a responsibility for a very narrow function.

Work team members have more resources, a wider range of cross-functional skills, much greater decision-making authority, and better access to the information needed for sound decisions. They plan, set priorities, organize, coordinate with others, measure, solve problems, schedule work, and handle absenteeism, team selection, and evaluation. Each team member receives extensive training in administrative, interpersonal, and technical skills necessary for a self-managed group.

Origins of SDWT

Strange as it may seem, SDWT started with the coal miners and their mentors in south Yorkshire, England in 1949—more than fifty years ago. It then spread to the United States, Sweden, and India. In the past twenty years, it has been embraced not only by industry giants such as Corning, Xerox, GE, TRW, and others, but also by service industries and not-for-profit institutions.

The Benefits

There is a significant change, as shown in Table 9-4, from the typical organization to the self-directed team organization. The result is that self-directed work teams have distinct benefits, both for the company and its workers:

- ❖ A productivity increase of 20–40 percent and up to 250 percent.
- ❖ A reduction in set-up time of over 1,000 percent—from 1½ days to 1½ hours.
- ❖ Flattening the organizational pyramid—transforming supervisors into facilitators.
- ❖ Paperwork reduction —elimination, if not useful to the teams.
- ❖ Flexibility—with the workflow more suited to focus factories and mass customization.
- ❖ Quality improvements —with an enthusiasm for problem solving.
- ❖ Greater commitment to corporate goals.
- ❖ Improved customer satisfaction and loyalty.
- ❖ Most important of all, the releasing of the creative juices of people, with a reduction—almost elimination—of alienation.

From the Bureaucratic Organization to SDWT

Downloading, Not Downsizing

A natural concern of companies contemplating self-directed work teams is that it will lead to downsizing and layoffs. However, successful companies use self-directed teams for downloading, not downsizing. That means top management can concentrate on strategic planning, downloading operational matters to middle management. Middle managers now have time to

Table 9-4. The typical bureaucratic versus the self-directed team organization.

Organizational Level	Bureaucratic Organization	SDWT Organization
Executives	Tactical Decisions	Strategic Decisions
Managers	Control	Coaching, Guiding
Supervisors	Operational Decisions	Facilitators
Operators		
❖ Job Categories	Many, narrow categories	One or two broad categories
❖ Authority	By supervisor	Through group decisions
❖ Output	Minimum work to meet external standards	Exceed self-made standards
Reward System	Tied to individual job, individual performance, and seniority	Tied to team performance and individual breadth of skills

coach the teams, champion innovative ideas, concentrate on technology, find resources, and interface between the teams and the larger organization. Supervisors, in turn, no longer have to chase parts and paper, but can concentrate on their natural roles of teachers, coaches, and guides.

Training—The Engine for Transition to SDWT

Training is an absolute prerequisite to the success of self-directed work teams. It can take two to five years for a team to reach peak effectiveness. Without proper training, teams can flounder, get frustrated, and suffer "give-up-itis." Training is essential in three areas: technical skills (which can be taught just as easily to line workers as they can be to technical types), administrative skills, and interpersonal skills.

Technical

❖ Proficiency in (the powerful tools of the twenty-first century, described in Chapter 11 can just as easily be taught to line workers as they can be to technical types):

☐ Design of experiments, positrol, process certification, and precontrol for problem-solving and prevention

☐ Poka-yoke for eliminating operator controllable errors

☐ Total value engineering to reduce cost and improve customer enthusiasm

☐ Cycle-time reduction: TPM, focus factories, pull systems, setup time reductions

Administrative Skills

☐ Benchmarking

☐ Next Operation as Customer (NOAC), flowcharting, and force field analysis

☐ Basics of P&L statements and balance sheets

☐ Other financials

☐ Record-keeping, reporting, and budgeting

☐ Scheduling and Kanban

☐ Job redesign

☐ Hiring and evaluating team members

Interpersonal Skills

☐ Communications (e.g., listening, making presentations)

☐ Team dynamics

☐ Motivation theories and practices
☐ Handling interpersonal conflicts

The Five-Stage Journey to Self-Directed Teams

There are five stages in the progress toward mature self-direction:[8]

Stage 1. Start-Up

❖ Executive steering committee is formed and the mission statement is developed.
❖ A multilevel team is created, the plan outlined, and pilot sites selected.
❖ Prework begins with mid-level managers and employees.
❖ Intensive training begins for supervisors and managers.

Stage 2. Confusion

❖ With supervisor authority fading, teams may have difficulty in reaching decisions.
❖ Job security becomes a major concern, and there are speculations about the "real" corporate intentions.
❖ Managers and supervisors may be apprehensive of shrinking roles.

Stage 3. Leader-Centered Teams

❖ Positive signs emerge. Support groups begin to respond to team requests.
❖ A natural leader emerges within the team.
❖ Conflicts decline between teams and norms evolve on meetings and assignments.
❖ Managers gradually withdraw from daily operations to work on external matters.

Stage 4. Tightly Formed Teams

❖ Loyalty and protectionism emerge within teams, and there is fierce competition between teams.
❖ Managers establish councils of elected team members to review issues of mutual concern.

Stage 5. Self-Directed Teams

❖ Commitment to corporate and team goals comes to the forefront.
❖ All team members acquire new skills, reach for new technical tasks, seek and respond to internal customer needs, handle administrative duties, refine work processes, and focus on external customers and competition.

Figure 9-4 depicts the gradual transfer of operational decision-making authority from managers to teams in each of the above five stages.

SDWT Success Stories

A few examples of companies incorporating self-directed work teams will suffice to underline their effectiveness:

❖ Procter & Gamble gets 30–40 percent higher productivity in its eighteen team-based plants. Until recently, it considered work teams so vital to its success that it deliberately shunned all publicity about them.

❖ Xerox plants incorporating work teams are at least 30 percent more productive than conventionally organized plants. The same is true for General Motors.

❖ Tektronix reports one self-directed work team turns out as many products in three days as it once did with an entire assembly line in fourteen days.

❖ General Mills reports that work teams schedule, operate, and maintain machines so effectively that the night shift operates with zero managers.

❖ Aid Association for Lutherans (AAL) increased productivity by 20 percent and cut processing time by 75 percent.

❖ Shenandoah Life processes 50 percent more applications and customer services using work teams with 10 percent fewer people.

Figure 9-4. The transfer of authority from manager to teams.

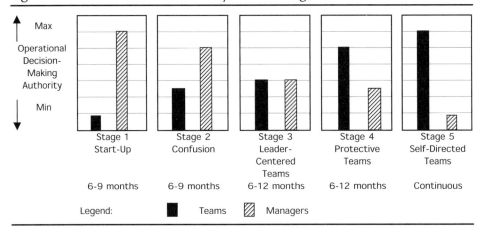

The Minicompany: Make People Before You Make Product!

If the teaching of financials is the starting point of open book management, and the breakup of bureaucracy that of self-directed work teams, then releasing the wisdom and creativity of all employees is the hallmark feature of the minicompany.[9] The old-line company transforms from "brick wall" management to a "no wall" management, from secrecy to transparency.

The Principles of Minicompanies

The principles behind the minicompany concept are simple but powerful:

- ❖ Build people before you build product.
- ❖ Provide people with information they need, problem-solving tools, and an educational environment to help people grow, just as children grow under their parents' care.
- ❖ Use the collective wisdom people have gained and harness their natural energies once they are released.
- ❖ View every individual in the organization as the "president" of his area of responsibility, providing products or services to satisfy customers. Using the Next Operation as Customer (NOAC) concept (discussed in more detail in Chapter 16), managers see shop floor people as their customers in the new tradition of the inverted organizational pyramid.
- ❖ Extend this concept one step further: Minicompanies are the "owners" of the little companies within a large company.
- ❖ Create total visibility of progress, with scoreboards established and updated by the minicompany "owners" themselves.
- ❖ Recognize contributions with tangible rewards.

The Benefits of Running a Minicompany

The overall benefit of the minicompany is that it truly unleashes the human spirit. There are additional benefits, as well:

- ❖ People develop a sense of ownership.
- ❖ By focusing on joint ownership, people learn to work as an effective team.
- ❖ The development of a mission statement and customer-supplier relationship charts will help people focus on clear objectives.
- ❖ Barriers between units of organizations are reduced.
- ❖ The total organization moves toward goal congruence as each minicompany's plans and progress are shared.

❖ Because the minicompany framework is the same at all levels, a comprehensive management framework is developed, tying the lowest level of the organization to the highest level.

The Basic Organizational Building Block of a Minicompany

One of the stated principles of a minicompany is that each individual in the minicompany can be considered a manager, and the team leader (or old supervisor) as the president. The team has internal customers, whose requirements must be met, and suppliers, whose outputs must meet the team's requirements. These suppliers include services provided by white-collar support operations. Managers now are seen as "bankers," share-holders, or even venture capitalists who provide the minicompany with resources—people, equipment, time, and money—to get the job done. Figure 9-5 is a schematic representation of a minicompany's organization chart.

The Wide Application of Minicompanies

As is the case with open book management and self-directed work teams, there are a large number of companies at various levels of minicompany implementation. They include Alps, BorgWarner, Robert Bosch, Ford Motor Co., General Motors, McDonald's, Motorola, Toledo Scale, Xerox Corp., among others. Each has shown encouraging results of quality, cost, delivery, safety, and morale improvements. The numbers are not as quantified or publicized as with the other two innovations, but one thing is clear. With any of these techniques, the genie of the irresistible human spirit is out of the bottle, and no amount of backsliding to Taylorism will ever lock it up again.

Figure 9-5. Organizational chart of a minicompany.

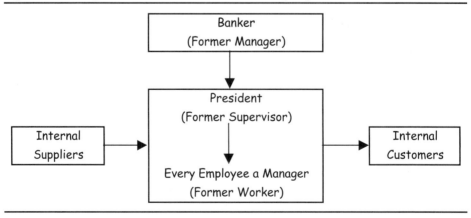

The Ten Stages of Empowerment

Table 9-5 is an evolution—in ten stages—of the march from Taylorism to unleashing the human spirit and the capabilities of workers to rise to their ever-increasing potential. Even in this age of enlightenment:

❖ Forty percent of companies are still in Stage 1 or 2—bureaucratic, authoritarian, and ruling by fear.

Table 9-5. Ten stages of empowerment leading to industrial democracy.

Stage	Action	Outcome
1. Bureaucratic	Managers make all decisions; workers are seen as a pair of hands.	Passivity
2. Information Sharing	Managers decide, then inform workers.	Conformity
3. Dialogue	Managers seek employee inputs, then decide.	Acceptance
4. Intragroup Problem Solving	Groups within each department meet to solve problems.	Commitment
5. Intergroup Problem Solving	Cross-functional teams solve larger company problems.	Cooperation
6. Start of Self-Direction	Teams improve quality, cost, cycle time; help determine flexible working hours, overtime, and redesign of work areas and business processing. Exposure to financials and business metrics gives teams a sense of ownership.	Taste of Management
7. Limited Self-Direction	Teams work with suppliers and customers with limited budget authority. Teams make contributions to company financials.	Limited Operations Management
8. Self-Direction	Worker committees take over peripheral areas such as cafeteria, recreation, and insurance.	Practice Management
9. Start of Ownership	Peer evaluation, compensation, hiring and firing of team members. Job redesign, start of minicompanies.	Empowerment
10. Ownership	Workers elect leaders, determine promotions, share in profits and losses, with bonuses and/or employee stock option plans.	Industrial Democracy

❖ Another 10 percent of companies have reached Stage 3.

❖ A further 40 percent of companies are in Stage 4 or 5, having formed teams with a focus on problem solving.

❖ About 8 percent of companies have given their teams the first taste of management—Stage 6—although coaching on business financials is an exception, rather than the rule.

❖ One- and one-half percent of the companies have possibly entered Stage 7, with a measurable team contribution to the company bottom line.

❖ Perhaps 0.5 percent have had teams take over noncore operations within their companies—reaching Stage 8.

❖ It is estimated that there are no more than eighty to 100 companies—all told—in Stage 9 of true empowerment.

❖ Only one company—Semco—can possibly have reached Stage 10.[10]

CASE STUDY

Semco—A Pinnacle of Industrial Democracy

There is only one company—Semco, located near São Paulo in Brazil—that can lay claim to having reached Stage 10 or total industrial democracy. No other company comes within a mile! Companies from all over the world have made pilgrimages to Semco to learn first-hand about its success. Its president, Ricardo Semler, has captured its spirit in his book *Maverick* that is a blockbuster.[11] Semler has been most generous with his time, lecturing in many parts of the globe on unleashing—in fact, not just in theory—the human spirit. Table 9-6 lists the innovations Semco developed for its workers and teams:

❖ Innovations common to other well-known empowering companies

❖ Innovations unique to Semco, making it the capital of industrial democracy

In terms of financial results, Semco has increased its profits seven times in less than ten years and has a waiting list of job applicants six years long!

Table 9-6. Innovations at Semco to unleash the human spirit.

Common to Other Empowering Companies	*Unique to Semco: Revolution in Industrial Democracy*
❖ Uncompromising ethics ❖ Company communications honest, even to the media ❖ No corporate debt ❖ "Skunk works" a way of life ❖ Autonomous focus factories ❖ No mumbo-jumbo of job titles ❖ Only three layers between CEO and workers ❖ Manufacturing cells, not assembly lines ❖ Pay based on a "basket of skills" ❖ Visible, transparent scoreboards ❖ Self-run cafeterias with sliding-scale rates ❖ No executive dining room ❖ No reserved parking ❖ Executives share secretaries ❖ No office cubicles ❖ No office furniture based on rank ❖ No formal dress code ❖ Regular rap sessions ❖ Free daycare centers ❖ Flextime ❖ Computers deemphasized ❖ Reducing dead-end jobs (e.g. clerks) ❖ Encouraging work at home ❖ Single-page memos ❖ No time clocks or searches at guard post	❖ Teams set their own salaries ❖ Salary incentives/penalties based on company performance ❖ Help in determining profit-sharing splits among teams ❖ Help in determining plant locations ❖ CEO rotational, not permanent ❖ Employees choose their leaders ❖ Employees decide on promotions ❖ Growth for the sake of bigness is no virtue ❖ Laid-off workers encouraged/guided to become suppliers ❖ Job openings: First preference for current employees, next for their friends ❖ Worker committees: Identify surplus managers; challenge expenses; decide on layoffs or lower wages ❖ Plant sizes kept to a maximum of 150 people ❖ Hiring/firing of their own team members ❖ Policy manuals to the shredder ❖ Job descriptions based on what workers want to do ❖ Job rotation at manager level ❖ Sabbaticals for managers ❖ Worker rating of supervisors ❖ Company health insurance, but employees decide how spent

CASE STUDY

Springfield ReManufacturing Corp. (SRC)—A Benchmark Company in the Area of Employee Empowerment[12]

A U.S. benchmark company and the one that started the open book management revolution in America is the Springfield ReManufacturing Corp. (SRC) in Springfield, Missouri. Its business is remanufacturing engines and engine components by removing worn-out engines in cars, bulldozers, and eighteen-wheelers and rebuilding them.

Its Shaky Start

SRC was a small department in the huge International Harvester plant in Melrose Park, Illinois. The plant was among the worst in America marred by racial incidents, death threats, burning of effigies, bombings, shootings, and aggravated assaults. Workers and managers were constantly at each other's throats. Things progressed from bad to worse until International Harvester sold the operation to thirty-six of its managers for $9 million. These managers were only able to scrape up $100,000 and had to borrow the rest—a debt-to-equity ratio of 89:1.

Open Book Management to the Rescue

With his back to the wall, Jack Stack, former manager and now CEO of SRC, developed the theory, concepts, and practice of open book management. He cites two preconditions that are essential before open book management can be launched:

1. *Management must have credibility.* It must earn a minimum level of mutual trust and respect. When managers flaunt what they have, when they intimidate, when they treat people badly, they forfeit their power.
2. *Employees should have some fire in their eyes.* They should not be made to feel that they are losers. For people to feel like winners, they must have pride in themselves and in what they do. There is no winning without pride; no ownership without pride.

There are all kinds of techniques to build credibility and light fires in peoples' eyes—listening to workers and their ideas, encouraging them to dream, creating small wins, and making a big deal out of little victories—and, in the process, having fun!

Rearranging the Financials in Terms Meaningful to Each Employee

Open book management uses the same two metrics common to all businesses—the balance sheet (which SRC calls the thermometer, to measure if it's healthy or not) and the income statement (which tells how the company got that way and what you can do about it). But these are not the typical financial statements of the CPA world. The detail is broken down to show how each employee affects the income statement and the balance sheet. Categories where the most money is spent are highlighted. Cost-of-goods-sold is broken down into its basic elements—material, labor, and overhead. Every element of the company is quantified—from the percentage of the budget spent on receptionists' notepads to the amount of overhead absorbed each hour that a machinist uses in grinding crankshafts. Sales numbers are posted daily: who's buying, what they're buying, how they're buying. The numbers are broken down, not just by customer but by product.

"Skip the Praise, Give Us the Raise!"

The real outcome for the week-by-week knowledge of the financials is SRC's bonus program, nicknamed "Skip the Praise—Give Us the Raise" (shortened to STP-GUTR and pronounced "stop-gooter"). Bonus programs and gain sharing are, by no means, new to industry. Furthermore, SRC's bonus of 13 percent of annual pay is not in the upper tenth percentile of incentives in other companies. There are, however, some unique features that makes STP-GUTR at SRC so cherished by its employees:

❖ All employees are in the same boat—from CEO to line operator. The ceiling bonus for managers and professionals is 18 percent of annual pay; for everyone else it is 13 percent. This 5 percent differential is the lowest and most egalitarian among all prevalent incentive systems.

❖ The "floor" for opening the bonus gate is a modest profit of 5 percent before taxes or 3.3 percent after-tax profit—one of the lowest in industry. This enables bonuses to be paid even during business down-cycles.

❖ As profit margins increase, the bonus percentages are also ratcheted up.

❖ There is a tie-in between the profit-on-sales target and the balance sheet goal, to assure adequate cash flow and liquidity.

❖ The payouts are made every quarter as opposed to the usual practice in industry of annual payouts. Annual payouts lose their impact on workers, who might not correlate their individual contributions with the delayed benefits.

❖ SRC goes to great lengths to make sure its people understand how the bonus program works. Bad communications is the main reason most bonus systems fail.

❖ SRC closely tracks all financials and announces bonus results once a week. It holds meetings, sets up an electronic ticker tape in the cafeteria, and flashes the score at lunch.

❖ The company believes that the real power of its bonus program lies in its ability to educate its people about business.

Results at SRC in Adopting Open Book Management

The results speak for themselves:

❖ At the end of its first year of operation, SRC had a loss of $670,488 on $16 million in sales.

❖ Ten years later, SRC had the pretax earnings of $27 million on sales of $83 million.

❖ Its stock, worth 10 cents a share at the time of its buyout, was worth $18.36 in ten years, an increase of 18,200 percent.

❖ Despite lean years and a recession, it never laid off any of its workers and has never missed a bonus.

SELF-ASSESSMENT/AUDIT ON EMPLOYEE EMPOWERMENT

Table 9-7 is a self-assessment/audit that a company can conduct to score its outlook on employees. It has six key characteristics and fifteen success factors, each worth five points for a maximum score of seventy-five points.

Table 9-7. Employees: key characteristics and success factors (75 points).

Key Characteristic	Success Factors	Rating				
		1	2	3	4	5
4.1 Motivation: From "How to" to "Want to"	1. The company has recognized that motivation cannot be extrinsic to the workers; it must be intrinsic—stemming from within them. 2. The company has recognized that this inner motivation springs from: the challenge of the job itself, achievement, recognition, responsibility, and advancement. 3. The company has redesigned worker jobs by combining tasks; forming natural work units; establishing client relationships; and opening feedback channels as important in achieving high internal motivation in the worker. 4. The company has made vertical job enrichment the centerpiece of job redesign and applied it to transform dull, boring, repetitive jobs and create fun in the workplace.					
4.2 Job Security	1. While no company can provide iron-clad guarantees of jobs for workers, the company minimizes, to the greatest extent possible, layoffs through under-staffing and overtime; retraining; job rotation; and temporary insourcing.					
4.3 Empowerment Readiness	1. The company has undertaken a survey, before launching an employee empowerment process, to determine if its management and workers are truly ready for empowerment.					
4.4 Team Competition	1. The company sponsors the formation of numerous teams—generally cross-functional—to establish their own goals (e.g. quality, cost, cycle time, innovation) and encourages friendly competition among them. It gives these teams time to work on their projects and celebrates the winning teams.					

(continues)

Table 9-7. (Continued).

Key Characteristic	Success Factors	Rating				
		1	2	3	4	5
4.5 Empowerment Systems	1. The open book management of empowering people has been adopted, with its unique feature of opening and teaching all financial information in such a way that all employees know how they contribute to the income statement and the balance sheet.					
	2. A fair, egalitarian, and motivational bonus system has been designed by the company for all employees to acquire a piece of the company ownership.					
	3. The company has created self-directed work teams to lift and remove the oppressive weight of bureaucracy from its people.					
	4. There is a heavy emphasis on training—in technical, administrative, and interpersonal skills—to prepare people for self-directed work teams.					
	5. The company's "brick wall" management has been transformed into "no wall" management—from secrecy and withholding information to transparency of all information and company progress, with employees fully informed.					
	6. The company emphasizes the concept of the "next operation as customer" in all areas of the company.					
	7. The company has created minicompanies as the "owners" of the little companies within the large company.					
4.6 Empowerment Stage	1. The company is constantly escalating its empowerment stage from the low stages and is moving toward industrial democracy.					

Ten

FROM TRADITIONAL INDICATORS TO ROBUST METRICS

❖ ❖ ❖

When you can measure
what you're speaking about
and express it in numbers,
you know something about
it, but when you cannot
measure it, when you
cannot express it in
numbers, your knowledge is
of a meager and
unsatisfactory kind. It may
be the beginning of
knowledge, but you have
scarcely advanced to the
stage of science.

—LORD KELVIN

Measurement Axioms

To add to the great scientist Lord Kelvin's profound statement, there are two axioms regarding measurement:

1. That which is not measured is not managed.
2. Too many measurements end up with no measurement.

The first can be illustrated with an analogy. What would happen to a sport—any sport for that matter—without a score? Imagine the game of bowling without a score. If prevailing industry rules were applied to the game, the bowling alley would be draped to prevent score keeping. Coaches would critique the bowlers' performance, telling them to do better but not saying how. The rules of the game would change without reason. Social interaction among bowlers would be discouraged and instead the threat of expulsion from the game would be the way to keep bowlers at play.[1] It is also known that workers would only exert themselves to fulfill the measurements important to management. If management tracks shipments hourly but only mouths quality platitudes in vague terms, the workers will focus on the former and allow quality to twist slowly in the wind.

The mirror statement of axiom 1 is also true, "What gets measured, gets accomplished."

The second axiom is well known to behavioral scientists, especially in the formulation of goals. Workers cannot concentrate on too many measurements and too many goals, neither can corporations. Many companies, as part of their management by objectives (MBO) drive, formulate thirty, forty, fifty goals each year—cafeteria style—and expect their workers to juggle their limited time and resources to achieve them. Workers, as a consequence, become harried, unfocused, and demoralized and the results are mediocre at best.

There should be no more than three or four key goals and their associated measurements that companies, divisions, businesses, teams, and departments should focus on in order to maximize results.

Measurement Principles

Several important principles constitute the foundation of a robust Ultimate Six Sigma measurement system:

1. Any department or team can and must be measured for effectiveness.
2. The measurements must be simple, meaningful, fair, and easy to

apply, with little calibration error. (Sometimes two or more measurements may be needed to afford a true perspective on the situation under consideration.)

3. Employees must be involved in the formulation of the measurements.

4. When employees have operational "ownership," they set higher goals than management.

5. Team measurements should be transparent and prominently visible to all—employees, managers, and customers.

6. Each team should be measured against itself, using time for comparison. Pitting one team against another should be discouraged.

7. The benefit-to-cost ratio of any measurement parameter must be at least 5:1 and preferably 10:1. Nothing is worse than measuring efficiently a parameter that adds little value.

8. Leading indicators should be given much greater attention than lagging indicators.

9. Traditional and weak indicators should be substituted with twenty-first century, robust indicators.

10. Quantitative indicators should be reinforced with qualitative indicators.

Universality of Measurements

In manufacturing industries, production employees are measured constantly, with resultant high rates of productivity. But white-collar operations in these industries have historically resisted the scrutiny of measurement and accountability. To them, measurement is for those "second-class" citizens in manufacturing. "We work intellectually," say the white-collar operations. "How can we put a number on our cerebral thinking?" Further, if any chance, a measurement does get introduced, they start guerrilla warfare in the trenches to torpedo it.

The same resistance is encountered in the service sector in commercial services, schools, hospitals, nonprofit institutions, and the government—local, state, and federal. (When I served as president of the board of trustees of New Trier Township on Chicago's North Shore, we had the task of funding social service agencies through tax levies. To determine their effectiveness, I suggested the agencies measure the satisfaction of their clients. This brought a howl of protest, but the power of the township's purse strings forced an "attitude adjustment.") Yet two parameters—customer (internal and/or external) evaluations and cycle time—are effective measures for any support service. These will be explained later in this chapter.

Measurement Simplicity and Visibility

Complex formulas for calculating a measurement parameter generate confusion for the average worker. As an example, the calculation of team bonuses based on achievement, when diluted or negated by profit and return-on-investment "gates," can cause suspicions among team members unless the metrics are clearly explained to them.

Performance indicators cannot be buried in a computer terminal or printout, either. They must be visible for all to see and honest to have credibility.

The Scoreboard

An example of this visibility and total transparency of measurements is a prominently displayed scoreboard, a technique used by minicompanies (see Chapter 9) to register its progress. The scoreboard includes the following information:

- ❖ The minicompany name, team members, mission, and goals
- ❖ The internal customer-supplier relationship chart (see Chapter 16)
- ❖ Progress on quality, cost, cycle time, delivery, innovation, service, and safety
- ❖ Customer feedback
- ❖ A matrix of team members comparing the type and number of skills they've acquired
- ❖ A defined focus for the month
- ❖ The minicompany's contribution to the company's financials

In short, the scoreboard is viewed by employees, managers, and customers alike, in keeping with an "open kimono" policy that is far more open and transparent than conventional policies that have company information passed on to financial analysts.

Leading Versus Lagging Indicators

Several parameters measure only results. They are important, but too late to improve or correct. It is like steering a car by looking at the rearview mirror. These indicators are lagging, as contrasted with leading indicators, which are sensitive barometers of the financial weather that will follow. For example:

- ❖ Profit, return on investment, and asset turns are lagging indicators. They are results not causes, outputs not inputs. Actions are confined to post-mortems. By contrast, customer loyalty, retention, and longevity (Chapter 6) are leading indicators. You can work on them, in advance, to influence the lagging indicators.

❖ Market share is equally a lagging, and often misleading, indicator. Increasing market share may give you a false indication that all is well when, in reality, customer defection may be slowly rising to eventually torpedo the business.

❖ Inventory can be both a leading indicator and a lagging indicator. It is a leading indicator relative to return on assets, since it directly influences the latter (along with fixed assets and receivables). But, in a larger sense, it is a lagging indicator—a result. You cannot actually work on inventory unless you work on its leading indicator twin—cycle time: in raw materials, work-in-process (WIP), and finished goods.

Traditional (Weak) Versus Robust Indicators

Companies, led by the nose by accountants (who have a penchant for being exactly wrong rather than approximately right!), have blindly followed traditional metrics that are wrong, weak, or obsolete, rather than develop corresponding metrics that are right, robust, and in tune with the twenty-first century (see Table 10-1).

Because several of these metrics listed in Table 10-1, especially the newer robust ones, are not well known, further explanation is in order.

1. *Financial Statements.* The current system of financial reporting is similar to a shop selling coats in just one size. The customers of financial information no longer want to be prisoners of "the one size fits all" accounting system. The recommendations for changes in financial reporting are contained in a committee report, "The Information Needs of Investors and Creditors" issued by the American Institute of Certified Public Accountants (AICPA) in response to mounting litigation on the opaqueness of financial statements.[2]

2. *Shareholder Value Versus Stakeholder Value.* Increasing shareholder value is the sacred mantra of public corporations, as measured by the earnings per share, quarter by quarter. While this is important, the measurement of the other stakeholders' satisfaction—customers, employees, suppliers, distributors, and dealers—is much more important. But how do you measure stakeholder value? One measurement device that can convert a subjective measurement into an objective one is an opinion scale, from one to ten (with one being the worst and ten being the best), determined by *core* stakeholders. This is the same technique by which an attribute (e.g., pass/fail, accept/reject) can be transformed into a variable. (We call

Table 10-1. Traditional versus robust indicators.

Traditional	Robust

1. Financial Statements
  ❖ Historical cost-based systems
 	
  ❖ Nonrecurring losses given line-item treatment while gains treated as ordinary income
  ❖ Disaggregation disclosures inadequate to predict earnings and cash flow
  ❖ Estimates, assumptions, and off-balance sheet risks are sketchy
 	
  ❖ Little information on company's success factors and nonfinancial performance measures
 	
  ❖ Forward-looking information missing
	1. Financial Statements
	  ❖ Need for fair value information on long-term assets and liabilities
	  ❖ Need for flagging stable, sustainable earnings
	
	  ❖ Need for details and trends business by business
	
	  ❖ Need for more qualitative and quantitative information on risks associated with financial instruments
	  ❖ Need for disclosure on company's success factors (e.g., customer retention, quality, innovation, empowerment, etc.)
	  ❖ Need for information on near-term opportunities and risks that are quantifiable

2. Emphasis on shareholder value	2. Emphasis on measuring value of all stakeholders
3. Profit on sales	3. Return on assets
4. Maximizing number of customers	4. Concentration on core customers
5. Overhead cost allocations based on direct labor	5. Activity-based cost (ABC) accounting
6. Quality determined by defect levels	6. Quality measured by the cost of poor quality (COPQ)
7. Warranty costs	7. Lifetime costs of a defecting customer
8. Large supplier base, distributor/dealer/servicer base	8. Sharply reduced, supplier partnership base, distributor/dealer/servicer base
9. Large part number base	9. Sharply reduced part number base
10. Minimize product options through standardization	10. Maximize product options through mass customization

this a ''Bo Derek scale,'' referring to the movie *Ten*, where Bo Derek was rated a perfect ten.)

3. *Profit on Sales Versus Return on Assets*. Companies and their stockholders report and focus on profits as a percentage of sales. But the return on assets is a metric worthy of much greater pursuit. As an example, a company with 5 percent profit on sales and two asset turns achieves only

a 10 percent return on assets. By contrast, a company with just a 2 percent profit on sales and ten asset turns achieves a 20 percent return on assets— twice the return with less than half the profit. Benchmark companies pursuing this strategy have achieved 100 percent return on assets!

4. *Maximizing Number of Customers Versus Reducing the Customer Base.* This topic was discussed in depth in Chapter 6. There is no mileage in expanding your customer base. On the contrary, send your worst customers to your competition.

5. *Overhead Cost Allocations, Direct Labor–based Versus ABC-based.* Allocating overhead costs based on a direct or labor base is a practice whose time has long gone. It is obsolete, inaccurate, and can lead to wrong business decisions. Allocations based on material costs are better; in the last ten to fifteen years, activity-based costing (ABC) has come into its own as a realistic and effective way to allocate overhead costs.

6. *Quality Defect Levels Versus the Cost of Poor Quality.* Dr. Joseph Juran, one of the quality gurus of the world, stressed that for the quality movement to be credible, it must convert quality from the language of defect levels to the management language of money. Quality was traditionally measured in terms of defect levels, such as acceptable quality levels (AQL), average outgoing quality levels (AOQL), and lot tolerance percent defective (LTPD). Later, as defect levels dropped dramatically, they were measured in parts per million (ppm), parts per billion (ppb), total defects per unit (TDPU), c_p (process capability) and c_{p_K} (noncentered process capability). Armand Feigenbaum of General Electric introduced the concept of the cost of quality over forty years ago in his landmark book *Total Quality Control*. It has four categories: external failure costs, internal failure costs, appraisal costs, and preventive costs. Of these, only the last, preventive costs, is desirable. The other three are undesirable and constitute the cost of poor quality (COPQ). The most easily gathered elements of COPQ are shown in Table 10-2.

For a company that has not begun the quality revolution, this cost of poor quality ranges from 10–20 percent of the sales dollar. That is pure waste. And when it is realized that the average profit on sales for a company is only 5 percent, the COPQ loss is two to four times the profit. Here is one of industry's greatest money-making machines, next only to reducing customer defections. A 50 percent reduction in COPQ can in-

Table 10-2. Easily gathered elements of cost of poor quality (COPQ).

External Failure Costs	Internal Failure Costs	Appraisal Costs
Warranty	Scrap	Incoming inspection
Product recalls	Analyzing	In-process inspection
Liability lawsuits	Repair/rework	Outgoing checks
	Inventory carrying costs	Test

crease profit 100 percent! Yet the accounting system is blissfully ignorant of its profit potential, just as it is ignorant of customer defection reduction, the cumulative impact of one truly dissatisfied customer over a lifetime, ABC-based allocations for overhead costs, and the impact of inventory, which accounting views as an asset instead of a liability.

I have made a parallel study of the cost of poor quality for a company that has not addressed the profit potential of quality. That figure per employee per day is $100 to $200. (One enterprising company dramatized this cost to its employees by smashing a brand-new Mercedes in the parking lot. That is how much, the company said, poor quality cost each day!) If COPQ can be cut to one-third, from $150 per day to $50 per day, the total savings for the United States as a whole could be $500 billion per year (assuming 20 million people in manufacturing alone and 250 working days per year). Even if 50 percent of these savings are passed on to customers, the positive impact on tax receipts to the U.S. Treasury would be $75 billion.

The COPQ elements in Table 10-2 are just the tip of the iceberg when it comes to COPQ costs. Figure 10-1 depicts the hidden cost of poor quality—those far more insidious elements below the waterline that accounting is largely incapable of even defining, much less measuring. It is estimated that if these hidden COPQ costs were capable of being mea-

Figure 10-1. The hidden costs of poor quality.

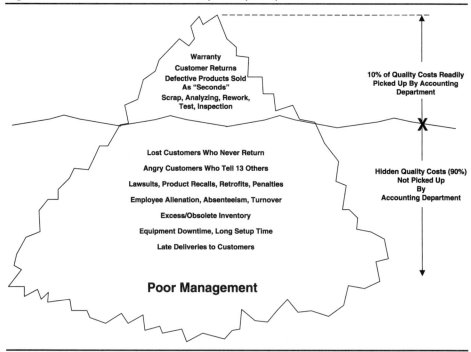

sured, the true and total quality loss to a company could be as high as 50 percent of the sales dollar.

7. *Warranty Costs Versus Lifetime Costs of a Defecting Customer.* Most companies are unaware of the multiplier effect in sale losses caused by a defecting customer. Worse, their nineteenth-century accounting departments do not know how to even begin estimating that loss. Typically, warranty costs run between 0.5–2 percent of the sales dollar. CEOs are not phased by these apparently small percentages, which are just the tip of the iceberg that can sink their companies.

What is the real cost of a customer who is so frustrated with a company's products that he will never, ever buy from that company again? The automotive industry has estimated that cost as not that of one car sale— say, $30,000—but a cost ranging from $500,000 to $1.5 million. How? Let us assume that:

❖ The customer, during his adult life, would buy ten cars.

❖ He will tell twenty other customers of his extremely poor experience with Car Company A.

❖ Three of these other customers will be influenced and not buy Car Company A's cars.

The loss, therefore, is not $30,000 but $300,000 times three or $900,000. Add to this the loss to Car Company A of out-of-warranty service, out-of-warranty parts, and financing costs—say, another $20,000 in four years. The total cost: $920,000.

8. *Large Supplier Base Versus a Sharply Reduced Base of Partnership Suppliers.* In Chapter 13, the rationale for a reduced customer base was clearly made. (It is just as desirable to reduce the distributor, dealer, and servicer bases.) Similarly, there is a compelling need to reduce the large and unwieldy number of suppliers for a company—ranging from a few hundred suppliers for a small company to thousands for large corporations. Companies do not have the time and resources to manage and help such a large supplier base. Instead, by concentrating on a few suppliers with whom they can enter into a true partnership relationship (the topic of Chapter 13), a company can truly improve supplier quality, cost, and cycle time. Therefore, one of the metrics used in supply management is to reduce a company's supplier base by factors of 10:1 and even 20:1.

9. *Large Part Number Base Versus a Sharply Reduced Part Number Base.* It makes no sense to reduce the supplier base if, at the same time, the part number base is increased by a proliferation of new parts. The economic gains of the former are neutralized by the loss of the latter. It costs $1,300 to $12,000 for the addition of one part number to a system, according to an excellent study published in the *Harvard Business Review*.[3] This cost in-

cludes the expenses for design, planning and control, inventory, and record-keeping. Therefore, a parallel metric of reducing the supplier base of a company is reducing its part number base.

10. *Standardization Versus Mass Customization.* In an effort to cut costs, companies used to severely limit the number of options associated with a product. The classical example was the old Henry Ford chestnut that his customers could have any color car they wanted as long as it was black. General Motors seized on that weakness to offer customers any color they wanted and went on to eclipse Ford Motor Co. at the marketplace. In a sense, that was the start of mass customization. Today, with mass customization becoming the rage with customers, the number of permutations and combinations in options can number more than 100,000. It becomes important to limit the number of such permutations by determining the benefit-to-cost ratio of the less significant options.

Types of Measurement

While a number of parameters of measurement are listed in the following categories, great care must be exercised in selecting only a few of those parameters that are meaningful to an organization, its values, and its culture. Any department's or group's performance can be assessed by its:

❖ Quality effectiveness
❖ Cost effectiveness
❖ Cycle-time effectiveness
❖ Service effectiveness
❖ Innovation effectiveness
❖ Empowerment effectiveness

Of these, the last three—service, innovation, and empowerment—are generic to all departments/teams, whereas the first three—quality, cost, and cycle time—can be tailored to each function/team.

Generic Measures of Service Effectiveness

As measured by external or internal customers, on a scale of one to ten, these measures include:

Dependability
❖ Trustworthiness
❖ Consistency

❖ Credibility

❖ Promises kept

Cooperativeness

❖ Responsiveness

❖ Speed in solving problems/complaints/questions

❖ Flexibility

❖ Approachability

❖ Courtesy

Communications

❖ Time spent with core customers

❖ Time spent with customer-contact people

❖ Visits by CEOs and senior management to core customers

❖ Feed-forward information on problems

Generic Measures of Innovation Effectiveness

❖ Number of patents generated as a percentage of the technical population

❖ Number of patents commercialized as a percentage of the technical population

❖ Sales dollars from new products (over one to two years) as a percentage of total sales dollars

❖ Number of benchmark projects implemented as a percentage of sales

❖ Number of small product starts (with less than 25 percent change from their predecessor products) as a percentage of the R&D budget

❖ Frequency of a stream of new products (with less than 25 percent change from their predecessor products) as a percentage of the R&D budget

❖ Number of customized features (i.e., differentiators) added to a standard product

❖ Number of "skunk work" projects created as a percentage of the technical population

❖ Number of "skunk work" projects commercialized as a percentage of the technical population

❖ Number of ideas lifted from competition through reverse engineering (and other means) as a percentage of the R&D budget

❖ Number of suggestions as a percentage of the employee population

❖ Number of rules and regulations challenged and deleted as a percentage of sales

❖ Number of ideas from suppliers and the dollar value of ideas implemented

Generic Measures of Empowerment Effectiveness

❖ Ratio of total compensation of CEO to that of lowest-level worker

❖ Ratio of corporate staff to total number of employees

❖ Gain sharing as a percentage of base pay for the average worker

❖ Number of managers transformed into leaders as a percentage of the total management population

❖ Number of management layers reduced as a percentage of total number of layers between CEO and worker

❖ Number of supervisors eliminated as a percentage of the total supervisor population

❖ Percentage of workers in teams

❖ Number of team projects completed as a percentage of sales

❖ Team project savings as a percentage of sales

❖ Training cost as a percentage of total payroll

❖ Tangible savings from training as a percentage of sales

❖ Ratio of number of jobs redesigned to those not redesigned

❖ Ratio of managers to workers laid off during company restructuring/downsizing

❖ Average number of hours per week of management by walking around (MBWA)

❖ Longitudinal progress on employee attitude surveys

❖ Percentage of workers having near full trust in top management

❖ Percentage of workers who consider company has uncompromising integrity and ethics

Specific Measures of Quality, Cost, and Cycle Time by Function/Team

Table 10-3 lists the parameters that can be used to track quality, cost, and cycle-time effectiveness by function/team. Nine key functions are covered—marketing, sales, engineering/design/R&D, supply chain management (SCM), manufacturing, quality assurance (QA), accounting, human resources, and information technology (IT).

Again, it must be emphasized that although the list of measurements

(text continues on page 167)

Table 10-3. Tracking quality, cost, and cycle-time effectiveness by function/team.

Function/ Team	Measurement of Quality Effectiveness	Measurement of Cost Effectiveness	Measurement of Cycle Time Effectiveness
Marketing	❖ Accuracy in profiling customer requirements (actual vs. predicted) using: 1. Quality function deployment (QFD) and mass customization 2. Focus groups, clinics, panels ❖ Accuracy of projected sales volume and product pricing ❖ Accuracy of market demographics ❖ Sales rating of marketing effectiveness	❖ Ratio of value of market intelligence to the cost of marketing ❖ Ratio of cost of marketing to sales dollars ❖ Cost of proliferations in mass customization as percentage of product cost	❖ Speed and timeliness of market intelligence as percentage of product cost ❖ Cycle time in concept stage of new product development vs. previous comparable product
Sales	❖ Customer defections as a percentage of total number of customers ❖ Time spent with customer per sales win ❖ Customer rating of sales effectiveness ❖ Engineering (internal customer) rating of sales effectiveness ❖ Accuracy of sales forecasts (predicted vs. actual) ❖ Accuracy of sales order from customer to factory ❖ Number of changes in customer requirements during new product introduction as a percentage of product cost	❖ Ratio of sales cost, to sales dollars ❖ Ratio of sales bids to sales won ❖ Percent profit increase from percent customer retention increase ❖ Percent profit increase from years of customer longevity ❖ Number of noncore customers reduced as a percentage of total customer base ❖ Brand loyalty increases as dollar amount of total sales dollars	❖ Average longevity of customer retention ❖ Cycle time from customer order to factory order ❖ Average time from customer complaint to resolution ❖ Percent of delinquent orders

(continues)

Table 10-3. (Continued).

Function/ Team	Measurement of Quality Effectiveness	Measurement of Cost Effectiveness	Measurement of Cycle Time Effectiveness
Engineering/ Design/R&D	❖ Number of design changes after Job 1 as percentage of product cost ❖ Actual vs. target yield at Job 1 ❖ Number of c_{p_K} parameters greater than 2.0 as percent of total important parameters ❖ Field reliability, actual vs. target ❖ Warranty cost as percentage of sales dollar ❖ Mean time to diagnose (MTTD) ❖ Mean time to repair (MTTR) ❖ Built-in diagnostics as percentage of total critical features ❖ Ergonomics vs. competitor (customer perception) ❖ "Wow" features as percentage of total number of features ❖ Cost of product liability suits vs. historic ❖ Manufacturing's rating of engineering effectiveness ❖ Design quality (at Job 1) vs. earlier comparable design ❖ Number of design flaws prevented through Multiple Environment Over Stress Testing (MEOST) ❖ Number of parts certified as having realistic specifications and tolerances using variable search and scatter plots as percentage of total parts ❖ Number of parts with classification of characteristics as percentage of total parts	❖ Product cost vs. competition ❖ Design cost, actual vs. target ❖ Design for manufacturability score ❖ Early supplier involvement (ESI) savings as a percentage of product cost ❖ Ratio of cost targeted components to total number of components ❖ Total "black boxes" outsourced as a percentage of product cost ❖ Value engineering savings as a percentage of product cost ❖ Mass customization options as a percentage of product cost ❖ Design cost vs. earlier comparable design ❖ Part number reduction vs. historic ❖ Standard (common) parts as a percentage of total parts count	❖ Average time to resolve design problems ❖ Time (months) to recover costs (cash flow) of human inventory in design ❖ Design cycle time, actual vs. target ❖ Design cycle time vs. competition ❖ Tooling cycle time as a percentage of total design time ❖ Design cycle time vs. earlier comparable design ❖ Number of differentiators added to product every quarter

Supply Chain Management (SCM)*

- Supplier quality levels, as measured by customer vs. target
- Supplier reliability levels, as measured by customer vs. target
- Number of certified parts (no incoming inspection) as percentage of total number of parts
- Number of parts with c_{p_k} of 2.0 and more as percentage of total number of important parts
- Number of parts with mutually determined specifications (by supplier and engineering) as a percentage of total number of parts
- Number of precontrol charts as a percentage of important/critical parts
- Percent reduction of supplier base from historic
- Percent reduction of distributor/dealer/service bases from historic
- Manufacturing rating of SCM effectiveness
- Number of partnership suppliers extending active, concrete help to their subsuppliers as a percentage of the total supplier base

- Total "ownership" costs (price + COPQ + delinquency) as percentage of purchase costs on important/ critical parts
- Material inventory turns
- Total cost of supplier parts vs. target
- Procurement costs as a percentage of material costs
- Savings affected by commodity team and supplier as a percentage of material costs
- Profit margin increases gained by partnership suppliers vs. historic
- Larger volumes gained by partnership suppliers vs. historic volumes
- Longer-term contracts gained by partnership suppliers vs. historic
- Number of supplier costs negotiated as a ceiling vs. number negotiated as a floor
- Number of financial incentives/penalties established on key parts as a percentage of total number of key parts
- Savings to company and to a partnership supplier as a result of idea incentives
- Number of parts on volume variable pricing vs. total number of key parts

- Lead time improvements in important/ critical parts
- Number of partnership suppliers with lean manufacturing systems (e.g., Kanban) as a percentage of total suppliers
- Number of partnership suppliers with firm lead time reductions extended to their subsuppliers
- Total supplier lead time reductions vis-à-vis historic figures
- Cycle time for purchase orders, from factory orders to suppliers
- Number of partial authorizations for supplier for raw material as a percentage of total material dollars

(continues)

Table 10-3. (Continued).

Function/ Team	Measurement of Quality Effectiveness	Measurement of Cost Effectiveness	Measurement of Cycle Time Effectiveness
Manufacturing	❖ Number of processes characterized and optimized (using variables search & scatter plots) as a percentage of total number of processes ❖ Number of key process gains "frozen" with positrol as a percentage of total number of processes ❖ Number of key processes made "robust" through process certification as a percentage of total number of processes ❖ Comprehensive date retrieval system that is accurate, timely, visible, and transparent to all, based on employee surveys ❖ Number of important operator-controlled processes using poka-yoke as a percentage of total operator-controllable processes ❖ Number of multi-skilled operators (three or more skills) as a percentage of total number of operators ❖ Number of reliability failures uncovered in production through "field escape control" vs. reliability failures in the field not uncovered ❖ Number of design of experiment (DOE) projects completed and associated quality improvement as a	❖ Number of key processes with OEE greater than 85% as a percentage of total key processes ❖ Savings per year as a result of team projects as a percentage of total manufacturing population ❖ Number of suggestions/employee/year and associated savings ❖ Raw material, work in process (WIP), and finished goods inventory turns ❖ Number of stations in product flow vs. number in process flow ❖ Space reduction as a percentage of total manufacturing floor space	❖ Reductions in work in process (WIP) cycle time vs. historic ❖ Number of focus factories in a single plant as percentage of sales dollars ❖ Number of product flow stations (vs. process flow stations) as a percentage of total number of stations ❖ Setup time reductions vs. historic setup times ❖ Lot size reductions vs. historic lot sizes ❖ Reductions in master schedules vs. historic ❖ Percent of forecasts ignored ❖ Number of multiple skills per employee ❖ Reductions in lot sizes from customer vs. historic ❖ Increases in frequency of customer orders vs. historic ❖ Space savings through cycle-time reduction ❖ Sales rating of manufacturing effectiveness (percentage of orders delinquent)

percentage of total manufacturing population
- ❖ Accident/injury reductions per line worker
- ❖ Outgoing quality, actual vs. target
- ❖ Total defects per unit (TDPU), actual vs. target
- ❖ Cost of poor quality (COPQ), actual vs. target
- ❖ Field reliability, actual vs. target

| Quality Assurance (QA) | ❖ Number of reduced customer defections as a percentage of total number of customers
❖ COPQ as a percentage of sales dollars
❖ External customer satisfaction index (CSI)
❖ Number of problems solved/prevented, with quality as coach, as a percentage of sales dollars
❖ Internal customer evaluations (manufacturing and engineering) of QA
❖ Outgoing quality, TDPU, yield/cycle time
❖ c_{p_k} greater than 2.0 of important parameters as a percentage of all important parameters
❖ Customer satisfaction index (CSI) | ❖ Total quality costs as a percentage of sales dollars
❖ Cost of a single lifetime customer defections per product line
❖ Cost of QA department as a percentage of sales dollars
❖ Ratio of inspectors to total number of direct labor employees
❖ Savings from number of problems solved/prevented as a percentage of sales dollars | ❖ Cycle time of quality problems solved vs. not solved
❖ Overall equipment effectiveness of processes greater than 85% vs. all processes |

(continues)

Table 10-3. (Continued).

Function/ Team	Measurement of Quality Effectiveness	Measurement of Cost Effectiveness	Measurement of Cycle Time Effectiveness
Accounting	❖ Internal customer (e.g., management, manufacturing) evaluations of accounting effectiveness ❖ Accuracy of reports/paychecks ❖ Clarity of reports ❖ Number of leading measurement indicators vs. lagging indicators ❖ Trend analysis/warnings ❖ Explanation of financials to all workers ❖ Reforms in income statements and balance sheets	❖ Use of activity-based costing (ABC) system to allocate overheads rather than direct labor based ❖ Gathering and analysis of COPQ ❖ Internal customer (e.g., manufacturing, management) evaluation ❖ Number of reports reduced (those not useful to customer) as a percentage of total reports ❖ Ratio of value of reports to cost of reports ❖ Cost of customer defections as a percentage of sales dollars ❖ Percent profit increase resulting from percent customer retention increase ❖ Number of royalty payments granted to partnership suppliers even with no business awarded to them ❖ Accounting costs as a percentage of sales	❖ Timeliness of reports, actual vs. target ❖ Timeliness of year-end closing ❖ Number of accounts payable invoices paid within 30 days as a percentage of total invoices ❖ Number of accounts receivable items collected within 30 days as a percentage of total receivable items
Human Resources	❖ Voluntary personnel turnover as a percentage of total headcount ❖ Percent of employees not satisfied in attitude surveys ❖ Number of unfilled requisitions as a percentage of total requisitions ❖ Number of people hired on new criteria vs. old	❖ Training dollars as a percentage of payroll ❖ Number of pay increases based on performance vs. total pay increases ❖ Number of middle-level managers eliminated as a percentage of total middle-level managers ❖ Number of management layers reduced	❖ Cycle time to fill requisition ❖ Cycle time to pay medical claims ❖ Mean time between accidents

	❖ Number of performance appraisals based on new criteria vs. old ❖ Number of promotions based on new criteria vs. old ❖ Training effectiveness, dollars saved as a percentage of total sales ❖ Ratio of applicants approved, accepted ❖ Ratio of top management salaries to bottom worker salaries ❖ Ratio of bonuses/incentives to top management to bonuses for bottom workers	❖ Number of supervisors eliminated as a percentage of total supervisors ❖ Ratio of temporary, permanent employees ❖ Cost of downsizing/layoffs as a percentage of sales dollars ❖ Cost of fringe benefits per average worker ❖ Cost of accidents as a percentage of sales dollars	
Information Technology (IT)	❖ Errors per line code ❖ Cost of software errors as a percentage of sales dollars ❖ Internal customer (e.g., manufacturing, engineering, QA) evaluations of IT effectiveness	❖ Project costs, actual vs. target ❖ Ratio of IT costs to company sales dollars	❖ Cycle time from internal customer to delivery ❖ Cycle time of error corrections

*In the context of SCM, the term *supplier* refers both to 1) upstream first and second-tier suppliers and 2) downstream suppliers (i.e., distributors, dealers, and servicers).

is comprehensive and all-inclusive, a company should pick and choose only those few parameters that reflect its values and contribute to customer value/loyalty and financial results.

Company Effectiveness Index (CEI) Versus Competition as Measured by Customers

As important as measures are to assess team/department effectiveness, it is far more crucial to measure the customer's evaluation of the company's overall effectiveness and that of its competition.

An innovative practice is to develop a single overall index, or score—from one to 100—that integrates and quantifies the various elements by which a customer evaluates the company's effectiveness and compares that score to the comparable score attained by the company's competition.

Table 10-4 is a generic model for a customer's evaluation of a company's products. The first column lists the specific requirements the *core* customer considers essential (e.g., quality, durability, uptime, safety, price, delivery). These requirements can be derived from a quality function

Table 10-4. Company effective index (CEI), a generic model for products as measured by a core customer.

Requirement	Importance (I)	Company Rating (R)	Company Score (S)	Competitor Rating (CR)	Competitor Score (CS)
	Scale: 1–5	Scale: 1–5	(S) = (I) × (R)	Scale: 1–5	(CS) = (I) × (C)
Quality (upon receipt)					
Reliability (within warranty)					
Durability (lifetime)					
Serviceability					
Uptime (% use)					
Technical performance					
Features (that sell)					
Safety					
Human engineering					
Reputation					
Sales					
Cooperativeness					
Price					
Resale Price					
Delivery					
Total Score	Sum of (I) = (Y)		Sum of (S) = (T)	—	Sum of (CS) = (C)

Company effectiveness index (CEI) expressed as a percentage:
For Company = T/5Y × 100
For Competitor = U/5Y × 100

deployment (QFD) study or through mass customization. The second column depicts the importance (I) that the customer assigns to each requirement on a scale of one to five, with one as the least important and five the most important. In the third column, the customer rates the company's performance (R) for each requirement, using the same scale of one to five. The fourth column multiplies the figures in columns two and three to determine the company's score for each requirement: (S) = (I) × (R).

To determine the overall company effectiveness index (CEI), the importance numbers in column two are totaled (Y), as are the scores in column four (T). The overall company effectiveness index is expressed as a percentage: T/5Y × 100. (The five in the formula comes from the maximum rating of five for each requirement.)

The maximum CEI is 100 percent. A CEI rating below 20 percent would indicate a company that is terminally ill; between 20–40 percent it would be in intensive care; between 40–60 percent it would require hospitalization; between 60–80 percent it would need periodic checkups; and a rating above 80 percent would be a sign of robust health.

The same CEI can be expanded to determine how the customer compares the company against its best competitor. Column five is the customer's rating of the competitor's performance (CR) for each requirement on the same scale of one to five. The sixth column multiplies the numbers in columns two and five to determine a competitor's score for each requirement: (CS) = (I) × (CR). For the overall CEI of the competitor, the scores in column six (CS) are totaled (U). The overall competitor CEI is U/5Y × 100, expressed as a percentage.

The same type of index can be used for the service industry, for support services in the manufacturing industry, and for internal customer/supplier links. The requirements and format are shown in Table 10-5. This internal supplier effective index (ISEI) is developed further in Chapter 16.

Table 10-5. Internal supplier effective index (ISEI), a generic model for support services as measured by a core customer—external or internal.

Requirement	Importance (I) Scale: 1–5	Rating (R) Scale: 1–5	Score (S) (S) = (I) × (R)
1. *Quality* ❖ Completeness ❖ Accuracy ❖ Clarity ❖ Meaningfulness			
2. *Timeliness* ❖ On-time delivery ❖ Cycle time			
3. *Cost* (to customer)			
4. *Dependability* ❖ Promises kept ❖ Credibility ❖ Trustworthiness			
5. *Cooperativeness* ❖ Responsiveness ❖ Flexibility ❖ Approachability ❖ Courtesy			
6. *Communication* ❖ Listening ❖ Feed forward information			
Total Score	Total (Y) =	—	Total (T)

ISEI = T/5Y × 100

The elegance of the company effectiveness index is the remarkable way in which it simultaneously analyzes:

❖ The relative importance customers attach to their priority requirements

❖ The strength and weakness of the company for each requirement, as determined by the customer

❖ The strength and weakness of the company for each requirement vis-à-vis its best competitor, again as determined by the customer

❖ An internal customer's quantitative evaluation of an internal supplier

CASE STUDY

FedEx Corp.—A Benchmark Company in the Area of Measurement[4]

The service sector has a long notoriety for defective work, price gouging, late responses, indifference, and even rudeness. Often, even answering a simple phone call requires punching in more than thirty digits and waiting ten to fifteen minutes to get a human voice at the other end.

An honorable exception to this shoddy service is Federal Express. Despite competition from e-mail, faxes, and other carriers, it is highly profitable, with a universal reputation for service excellence. This case study examines its simple yet highly effective measurement system. The company's two ambitious goals are straightforward and unequivocal:

❖ One hundred percent customer satisfaction after every interaction and transaction

❖ One hundred percent service performance on every package handled

The key components of the Federal Express measurement and service performance are:

❖ Define service quality from the perspective of the customer, not by internal standards.

❖ Develop measurement of actual service failures, not just percentages of service achievement.

❖ Weight each category of service failure by its importance to the customer.

❖ Measure performance against the 100 percent standard.

❖ Provide accurate, immediate feedback for employee action and innovation toward the 100 percent customer satisfaction and service performance goals.

Service Quality Indicator (SQI)

The company developed a twelve-item statistical measure of customer satisfaction, with relative weights (see Table 10-6). Federal Express multiples the actual failures within each indicator by the weight factor, then totals the score. The goal of a 90 percent reduction in the total SQI was achieved in less than five years.

SELF-ASSESSMENT/AUDIT ON MEASUREMENT EFFECTIVENESS

Table 10-7 is a company self-assessment audit of its measurement effectiveness. It has six key characteristics and fifteen success factors, each with five points for a maximum score of seventy-five points.

Table 10-6. Federal Express service quality indicators (SQI).

Indicator	Weight
Abandoned calls	1
Complaints reopened	5
Damaged packages	10
International	1
Invoice adjustment requested	1
Lost packages	10
Missed pickups	10
Missing proof of delivery	1
Lost and found (identifying labels missing)	1
Right day late deliveries	1
Traces (proof of performance)	1
Wrong day late deliveries	5

Table 10-7. Measurement effectiveness: key characteristics and success factors (75 points).

Key Characteristic	Success Factors	Rating				
		1	*2*	*3*	*4*	*5*
5.1 Measurement Axioms	1. The number of measurements for a company, division, business or team are limited to three or four key metrics that can be focused on with laserlike intensity.					
5.2 Measurement Principles	1. Measurements are simple, meaningful, fair, and easy to apply with little calibration error. 2. Team measurements are transparent and prominently visible to all employees, managers, and customers. 3. Each team is measured against itself, using time for comparison. Pitting teams against one another is severely discouraged. 4. The benefit-to-cost ratio of all measurements are over 5:1. 5. Leading indicators are given much greater attention than lagging indicators. 6. Scoreboards, updated weekly, measure the progress of self-directed work teams, or minicompanies, in all key metrics.					
5.3 Financial Statements	1. Traditional financial statements are replaced by: ❖ Fair value information on long-term assets and liabilities ❖ Details and trends, business by business ❖ Qualitative and quantitative information on risks and opportunities associated with financial instruments ❖ Disclosure on company's success factors (e.g., customer retention, innovation, quality, empowerment, etc.)					

5.4 Core Customers	1. The company concentrates on its core customers—platinum and gold customers (see Chapter 6) and reduces and/or eliminates its bronze and tin customers. 2. The company measures the retention and longevity of its core customers.					
5.5 Generic Measurements	1. The company has switched to activity-based costing (ABC) accounting rather than using direct labor to allocate overhead costs. 2. The company gathers, analyzes, and systematically reduces the cost of poor quality (COPQ). 3. The company estimates the lifetime losses of defecting customers instead of looking at the tip of the cost iceberg—warranty costs alone. 4. The company uses a number of parameters to measure service effectiveness, innovation effectiveness, and empowerment effectiveness.					
5.6 Team/ Department Measurements	1. The company uses specific measures of quality, cost, and cycle time to assess the effectiveness of specific teams/ departments (e.g., marketing and sales, engineering, supply chain management, manufacturing, QA).					

Eleven

From Obsolete Tools of the Twentieth Century to the Powerful Tools of the Twenty-First Century

❖   ❖   ❖

In a Chinese village,
swarms of locusts were
devouring crops and ruining
its fragile economy. The
villagers went out en masse
to kill the locusts by hand.
They labored for seven
whole days. Hundreds of

174

> *thousands of locusts lay*
> *dead. . . . In the U.S., a few*
> *bags of insecticide would*
> *have done the job in one*
> *hour! The silent power of*
> *tools!*
>
> —Keki R. Bhote

Ten Powerful Tools of the Twenty-First Century

A company, aspiring to achieve the appellation of Ultimate Six Sigma, needs a triad of excellence: excellence in leadership, excellence in its workforce, and excellence in tools. Chapter 7 focused on the characteristics required for visionary, inspiring leadership. Chapter 9 discussed the characteristics required for happy, productive employees. This chapter emphasizes a whole array of new tools needed for a company to survive in the fierce winds of global competition in the twenty-first century. The old, ineffective, and obsolete tools, critiqued in Chapter 2, should be replaced by the following new tools. In the 1980s, I introduced or harnessed these tools at Motorola in order to achieve our breakthrough, not only in the 10:1, 100:1, and 1,000:1 quality improvement, but also in corresponding reductions in cost and cycle time. These same tools can help any corporation achieve at least an order—or two orders—of magnitude in improving quality, cost, and cycle time:

- ❖ *Design of Experiments (DOE):* The Shainin/Bhote DOE (not classical or Taguchi methods) is a problem-solving tool par excellence.
- ❖ *Multiple Environment Over Stress Testing (MEOST):* The drive for zero field failures.
- ❖ *Mass Customization and Quality Function Deployment (QFD):* Captures the voice of the customer.
- ❖ *Total Productive Maintenance (TPM):* Maximizes equipment/machinery productivity.
- ❖ *Benchmarking:* Closes the gap between you and the best-in-class company.

❖ *Poka-Yoke:* Prevents operator-controllable errors.

❖ *Next Operation as Customer (NOAC) and Business Process Reengineering (BPR):* Improves white-collar quality, cost, and cycle time.

❖ *Total Value Engineering:* Maximizes customer loyalty at minimum cost.

❖ *Supply Chain Optimization:* Maximizes profit enhancement for both customer and supplier

❖ *Lean Manufacturing/Inventory and Cycle-Time Reduction:* Maximizes inventory turns with these techniques.

In the remainder of this chapter, each of these tools is capsulized, with a brief description of its need, objectives, and benefits. Next, the methodology for achieving success in each tool is charted in somewhat greater detail. Nevertheless, it is beyond the scope of this book to go into much greater depth. Each tool would require a textbook in itself. Instead, there are references to appropriate texts for each tool. In addition, caveats to avoid pitfalls and notes of caution are outlined for each tool.

Design of Experiments (DOE)—Shainin/Bhote Not Classical or Taguchi: Problem-Solving Tool Par Excellence

History

More than 90 percent of companies do not know how to solve chronic quality problems. The emphasis is on the word *chronic*. And almost 90 percent of engineering specifications are wrong. Historically, design of experiments has been in existence for over seventy-five years. Its inventor was Sir Ronald Fisher—the real father of modern quality control (not Dr. Walter Shewhart, who was the originator of control charts). Fisher only used a pure but complex form of DOE called the Full Factorial. His successors—the advocates of classical DOE—modified the Full Factorial into a Fractional Factorial approach and weakened the purity of DOE by confounding main effects with interaction effects. Dr. Genichi Taguchi of Japan—in the name of simplicity—further weakened the Fractional Factorial method with his Orthogonal Array.

Since the 1930s, the late Dorian Shainin (who was more effective than Dr. W. Edwards Deming and Dr. Joseph M. Juran in problem solving) introduced a series of DOE techniques over the last fifty years that are simple, easy to understand and implement, cost-effective—and, above all—statistically powerful. As a disciple of Shainin for the last forty years, I enhanced and honed his techniques at Motorola, especially since 1982, when Chairman Bob Galvin challenged us to achieve a 1,000:1 quality improvement in ten years. Without our DOE, Motorola would not have achieved the spectacular results that accorded it worldwide fame. The record is captured in my two landmark books. The first, published in 1991, has sold over 100,000 copies and has been translated into four European languages.[1] The second edition, published in 2000, is even more comprehensive—with additional techniques, case studies, and workshop exercises.[2]

Weaknesses of Classical/Taguchi DOE and Strengths of Shainin/Bhote DOE

Although all three approaches to DOE—classical/Taguchi and Shainin/Bhote—are superior to all other problem-solving techniques, the classical/Taguchi methods have fundamental, philosophical, statistical, cost, and effectiveness weaknesses that have rendered their appeal an illusionary mirage. Table 11-1 lists the weaknesses of Classical/Taguchi DOE versus the Shainin/Bhote DOE strengths in a number of important characteristics.

With this brief background on the three approaches to DOE, the remainder of this section focuses on the Shainin/Bhote DOE. Table 11-2 tabulates this vital tool's needs, objectives, and benefits. Figure 11-1 graphically portrays the very significant contributions of DOE to business excellence in the form of a "spider chart."

DOE aids so profoundly in so many areas—problem solving, profit/return on investment, customer loyalty/retention, overall quality breakthrough, reliability, cost reduction, cycle-time reduction, space reduction, robust design, total productive maintenance (TPM), supplier chain optimization, and employee morale—that no company can afford to do without it.

Method of Achieving Success

The versatility of the twelve Shainin/Bhote DOE techniques are shown in block diagram form in Figure 11-2 and listed here.[2] (The first four are the clue generation tools. They talk to the parts—a much better technique than engineering guesses, hunches, opinions, biases, and theories! They are detailed in Chapter 14.) The objective of each technique is summarized as follows:

Technique	*Objective*
1. Multi-Vari[3]	To reduce a large number of unmanageable variables to a much smaller family of related variables containing the root cause of a problem (referred to as Red X). (The concentration chart is one of these smaller families.)

Table 11-1. Classical/Taguchi DOE weaknesses and Shainin/Bhote DOE strengths.

Characteristic	Classical/Taguchi Weaknesses	Shainin/Bhote Strengths
Philosophy	Causal factors are determined by guesses, opinions, hunches, and theories.	Causal factors are determined by "talking to the parts. The parts are smarter than the engineers."
Statistical Power	Weak: All second-order and higher-order interactions are confounded with main effects.	Excellent: Accomplishes separation and quantification of main and interaction effects.
Versatility	Poor: Only 1 or 2 techniques.	Excellent: 12 techniques.
Comprehension/ Implementation	Difficult for engineers to understand, much less use.	Easy, even for direct labor to understand and implement.
Training Time	1 to 3 weeks.	2 days, 1 day instruction, and 1 day of hands-on practice.
Cost	High: There are as many as 5–10 trials per problem because of wrong guesses.	Low: Clue generation tools reduce number of trials to 1 or 2.
Results	Modest: 2:1 to 3:1 improvement at best.	High: 5:1 to greater than 100:1 improvement. Average: Over 10:1.
Worker Involvement	Black belts only: Involves one out of every 100 workers.	Converts a whole factory into black belt performers.
Climate	Passivity among most employees.	Joy in the workplace. DOE success begets more success.

2. Components Search[4]   To swap parts and subassemblies between a best and a worst product to quickly and neatly identify the root cause (Red X).

3. Paired Comparisons[5]   To compare eight best and eight worst products; to separate the important quality characteristics of the product (with a minimum 90 percent confidence) from the unimportant ones.

4. Product/Process[6] Search   To separate the important process parameters that produce good and bad products from the unimportant ones (with a minimum 90 percent confidence).

Table 11-2. Shainin/Bhote DOE: need, objective, and benefits.

Need	Objectives	Benefits
❖ 90 percent of industry does not know how to solve chronic quality problems. ❖ Most specifications are wrong or arbitrary. ❖ Defects and variation considered inevitable.	❖ Separate important factors from unimportant ones. ❖ Open up the tolerances of the unimportant factors to reduce costs. ❖ Optimize the important factors to establish realistic specifications and realistic tolerances. ❖ Freeze the gains with positrol. ❖ Reduce the impact of uncontrollable noise factors to achieve a robust design and increase the signal-to-noise ratio.	❖ Quality improvements from 5:1 to >100:1. ❖ Cost of poor quality reduced by 10:1. ❖ Better, faster designs. ❖ Greater customer satisfaction and loyalty. ❖ Better suppliers. ❖ Bottom-line improvements.

5. Variables Search[7]   To pinpoint the important variables, quantify them and their interaction effects, and separate them from the unimportant variables, then open up the tolerances of the latter unimportant variables to reduce costs. Because variables search is absolutely essential in design, it is explained in detail using a case study in Chapter 12.

6. Full Factorial[8]   Same objective as variables search, however, the Full Factorial is used if there are four or fewer variables to evaluate, the variable search if there are five or more.

7. B versus C[9]   To verify that a better (B) product/process, where an improvement has been made over a current (C) product/process, does indeed constitute a permanent improvement with 95 percent confidence. B versus C is briefly explained as

Figure 11-1. Contributions of DOE to business excellence: a spider chart.

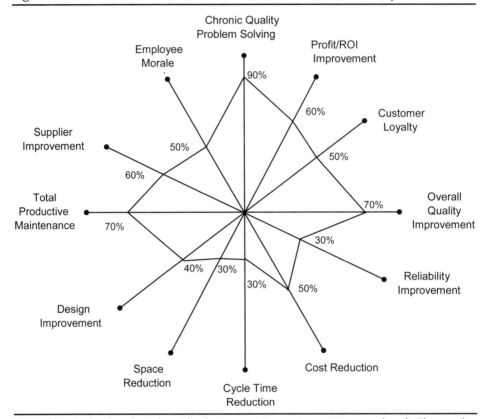

The total length of each spoke in the spider chart represents a maximum 100 percent benefit. The contribution of a particular technique (in this case DOE) is shown as a percentage of overall benefit.

	a verification test in design in Chapter 12.
8. Scatter Plot[10]	To determine realistic specifications and realistic tolerances of important variables (because 90 percent of specifications are wrong).
9. Response Surface Methodology (RSM)[11]	Same objective as scatter plot, but RSM is used where there are significant interaction effects between two or more input variables.
10. Positrol[12]	To monitor and control important process parameters in ongoing production by determining who, how, where, and when to monitor them. Chapter 12 explains positrol in some detail.

Figure 11-2. Problem-solving/variation reduction block diagram.

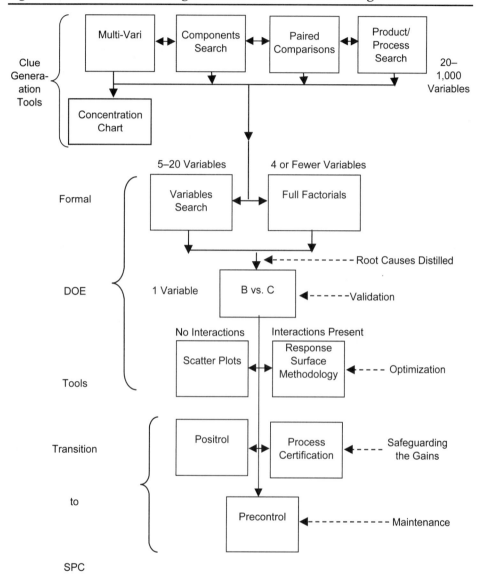

11. Process
    Certification[13]

To remove as many noise factors (e.g.,
poor maintenance, poor management/
supervision, poor environmental control, poor metrology, and poor worker
disciplines) as possible in order to increase the signal-to-noise ratio in any
DOE study and give it a chance to suc-

ceed. (For a fuller treatment of process certification see Chapter 12.)

12. Precontrol[14]   To monitor that good product quality, achieved through DOE, is continuously maintained in production. Precontrol is simpler, less expensive, and statistically more powerful than older, more complex and statistically weaker control charts still widely used in industry. (Chapter 14 gives a somewhat fuller treatment of precontrol.)

A more detailed framework connecting the twelve techniques of DOE is shown in Table 11-3. It is a ten-step surefire approach—far better than plan, do, check, act (PDCA) or its companion, the Ford 8-D methodology. Both of these methods do not even remotely begin to tell how to solve a chronic problem. By contrast, the twelve techniques of the Shainin/Bhote

Table 11-3. A generic problem-solving framework: a ten-step surefire approach: an Umbrella of Techniques.

1. Define the problem (the Green Y).
2. Quantify and measure the Green Y:
   ❖ Measure scatter plot (rather than gage R & R).
   ❖ Use Likert scale to convert attributes into variables.
3. Define the problem history (e.g., problem age, defective rate, and cost).
4. Generate clues using:
   ❖ Multi-vari (including concentration charts)
   ❖ Components search
   ❖ Paired comparisons
   ❖ Product/process search
5. Implement formal design of experiments:
   ❖ Variables search
   ❖ Full factorials
   ❖ B versus C
6. Turn the problem on and off to ensure permanency of improvement using:
   ❖ B versus C
7. Establish realistic specifications and tolerances (optimize) using:
   ❖ Scatter plots (for no interaction effects)
   ❖ Response surface methodology (if there are strong interaction effects)
8. "Freeze" the process improvements using:
   ❖ Positrol
9. Certify the process, nailing down all peripheral quality issues.
10. Hold the gains with statistical process control (SPC) and:
   ❖ Precontrol

DOE cover the entire waterfront of problem solving. If one technique is not appropriate, there is a second or third to take its place.

In all my consultations, I have had 98 percent success in marshaling one or more of these tools. (The only problems that are difficult to solve are when a full 100 percent of the product is bad, indicating a poor design. Even in such cases, many of these designs can be corrected using our Rolls-Royce technique—variables search (explained in Chapter 12).

For more detailed guidance on the Shainin/Bhote DOE, consult my earlier books on the subject, *World Class Quality—Using Design of Experiments to Make it Happen* (New York: AMACOM, 1991; the second edition was published by AMACOM in 2000). These books contain detailed theory, examples, case studies, and workshop exercises to enable a newcomer to get up to speed regarding problem solving.

Pitfalls to Avoid and Notes of Caution

1. Do not attempt a solution before defining the problem (called the Green Y). Often, defining the problem is halfway to its solution.

2. Try to find an earlier, easier Green Y that is correlated to the final Green Y and can solve the problem with less effort.

3. Make sure that the accuracy of the measuring instrument is at least five times the accuracy of the product being measured. Sometimes, this step alone can solve a problem.

4. Make sure that process certification is used *before* and *after* a DOE study to assure a good signal-to-noise ratio in the DOE study and to enhance its success.

5. Do not guess at the problem (as is done in classical and Taguchi DOE). Talk to the parts using one or more of the clue generation tools—the multi-vari, components search, paired comparisons, or product/process search.

6. Remember that, often, the Green Y problem has a root cause (called the Red X) that, in the next experiment, becomes a new Green Y with an underlying Red X. Sometimes, the ultimate root cause (the ultimate Red X) may be buried three or four layers deep, requiring three or four DOE trials.

7. Once an improvement is made, flip back and forth between the old and new product to verify that the improvement is permanent using the B versus C technique.

8. Extend DOE to your line workers. Never underestimate their intelligence. Train, encourage, and support them to conduct DOE and solve their own problems. Bring joy and excitement into their otherwise dull and monotonous jobs.

9. Extend DOE to your partnership suppliers as the best way to help them reduce costs, defects, and cycle time.

10. Monitor the gains in quality, cost, and cycle time and recognize, celebrate, and reward the DOE teams.

Multiple Environment Over Stress Testing (MEOST): The Drive for Zero Field Failures[15]

Background

Reliability is much more important than quality. It has two dimensions that quality does not have—time and stress. In addition, it is more directly related to the customer's requirements. To quote one of our Motorola vice presidents: "Regardless of specifications, the product must work—in the hands of the customer!" The traditional tools to achieve reliability—such as reliability prediction studies, failure mode effects analysis (FMEA), fault tree analysis (FTA), and brute-force mass testing—are woefully inadequate. They are, for the most part, paper studies and based on guesses, hunches, and opinions.

Traditional reliability tests have been patterned after the military approach of throwing money at the problem. The convoluted logic says that since you do not have the time to test one unit for 10,000 hours (to calculate mean time between failures, or MTBF), why not test 10,000 units for one hour? In the 1970s, accelerated life tests (ALT) gained some currency. However, they only conduct one environment or stress at a time and completely miss the interaction effects between two or more stresses that combine to produce failures. In the 1980s, two improvements were made—highly accelerated life tests (HALT) and highly accelerated stress screening (HASS). HALT is used in prototype development. It does combine stresses but takes these to product destruction limits, creating artificial failures and wasting time correcting each artificial failure. HASS is used in production but is expensive since 100 percent of the product is stressed. Furthermore, 10 to 20 percent of its useful life is wasted in such 100 percent tests.

Multiple Environment Over Stress Testing (MEOST) was first introduced in the days of the Apollo missions at NASA in the 1960s when every piece of space hardware had failed, with the exception of the lunar module, which took the two astronauts up to the moon and back safely. The lunar module had been tested with MEOST. Had it failed, we would have had the moon overpopulated by two people.

MEOST Principles

We cannot permit the customer in the field ascertaining product reliability. It is too late, too expensive (in terms of potential customer defections), and a fundamental abdication of a company's responsibility. By the same token, we cannot wait for six months to a year to determine if product changes to correct field problems are effective. That is only one step better than rolling the dice in Las Vegas. The objective is to *eliminate failures,* not to measure unreliability. Only through failures, forced in the prototype stage of a product, can the weak links of design be smoked out. Failures can be smoked out in the laboratory, prior to production and way before the product reaches the field by:

❖ Combining environments/stresses that simultaneously impinge upon the product in the hands of the customer

❖ Going beyond design stress to a maximum practical over stress

❖ Accelerating the rate of stress increase

By so doing, failures can be forced out in two or three days—or even hours—instead of weeks, months, and years.

All parts have built-in stresses. But weak parts have induced stresses two and three times the stresses in strong parts. These weak parts can be forced out in a very short time, leaving the strong parts intact. MEOST testing must be able to reproduce the same failures—the same failure modes and the same failure mechanisms—as failures in the field on the same product or similar product. That is the litmus test of MEOST's effectiveness. Table 11-4 lists a brief description of the need, objectives, and benefits of MEOST.

Method of Achieving Success

The following methodology, which takes you through preliminary stages of a product life cycle to the final stage of cost reduction, will help to assure success when using MEOST as a tool:

Pretest Considerations

❖ Obtain a profile of all maximum environmental/stress levels likely to be seen in the field simultaneously (e.g., thermal cycling, humidity, vibration, voltage transients, and so on).

❖ Select five or six of the most important of these simultaneous environments/stresses.

❖ *Increase the derating* (i.e., using a factor of safety of at least 2:1 to assure that a part is used well below its rated stress level) on each important/critical part as one of the best ways to enhance field reliability.

❖ Design and procure test chambers that simulate these combined stresses in the design laboratory.

Table 11-4. Multiple Environment Over Stress Tests (MEOST): need, objectives, and benefits.

Need	Objectives	Benefits
❖ Reliability prediction studies, such as MIL-Handbook 217E, are inaccurate. ❖ Failure mode effects analysis (FMEA) is good only as a preliminary paper study. ❖ Military testing of hundreds of units to calculate reliability only adds cost with little value. ❖ Traditional approaches to reliability are too mathematical, confusing, and ineffective. ❖ HALT and HASS are better methods but not good enough.	❖ Eliminate unreliability, not quantify it. ❖ Anticipate and capture field failures in design. ❖ Significantly shorten reliability test time from months to hours. ❖ Evaluate design changes quickly for reliability impact. ❖ Solve chronic field problems with a combination of MEOST and DOE. ❖ Reduce product costs with a combination of MEOST and value engineering.	❖ Reductions in field failure rates of 10:1 to 1,000:1. ❖ With a six-month and one-year field exposure, a quantified figure for reliability can be extrapolated with 90 percent confidence. ❖ In design, reduction of cycle time, cost, space, test equipment, and power consumption. ❖ Fast new product introduction into the marketplace to beat competition. ❖ Greater customer loyalty and retention.

Source: Keki R. Bhote, *World Class Quality—Using Design of Experiments to Make It Happen,* 2nd ed. (New York: AMACOM), 2000.

❖ Sample sizes should be of a minimum of three at prototype stage, ten at pilot run and production stages—as compared to hundreds of units in traditional reliability tests.

❖ Test duration should generally be eight to twenty-four hours, as compared to months of traditional reliability tests.

The Eight Stages of MEOST

Stage 1: Preliminary Stage

❖ Gradually increase one stress at a time up to the design limit. There should not be a single failure. Institute immediate corrective action if there is such a failure.

Stage 2: Preliminary Stage

❖ Extend stage 1 to maximum practical over stress limits (MPOSL) but never up to maximum destruct limit, as advocated by HALT. In stage two of the preliminary stage, ignore a single failure of any one type. If there are two or more failures of the same failure mode/ failure mechanism, institute immediate corrective action.

Stage 3: Prototype Stage

❖ Repeat stage 2 with important environments combined.

Stage 4: Pilot Run Stage

❖ Repeat stage 3 to evaluate design changes and suppliers.

Stage 5: Production Stage

❖ Repeat stage 4, but with fewer stresses and lower stress levels. Never use 100 percent testing as is done in HASS.

Stage 6: Field Stage One

❖ Return good product from the field after six months and repeat stage three.

Stage 7: Field Stage Two

❖ Repeat stage 6, but with one-year exposure in the field.

Stage 8: Cost Reduction Stage

❖ Reduce costs of high cost parts with value engineering and DOE and repeat stage three. The failure rates should not be below stage three results.

Pitfalls to Avoid and Notes of Caution

The following is a list of important observations to keep in mind when using MEOST:

1. The MEOST test in stage 3 (prototype stage) must be able to replicate the same failure mode/mechanism as found after months and years of field use on the same or similar product. If these failures are not reproduced, the MEOST "recipe" is not captured. This implies one or more of the following shortcomings:

❖ The right combination of environments/stresses have not been selected.
❖ The stress levels have not been raised sufficiently beyond design stress.
❖ The rate of stress change has not been sufficiently accelerated.
❖ The stress time is insufficient.

2. Assure sufficient derating on critical/important parts. Engineers try to push parts to their maximum rating to save costs. This is totally false

economy, because the company will pay higher costs in terms of warranty and incalculably devastating costs in terms of customer defections.

3. Do not be discouraged with MEOST's inability to find failure at the first try. The experiments should continue until the MEOST recipe is captured.

4. Pay particular attention to the test plan—the range of each environment, the meshing of these environments in a logical sequence, and the number of steps needed to reach the maximum for each environment.

5. Never push each stress to its maximum destruct limit.

6. In production, never consider 100 percent testing. It is costly and uses up field life.

7. Do not ship MEOST units to the field.

8. Introduce known failures (called "seeded" failures) in the product to test the ability of MEOST to catch them.

Mass Customization and Quality Function Deployment: Capturing the Voice of the Customer

The subject of mass customization[16, 17] is discussed in Chapter 6. Here only the format—need, objectives, and benefits—is tabulated in Table 11-5 for your quick reference. The remainder of this section is therefore devoted to a detailed discussion of quality function deployment (QFD). QFD's need, objectives, and benefits are also listed in Table 11-5.

Quality Function Deployment (QFD)

Product designs have been governed too long by the voice of the engineer and the voice of management, rather than the voice of the customer. As a result, eight out of ten new products end up on the ash heap of the marketplace. As one of our Motorola vice presidents said, "You may have the best dog food in the world, but if the dogs don't eat it, what good is it?" QFD was first developed by the Japanese in 1970 to capture the voice of the customer in design, and was adopted by the United States—its automotive industry, in particular—in the mid-1980s.[18, 19]

The crucial difference between mass customization and QFD is that the former concentrates on a customer of one, with his unique wants, whereas QFD has to depend on a sizable group of customers with common requirements. As mass customization continues to grow in importance and numbers, it is the rising sun and QFD the setting sun. Nevertheless, QFD—along with its related disciplines, such as conjoint analysis, multiattribute evaluations, and value research—still packs a wallop. It can be a useful tool for engineers to design products in half the time with half the defects, half the costs, and half the human resources expended in previous designs. It acts as an umbrella for marshaling a whole set of tools in design

Table 11-5. Mass customization and quality function deployment (QFD): need, objectives, and benefits.

	Need	Objectives	Benefits
A. Mass Customization	❖ Mass marketing obsolete. ❖ Mass production obsolete. ❖ QFD useful only for groups of customers with common requirements. ❖ Trend for each individual customer is to want highly individualized features of products and/or services.	❖ Switch from measuring market share to measuring a customer's lifetime value. ❖ Production quantity of one becomes as economical as quantities over 1,000 with the aid of information technology, computer-integrated manufacturing, and lean manufacturing. ❖ Capture of the customer by competition is made more difficult.	❖ Democratization of goals and services. ❖ Individualized product/service features at mass production costs. ❖ Reductions in delivery time to customers.
B. Quality Function Deployment (QFD)	❖ The "voice of the engineer" has dictated product designs. As a result, 80 percent of new products fail at the marketplace. ❖ Poor designs in terms of quality, cost, and cycle time.	❖ Determine customer needs ahead of prototype design. ❖ Have customer rate each requirement in terms of importance and in comparison with competition. ❖ Concentrate on the important, difficult, and new aspects of a design. ❖ Deploy product specifications into part specs, process specs, and test specs. ❖ Provide an umbrella for a whole series of related tools and techniques.	❖ Design in half the time, with half the human resources, half the defects, and half the costs of older designs. ❖ Help in the transition from customer satisfaction to customer loyalty. ❖ Increased market share. ❖ Faster time to market. ❖ Reduced "learning time" for engineers.

and in production (see Figure 11-3). It can translate the "what" of customer requirements into the "how" of engineering specifications, the "what" of engineering specifications into the "how" of parts specifications, the "what" of parts specifications into the "how" of process specifications, and the "what" of process specifications into the "how" of test specifications.

Method of Achieving Success

In QFD, the workhorse is the "house of quality," depicted in Figure 11-4. On the far left are listed the customers' (collective) most relevant requirements (the what) prioritized in terms of importance. On the far right are the customers' rating of the company's performance versus competition, for each requirement. In the middle is a relationship matrix comparing the link (strong or medium) between each customer requirement and each

Figure 11-3. Quality function deployment (QFD).

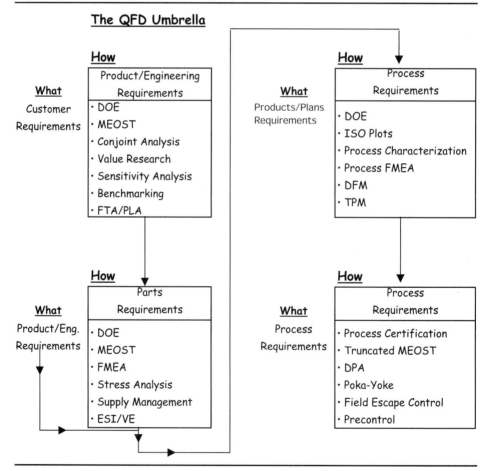

Figure 11-4. QFD house of quality.

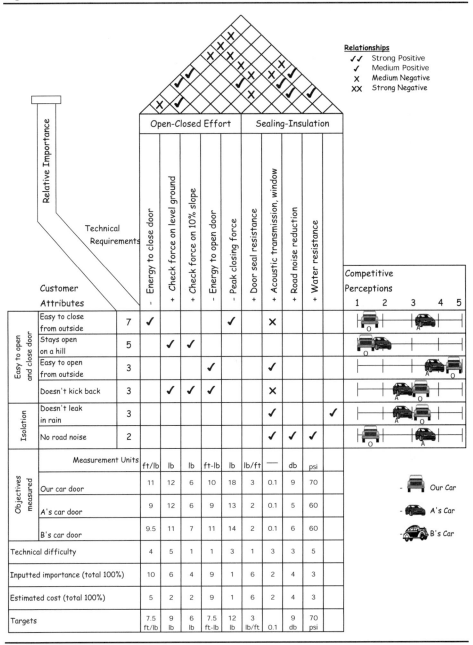

design specification (the how). A simple calculation then pinpoints the engineering specifications that must be concentrated on to meet customer requirements and assure competitive strength. At the bottom, there is a comparison of each specification against a target value and against competition, based on reverse engineering (i.e., an actual competitive analysis). On the roof is a correlation matrix to see if each pair of engineering specifications has a reinforcing correlation or a conflicting correlation.

Similar "house of quality" matrices are developed to deploy the engineering specifications to part specifications, process specifications, and test specifications.

Pitfalls to Avoid and Notes of Caution

A central weakness of QFD is its method of determining the commonality of customer requirements, given the trend that each customer has different needs and wants highly individualized products. This is the reason that mass customization has overtaken QFD in the last ten years.

Another weakness is the way customer requirements are gathered. Trade shows and questionnaires (via mail, telephone, or even person-to-person) are poor survey instruments. Focus groups, clinics, and panels are better but can bias the results, based on customer sampling, group dynamics, and the specific techniques used to stimulate discussion.

The following is a list of other important observations to keep in mind when using QFD:

1. Most QFD studies concentrate on customers' performance requirements, ignoring a higher priority that customers may attach to reliability, price, delivery, service, and so on.

2. Too many QFD studies draw up a long and unmanageable list of forty to seventy customer requirements, losing sight of the forest for the trees. The list should not exceed twenty.

3. There is frequently more than one customer group (e.g., consumers, distributors, dealers, OEM manufacturers, and so on). Their conflicting requirements necessitate separate QFD studies for each group.

4. The mechanics of filling in the house of quality matrix, though simple after a few practice rounds, can be a turnoff for first-time QFD practitioners, who might be discouraged by its seeming complexity.

5. Ninety percent of QFD studies stop at the first cascade—translating the "what" of the customer requirements into the "how" of engineering specifications. The remaining cascades—parts, process, and tests—should be deployed with similar QFD studies.

6. Most QFD studies are one-time events at the concept stage of design. However, changing customer requirements and expectations require QFD follow-up studies at the pilot run, production run, and field stages of product life.

Total Productive Maintenance (TPM): Maximizing Equipment/ Machine Productivity

Background

Next only to customer retention and the cost of poor quality (covered in Chapter 10), total productive maintenance (TPM)[20, 21] can be the greatest contributor to a company's profits. With a 4–8 percent cost of sales, it can improve corporate profits by 50–100 percent. The old industrial philosophy on plant equipment and maintenance was: "If it ain't broke, don't fix it!" This shortsightedness led to the perpetuation of quality/yield problems, cost overruns, and production stoppages. The first major reform was preventive maintenance, introduced in the United States in the 1950s. Nippondenso, a first-tier supplier to Toyota, went further to develop TPM in 1969. Today, the TPM prize in Japan is coveted almost as much as the Deming Prize, with an overall equipment effectiveness (OEE) level of 85 percent as a minimum required to even apply for the prize. Eastman Chemical Company's Tennessee Eastman division (a former unit of Eastman Kodak) was the first U.S. company to begin TPM in 1987. With 110 TPM teams, it has registered an 800 percent return on its investment. Unfortunately, most U.S. companies are not even aware of total productive maintenance, and the few that use it average a dismal OEE level below 50 percent.

This key metric for TPM is called overall equipment effectiveness (OEE), defined as the product of three percentages:

1. Yield
2. Uptime (the reciprocal of downtime)
3. Machine efficiency as a percentage, where:

Machine efficiency = Theoretical runtime/(Actual runtime + Setup time)

For a machine/process to achieve an OEE of 85 percent as a minimum, each of these three percentages must generally exceed 95 percent. Table 11-6 summarizes the need, objectives, and benefits of TPM.

Method of Achieving Success

Follow these steps to achieve success with TPM:

1. Train your line operators and maintenance technicians in simple pre-DOE techniques, such as Pareto charts, cause-and-effect diagrams, CEDAC, "the five whys."

Table 11-6. Total productive maintenance (TPM): need, objectives, and benefits.

Need	Objectives	Benefits
❖ A powerful tool almost unknown in the West. ❖ Older preventive maintenance is used less than 20 percent of the time, as compared to correction after machine is stopped (used 80 percent of the time). ❖ "If it ain't broke, don't fix it" mentality. ❖ Maintenance costs on equipment/machines range from 8–14 percent of sales. ❖ Overall equipment effectiveness (OEE) averages less than 50 percent in Western companies.	❖ Move from preventive maintenance to TPM. ❖ Radically improve process/equipment quality and productivity. ❖ Improve plant throughput and reduce cycle time and inventories. ❖ Establish worker maintenance teams from preventing, not correcting, equipment problems. ❖ Reduce life cycle equipment costs. ❖ Characterize and optimize key process parameters. ❖ Freeze abovementioned parameters with positrol.	❖ Increased labor productivity by 40–80 percent. ❖ Reduced equipment breakdown by 90 percent. ❖ Increased line capacity by 25–40 percent. ❖ Reduced cost per unit of maintenance by 60 percent. ❖ Increased OEE to more than 95 percent. ❖ Improved company profits by 50–100 percent. ❖ Harnessing line worker involvement in maintenance and adding to job challenge.

2. Train line operators and maintenance technicians in process certification.

3. Train line operators and maintenance technicians in the clue generation techniques of DOE.

4. Form line operator/maintenance teams to improve yields and reduce machine downtime.

5. Reduce setup/changeover times[22] by factors of 50:1 and more with techniques such as flowcharting and videotaping intensive practice runs (as done by pit crews in auto racing).

6. Give line operators responsibility for all routine preventive maintenance, such as cleaning, lubrication, record-keeping, tool storage/retrieval, positrol, and precontrol.

7. Utilize predictive maintenance with diagnostics and alarm signals that can monitor key process variables such as temperature, vibration, noise, and lubrication.

Pitfalls to Avoid and Notes of Caution in TPM

1. Do not underestimate the intelligence of line workers and their ability to conduct preventive maintenance and DOE studies.

2. In the long run, free up the maintenance technician to do process design work such as building machines and robots.

3. Encourage line worker/maintenance teams to practice, practice, practice in order to reduce setup/changeover times.

4. Give every team member a clear idea of the benefits—emotional, administrative, and financial—that he will derive from TPM team progress.

5. Work with suppliers of new equipment/processes in running DOE studies on the factory floor, and do not pay them until very high yields are achieved before the start of full production.

6. Start process development at least six months ahead of product development.

Benchmarking: Closing the Gap Between You and the Best-in-Class Company

Benchmarking is the continuous process of measuring products, services, and practices against the best competitors or industry leaders, closing that performance gap and leapfrogging competition.[23, 24]

History

In 500 B.C., a famous Chinese general Sun-Tzu wrote, "If you know your enemy and know yourself, you need not fear the result of a hundred battles." Benchmarking is 2,500 years old. The Japanese have practiced benchmarking, *dandotsu* (competitive evaluation) without labeling it as such. Xerox Corp. started modern benchmarking in 1979. After learning from Xerox, I introduced benchmarking at Motorola in 1985.

Method of Achieving Success

Internal progressive targets, established by companies year by year, are inadequate. They create a "king in the well" mentality. Benchmarking provides an external stimulus to learn about the best methods in other companies—both competitors and businesses in noncompetitive industries—to measure the gap, close it, and become the benchmark company. Even though critics label it "legal espionage," benchmarking is a perfectly ethical undertaking, provided the benchmarking company offers reciprocity to the benchmarked company in terms of practices and results.

Table 11-7 is a time-tested road map for a company to conduct a comprehensive benchmarking process. Table 11-8 lists the need, objectives, and benefits of benchmarking.

Pitfalls to Avoid and Notes of Caution in Benchmarking

1. There should be a corporate steering committee and a benchmarking czar. Without them, the effort could lack focus and there could be organizational "disconnects."

Table 11-7. The benchmarking road map.

1. Define why and what to benchmark.
2. Establish your own company's baseline.
3. Perform pilot runs in your company and adjacent companies.
4. Identify whom to benchmark.
5. Visit benchmark companies.
6. Determine the performance gap between your company and benchmark company.
7. Secure top management commitment.
8. Establish goals and action plans.
9. Implement plans and monitor results.
10. Recalibrate/recycle the process.

2. There must be linkage between benchmarking and key business strategies and outcomes.

3. The internal customer of the benchmarking study must be involved. Secure cooperation and help from support services.

4. In benchmarking, on-the-job training is more important than classroom training for team members.

5. It is mandatory to know your own company's baseline before benchmarking others.

6. Careful research is needed on which companies to benchmark. Noncompetitors are preferred, but do not hesitate to benchmark the competition as well.

7. A questionnaire should be pilot-tested and telephone interviews should precede on-site visits.

8. Make sure that the right people are interviewed at the benchmark company. Pay attention to their failures as well as their success factors.

9. Recycle the benchmarking process at least every one or two years.

10. Use the benchmark output as a springboard for further research.

Table 11-8. Benchmarking: need, objectives, and benefits.

Need	Objectives	Benefits
1. Internal target setting is inadequate.	1. Tie in with key corporate strategies.	1. Competition is leap-frogged.
2. Benchmarking is viewed by critics as "legal spying," which it is not. Hence, it is more often not utilized.	2. Close the gap between competitors in a key function, discipline, or product.	2. Cycle time in learning a discipline is shortened.
3. Without benchmarking, company progress is slow.	3. Become best in class in a key function, discipline, or product.	3. Customer requirements are better met.
	4. Institutionalize benchmarking as a way of life.	4. Quality, cost, and cycle time in products and services are improved.
		5. Tunnel vision is converted into a global outlook.

Use "out of the box" thinking to go beyond the benchmark company's accomplishments.

Poka-Yoke: Prevention of Operator-Controlled Errors

Generally, management lays most of the blame for quality defects on the line operator, even though it is well known that 85 percent and more of quality problems are caused by management, while 15 percent or less are caused by direct labor. But, assuming that a residue of quality defects is caused by direct labor, there is little merit in threatening or punishing them. Human beings make mistakes. Poka-yoke is a 1961 invention of the versatile Shigeo Shingo of Japan[25, 26]. It provides sensors—electrical, mechanical, or visual—that warn a line operator that a mistake has been made or, preferably, is about to be made. So warned, the operator can thus prevent a permanent defect. Poka-yoke is widely used in Japan in place of much less-effective control charts, but its introduction into the United States is barely fifteen years old and its use limited to just a handful of progressive companies.

Method of Achieving Success

A simple example can illustrate the principle of poka-yoke. A line operator inserting a microprocessor with twenty leads (ten on each side) into a printed circuit board can, by mistake, reverse its orientation. In that event, the electronic circuit would not work. Asking the operator to be more careful, or catching the mistake at an inspection station several positions down the assembly line, is too ineffective and too late. Instead, a simple design change of adding a dummy leg to the microprocessor—with ten leads on one side and eleven on the other and ten holes in the associated printed circuit on one side and eleven on the other—is a poka-yoke sensor. The operator, if attempting to put the part in backward, would be physically unable to insert the part into the printed circuit board. Such prevention is better than later correction down the line.

Most poka-yoke sensors are simple fixtures, often designed by the line operators themselves. The use of the sensors not only improves the quality of the product, but it also offers a challenge to the line operators in its design and construction, and generates job excitement when they succeed. Table 11-9 is a summary of the need, objectives, and benefits of poka-yoke.

Pitfalls to Avoid and Notes of Caution in Poka-Yoke

1. The poka-yoke sensors should be simple, not complex Rube Goldberg contraptions.
2. The sensors can be designed by a technician, skilled in improvising fixtures, and aided by the ingenuity of the line workers themselves.

Table 11-9. Poka-yoke: need, objectives, and benefits.

Need	Objectives	Benefits
❖ Human beings make mistakes, no matter how much they are threatened or how much they are paid. ❖ It is difficult to apprehend operator-controllable errors with sampling plans. ❖ Statistical process control (SPC) and control charts are ineffective in detecting operator-controlled defects. ❖ Management wrongly blames line workers for defects not of their making.	❖ Provide a feedback warning signal to the operator that an error is about to be made, and can therefore be prevented from turning into a "hard" defect. ❖ Eliminate operator-controllable errors. ❖ Save the costs of external—and most ineffective—inspection. ❖ Use poka-yoke as a more effective substitute for SPC and control charts.	❖ The costs of poor quality are reduced. ❖ External inspection time is reduced. ❖ Nonthreatening feedback to the operator is provided.

3. The best poka-yoke solutions are achieved through design of experiments (DOE) and design for manufacturability (DFM) at the design stage of the product.

Business Process Reengineering (BPR) and Next Operation as Customer (NOAC)

These twin and complementary tools are described in detail in Chapter 16 dealing with the service sector of industry and with white-collar operations within manufacturing companies.

Total Value Engineering: Maximizing Customer Loyalty at Minimum Cost

Value engineering (VE) differs from plain cost reduction, which retains the original design of the product or part but attempts to make it cheaper, with only slight modifications. Value engineering, by contrast, concentrates on the function of the product or part and provides that function in very different ways to improve quality, while reducing cost.

Invented by Larry Miles of General Electric in the 1940s,[27] VE was quickly championed by the U.S. Defense Department to eliminate "gold-plating"—that is, unnecessary specifications that add costs but little value. Unfortunately, outside of the defense industry, VE did not catch fire. Instead, the Japanese have embraced it with the vigor of Sinbad the sailor and use it to milk every yen of cost out of a product.

Total value engineering was developed by me as an enlargement of traditional value engineering.[28] Conventional VE concentrates on cost reduction while not deteriorating quality and reliability. Total value engineering goes beyond VE not only to enhance quality and reliability, but—more important—to maximize customer retention and loyalty while reducing costs. Table 11-10 differentiates traditional value engineering and total value engineering in ten distinct areas.

Method of Achieving Success

The need, objectives, and benefits of total value engineering are listed in Table 11-11. The methodology of total value engineering is, however, best

Table 11-10. Traditional value engineering versus total value engineering.

Area	Traditional VE	Total VE
Objective	Reduces cost without sacrificing quality.	Enhances all elements of customer satisfaction.
Scope	Generally confined to product.	Covers product, services, administration, organization, and people.
Customers	Addresses specifications only.	Determines customer requirements; measures customer satisfaction and excitement; turns it into customer enthusiasm.
Design	Improves design cost only.	Improves design quality, cost, cycle time, manufacturability, serviceability, and diagnostics.
Suppliers	Improves material cost only.	Improves supplier quality, cost, cycle time; reduces supplier base; creates partnership with suppliers.
Manufacturing	Improves production cost only.	Improves production cost, quality, cost, cycle time, and total productive maintenance.
Field	Negligible involvement.	Improves reliability, diagnostics, and serviceability.
Support Services	Negligible involvement.	Improves quality, cost, and cycle time in white-collar operations.
Management	Negligible involvement.	Improves motivation, stimulates job excitement, and facilitates gain sharing.
Organization	Employs cross-functional teams.	Employs cross-functional teams and a flat pyramid.

Table 11-11. Total value engineering: need, objectives, and benefits.

Need	Objectives	Benefits

❖ Customers want not just quality, but value.
❖ Cost reduction isn't enough. It does not change the basic design.
❖ Global competition is intensifying.
❖ Material, labor, and overhead costs are spiraling.
❖ Profits are anemic.
❖ Customer defections are not perceived, much less measured.
❖ There's poor employee morale.
	❖ Go beyond cost reduction and concentrate on the *function* of the product, then provide that function in new ways that substantially reduce costs while not decreasing quality/reliability.
	❖ Go beyond traditional value engineering (VE) and quality improvement to enhance all aspects of customer enthusiasm.
	❖ Strengthen all functions within a company:
	  1. From management to the worker
	  2. From manufacturing processes to business processes
	  3. From design to the field
	❖ Provide an organizational framework for continuous, never-ending improvement.
		❖ 25 percent lower product costs (average 10 percent minimum).
		❖ A 10:1 return on investment.
		❖ Minimum investment expense.
		❖ Improved quality and reliability.
		❖ Improvement on all elements of customer enthusiasm.
		❖ Higher customer retention and loyalty.
		❖ Royalties and incentives:
		  1. From customers
		  2. To suppliers
		❖ Higher employee morale.

illustrated by a case study. The project was the design of a flexible dust cover for an outdoor, portable, metal carrying case containing electronic checkout equipment. The case and its equipment were to be carried and operated at a missile site. The case was 18″ x 14″ x 10″ and contained thirty-seven pounds of instrumentation.

Provisions were to be made for storing the dust cover in the lid of the carrying case when the equipment was in operation. The material for the dust cover was specified as "Mil-Grey" (military gray) in color and qualified under the environmental requirements mandated for all ground support equipment.

The defense contractor set about designing the flexible dust cover for an estimated cost of $124 per unit. This cost included the product meeting all the environmental requirements imposed by the Defense Department. A total value engineering team then interjected three questions that prompted new answers to the problem:

Question 1: What is the function of the dust cover?

Answer: To protect the metal carrying case during transportation and storage. On the other hand, the metal carrying case protected the dust cover during operation at the missile site.

Question 2: What was the gold-plating in the military specifications (excess requirements)?

Answer: Why do we need the "Mil-Grey" (military gray) color? Why do we need environmental requirements when in actual operations the metal case is protecting the dust cover, not the other way around?

Question 3: What else will provide the function of protecting the metal case?

Answer: A paper bag, a laundry bag, a plastic zip-lock bag, even no bag.

The final total value engineering solution: Provide a rubber-seal gasket on the lid of the metal carrying case to protect it against dust and rain. Cost: $1.10—more than a 100:1 savings!

Pitfalls to Avoid and Notes of Caution in Total Value Engineering

1. While reducing cost, never sacrifice quality/reliability and customer "wow."
2. Don't use traditional thinking. Go outside the box. Start with the possibility of eliminating the function.
3. Test your idea with a B versus C approach (see Chapter 12) to verify improvement.
4. Give away credit to others.

Supply Chain Optimization and Lean Manufacturing

These last two of the ten powerful tools are also part of two important managerial disciplines. Therefore, supply chain optimization and lean manufacturing/inventory and cycle-time reduction are covered in detail in Chapters 13 and 14, respectively.

Awareness and Implementation of the Ten Powerful Tools

The good news is the sledgehammer power of these tools and the enormous benefits a company can derive from them. The bad news is the lack of awareness of these tools worldwide and the even more pathetic lack of implementation. Table 11-12 is my unscientific but reasonable estimate of the percentage awareness and percentage implementation of each of the ten tools among leading U.S. companies, as well as corresponding figures for the average U.S company. (In general, Japanese companies score somewhat better; European companies are even worse.) Industry has miles to travel!

Companies that are uncompetitive today, especially on the global scene, need these tools desperately to pull themselves up by their bootstraps. Companies that are doing well, in terms of the bottom line, also need these tools lest success-complacent success—becomes the father of failure.

CASE STUDY

TELCO—A Benchmark Company Deploying Powerful Tools of the Twenty-First Century

Two companies are in a class by themselves when it comes to the deployment of the twenty-first century tools detailed in this chapter. The first is my own company—Motorola—where many of these tools were forged in the white-heat furnace of

Table 11-12. Awareness and implementation of the ten tools.

Tool	Leading U.S. Companies		Average U.S. Company	
	Percent (%) Aware	Percent (%) Implementing	Percent (%) Aware	Percent (%) Implementing
1. Design of Experiments (Shainin/Bhote, not classical or Taguchi)	10	1	—	—
2. Multiple Environment Over Stress Testing	1	0.05	—	—
3. (a) Mass Customization (b) Quality Function Deployment	15 25	2 3	2 2	— —
4. Total Productive Maintenance	15	2	3	0.01
5. Benchmarking	60	20	10	1
6. Poka-Yoke	15	1	2	—
7. (a) Business Process Reengineering (b) Next Operation as Customer	50 10	5 0.5	10 1	0.01 —
8. Total Value Engineering	50	2	5	0.01
9. Supply Chain Optimization	60	10	10	0.5
10. Lean Manufacturing/Cycle Time Reduction	50	10	10	1

global competition. The span of time was a ten-year period—from the early 1980s to the early 1990s—and even though we saved billions of dollars by using these tools, it is difficult (at this late juncture, ten years later) to precisely quantify the results.

For this reason, I have chosen a different company, in a difficult business, in a different country, and in a different culture as the benchmark. The company is the Tata Engineering and Locomotive Company Ltd. (TELCO), which is part of the Tata Group of companies—the largest conglomerate in India. TELCO makes trucks, commercial vehicles, and passenger cars. At the start of 1999, it launched a passenger car—Indica—with 100 percent of the parts made in India, an audacious project for a so-called underdeveloped country. I was brought into TELCO by the Tata Group chairman—Ratan Tata—as its consultant, at about the time of the Indica launch. The overall framework of the consultation was the Ultimate Six Sigma, and the initial primary concentration was the twenty-first century tools. I taught these tools to more than 200 of TELCO's senior executives, managers, and professionals. The tutorials were followed up by hands-on coaching. The results (over a span of eighteen months) have been nothing short of spectacular. They are summarized in Table 11-13. While other companies (including Motorola) have gone further, no company anywhere in the world has shown such a remarkable grasp of the tools so fast or has achieved these amazing results in so short a time as TELCO.

SELF-ASSESSMENT/AUDIT ON EFFECTIVE DEPLOYMENT OF TWENTY-FIRST CENTURY TOOLS

Table 11-14 is a self-assessment/audit on the powerful tools of the twenty-first century that a company or external Ultimate Six Sigma audit can use to gauge its progress. A company (or external Ultimate Six Sigma auditors) can conduct the assessment to gauge how well it fares in the area of powerful tools. The table lists ten key characteristics and fifteen success factors, each worth five points, for a total of seventy-five points.

Table 11-13. Results of application of twenty-first century tools: TELCO of Pune, India.

Tool Implemented	Time Period	Results
1. General ❖ Cost of poor quality (COPQ) ❖ Profit	6 months	 ❖ 35 percent reduction in COPQ ❖ 33 percent profit increase
2. Design of Experiments (DOE) ❖ Number of people trained ❖ Number of DOE projects ❖ Quality improvement	15 months	 ❖ 1,747 (direct labor: 116, inspectors: 153, supervisors: 686, tech staff: 741, executives: 51) ❖ Started projects: 428, completed projects: 268 (warranty/customer improvements: 121, c_{p_k} improvement: 34, reject/rework reduction: 91, miscellaneous: 22) ❖ From 3:1 to more than 1,000:1
3. Multiple Environments Over Stress Testing (MEOST) ❖ Number of projects ❖ Field failure rates ❖ Warranty costs	6 months	 ❖ Started: 23, completed: 12 ❖ Field failures reduced 4:1 to zero ❖ Reduced by 21.6 percent
4. Supply Management ❖ Commodities ❖ Savings	3 months	 ❖ 19 commodities ❖ $600,000 actual, $7.12 million projected
5. New Product Introduction	1 month	Activity in quality, cost, cycle time just started
6. Variation Reduction ❖ Number of projects ❖ Precontrol charts	9 months	 ❖ Total: 1,254 (409 with c_{p_k} greater than 2.0; 845 with c_{p_k} greater than 1.33) ❖ Total: 409 (233 complete, 176 in process)
7. Poka-yoke ❖ Number of projects	6 months	 ❖ Completed: 177

Table 11-14. Powerful tools: key characteristics and success factors (75 points).

Key Characteristic	Success Factors	Rating				
		1	2	3	4	5
6.1 Design of Experiments (DOE)	1. Shainin/Bhote DOE (not classical or Taguchi DOE) is used systematically in product/process design to prevent chronic quality problems from reaching production. 2. Shainin/Bhote DOE (not classical Taguchi or DOE) is used systematically in production, with suppliers and in services, to correct chronic quality problems. 3. Shainin/Bhote DOE is used to encourage and motivate line workers/direct labor to solve chronic quality problems in teams and thereby bring joy into the workplace.					
6.2 Multiple Environment Over Stress Testing (MEOST)	1. MEOST is used to simulate field failures early in design to smoke out weak links and move toward a goal of zero field failures. This is done: ❖ At the prototype and pilot run stages of design ❖ In production (with mini-MEOST) ❖ In the field, after 6 months and 12 months of field exposure 2. MEOST is used (stage 8 of the product life cycle) to reduce costs without any sacrifice of reliability.					
6.3 Mass Customization/Quality Function Deployment (QFD)	1. Mass customization is used to individualize the requirements of core customers and provide them with products/services they want, when they want them, where they want them, and at prices they want. This is done by: ❖ Keeping a finger on the pulse of the core customers at all times ❖ Postponing customization as long as possible in the supply chain ❖ Using a combination of IT, CIM, FMS, and automation to produce a quantity of 1 as economically as 1,000 to 10,000 2. QFD is used to determine largest volume of common customer requirements and employs the "house of quality" discipline to translate these into engineering specifications, part specifications, and process specifications.					

(continues)

Table 11-14. (Continued).

Key Characteristic	Success Factors	Rating				
		1	2	3	4	5
6.4 Total Productive Maintenance (TPM)	1. TPM is employed to improve overall equipment effectiveness (OEE) to a *minimum* of 85 percent on each important process/machine, using maintenance/line operator teams and the disciplines of DOE and setup time reduction.					
6.5 Benchmarking	1. Benchmarking is systematically and periodically employed to compare strategically important functions, disciplines, and processes of the company with best-in-class companies to: ❖ Determine and reduce the gap. ❖ Leapfrog the benchmark company with "out of the box" thinking and implementation.					
6.6 Poka-Yoke	1. Poka-yoke sensors are designed and used so that operator-controllable errors can be detected and prevented before such errors can be converted into actual defects.					
6.7 Business Process Reengineering (BPR)/Next Operation as Customer (NOAC)	1. BPR is used to transform an entire corporate culture by changing employee values and by changing management practices in the way employees are hired, trained, evaluated, compensated, and promoted. 2. NOAC is used in the service sector (including those in manufacturing industries) to improve quality, cost, and cycle time in all business processes. With cycle time as a metric, NOAC uses: ❖ Process mapping and elimination of nonvalue-added steps ❖ Radically new methods and "out of the box" creativity to achieve breakthrough improvements					
6.8 Total Value Engineering	1. Total value engineering principles are employed in all areas of a company to maximize customer loyalty and retention while reducing costs.					

6.9	Supply Chain Optimization	1. Principles of partnership with key suppliers, distributors, dealers, installers, and servicers are employed to mutually improve profits for all sides by rendering active and concrete help in quality, cost, and cycle time.					
6.10	Lean Manufacturing/Inventory and Cycle Time Reduction	1. The discipline of lean manufacturing is employed through: ❖ Pull versus push systems ❖ Emphasis on product/process quality with DOE and TPM ❖ Focus factories and linear flows/ schedules ❖ Small lots, short setup times ❖ Multi-skilled operators and manufacturing cells ❖ Extension of lean manufacturing to customers and suppliers					

FROM HISTORIC LEVELS TO DESIGNS IN HALF THE TIME WITH HALF THE DEFECTS, HALF THE COSTS, AND HALF THE MANPOWER

❖   ❖   ❖

Product life cycles today are shorter than design cycle times of yesterday. In this environment, the company with the shortest design cycle time will have the greatest competitive advantage. One of the most important

factors in reducing this

design cycle

time is first-time quality

success. . . .

—JIM SWARTZ,

MANAGER OF

BENCHMARKING,

DELCO ELECTRONICS

Today's Design Engineers—Not Quite Old Guard but Not Yet Avant Garde

If the difference between failure and success at the corporate level is the difference between management and leadership, the difference between failure and success at the product (or service) level is the difference between old line, isolation-booth engineers and the newer, customer-oriented, team-based designers. Their differences spill over into the areas of design quality, design costs, design cycle time, and design innovations with leadership (or management) of the design function playing a key role for better or for worse.

Poor Quality—The Bête Noir of Design

The corrosive influence of poor design quality is universal. Eighty percent of quality problems are unwittingly but nevertheless, actually designed into the product. For example:

- ❖ Customers see design problems not on just a few units, but often across the board.
- ❖ Reliability in design is more a quantitative false hope, given the engineer's inability to accurately predict, accurately test and accurately estimate field failure rates at the time of product launch.
- ❖ Suppliers often get blamed for poor product quality, even though over 50 percent of poor specifications are the engineer's fault.

❖ Manufacturing is in the unenviable position of having to catch engineering's half-baked product tossed over the wall, with the time bomb ticking and muddling through to shipment.

❖ Service receives the storm of complaints about a poor product that, ironically, it warned about earlier in the design cycle.

❖ Line workers become the scapegoats for poor product quality because engineering conveniently blames them for having "lost their pride of workmanship."

❖ And hovering over all these effects caused by poor engineering is the ever-present danger of product liability suits. Prosecution lawyers are especially ecstatic whenever they discover a design flaw that affects not just one but 100 percent of the units in the field.

Tough Design Costs—A Vicious Downward Spiral

It is well known that 70–75 percent of product costs are a function of the design. The large number of defects uncovered in the design cycle adds to these costs, making it difficult for a company to be competitive. Because of this competition, pressure is applied on the engineers to cut corners and reduce costs. The result is a further sacrifice of quality and reliability, producing, in turn, more costs, more delays, and more frustration—a veritable downward spiral.

A case in point is "derating," where a factor of safety should be applied to designed parts so that they are not subjected to their maximum stresses. Because derating a part generally entails higher costs, engineers are most reluctant to subscribe to derating. Instead, they push the stress on the part as high as possible in order to shave a few pennies from the bill of materials. They justify this by showing that the bill of material savings more than offsets warranty cost in the field. What they do not take into account, or even understand, is that the permanent loss of a disgruntled customer is more than 100 times larger than the paltry savings in the bill of materials.

Long Design Cycle Time—Giving the Keys to Your Competition

It is now a well-accepted axiom that the better the quality, the shorter is the cycle time; and the shorter the cycle time, the better is the quality. Design cycle times have been reduced in recent years, but they are still far too long, allowing nimble competitors to steal a march. Design cycle time today is the name of the game. A company that can design and launch a product into the marketplace much quicker than its competition will be the Charles Lindbergh of design flights to the marketplace. (Ever hear of the No. 2 flier to cross the Atlantic?)

Stunted Innovations

Finally, because designers are not given free rein to exercise their creative juices by an impatient management, many innovations remain unborn. The dollar amount of sales from new products as a percentage of total sales remains low; the frequency of a stream of new products, introduced to keep competition off-balance, is also dismally low; and "skunk work" projects dry up in the withering heat of management opposition.

The Ultimate Six Sigma Objective in Design

The main objective of a far-reaching engineering/development/R&D effort is to attain customer "wow" designs in half the time with half the costs, half the defects, and half the manpower of older designs. This requires a revolution in:

- ❖ The organization of the engineering function
- ❖ Management reinforcement
- ❖ Capturing the voice of the customer
- ❖ Breakthrough quality/reliability technologies
- ❖ Steeper cost reduction techniques
- ❖ Accelerated cycle-time reductions
- ❖ Releasing the natural but currently bottled-up creativity of the engineer

Organization of New Product Introduction (NPI)

Concurrent Engineering—a Team Effort

In the past fifteen years and more, companies have abandoned the traditional engineering role as a self-contained, isolated, departmental function. They have also, to a lesser extent, departed from matrix management with its solid and dotted lines that result in two bosses—a functional boss and a project boss—fighting for control.

Companies have embraced the principle of a team approach—concurrent engineering (also called simultaneous engineering). The first step was to team manufacturing with engineering at the start of design to overcome design weaknesses when the product reaches production. The second step is to expand the team concept by adding the disciplines of quality, purchasing, service, and finance at the start of design. The team is led by a program manager or project manager. Duties of the program team members are usually defined as follows:

❖ *Program Manager*. Fully responsible and accountable for all aspects of the program and for completing the program on schedule within the cost targets and in full accordance with the program scope.

❖ *Development Engineering Team Leader*. Responsible for all design aspects of the program.

❖ *Manufacturing Team Leader*. Responsible for all manufacturing disciplines involved within the program.

❖ *All Other Team Members*. They represent their respective departments and perform all functional duties of their disciplines for the program; they also coordinate any additional activities required from their departments to support the program.

Figure 12-1 is an example of a concurrent engineering team organization for a major design project in a large multinational company. (Smaller projects have smaller and simpler structures.) For best results, the team members should be colocated.

Unfortunately, the fine intent of concurrent engineering has been diluted in practice. The main reason is the turf warfare between the functional managers who loan their members to the project team and the program manager. Both functional managers and program managers demand allegiance to themselves, so team members are caught in the middle. Successful companies have solved this dilemma by granting full and unequivocal authority to the program manager once the team is formed and running. The program manager becomes the undisputed czar of the project—-from inception until it is launched into the field. In Japan, this authoritarian is the "shusa"—a veritable shogun—invested with great power, including the placement of team members after product launch is completed.

Another dilution is the part-time use of most of the team members. Part-time use ends up with part-time results. There are so many activities and duties in a product launch that a resourceful program manager can and should keep all team members busy with full-time, meaningful work. Even if it is outside their specialized disciplines, team members are adept at multiple skills and the design project moves forward in harmony and speed.

Milestone Chart and Sign-Off Authority

Once the project team is formed, an agreement—sometimes called the contract book—is drawn up between top management and the team members on their respective financing and capital equipment necessary to support the projects. The team, in turn, agrees to meet the quality, cost, and design cycle-time goals on the project. The contract book prevents finger-

Figure 12-1. A concurrent engineering team organization for new product introduction.

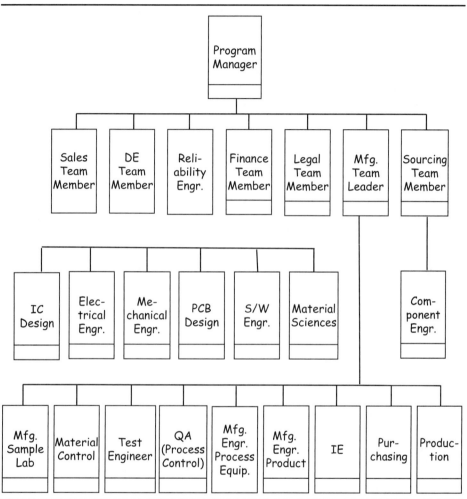

pointing and excuses on derailed projects because a "prior" agreement had not been made clear.

Table 12-1, which is a milestone chart for new product introduction, provides a detailed road map for the major steps and sequences as the design progresses. It shows:

❖ The team member with prime responsibility (P) for each step and those team members with contributing responsibility (C) for that step.

❖ Sign-off authority (S) at critical stages of the design's progress, which enables top management to either approve continuation of the project or scrub it.

Table 12-1. New product introduction: milestones and responsibilities chart.

No.	Milestone	Senior Management	Program Manager	Design Leader	Manufacturing Leader	Quality	Sales/Marketing	Service	Sourcing	Finance	Major Tools
A.	**Organization**										
1.	Program Team Kick-Off	S	P	C	C	C	C	C	C	C	Concurrent Engineering
B.	**Mgmt. Guidelines**										
1.	Max. 25% Redesign	S	P	P		C	C			C	Contract Book
2.	Stream of Rapid New Products	S	P	P		C	C			C	Lessons Learnt Log
C.	**Voice of Customer**										
1.	Elements of Customer "WOW"		P	P	C	C	C	C	C	C	"Bhote's Law"
2.	Customer Specifications		P	P	C	C	C	C	C	C	QFD, Mass Customization
D.	**Design Quality/Reliability**										
1.	Feasibility Study			P		C	C			C	CAD; Software Architecture
2.	Preliminary Design			P		C		C	C		De-rating, FMEA, FTA, PLA
3.	First Design Review		S	P	C	C	C	C	C	C	Checklist
4.	Prototype Design		C	P	C	C	C		C	C	DOE, MEOST, Metrology
5.	Quality Systems Audit		C	C	C	P	C	C	C		The Ultimate Six Sigma Assessment
6.	Second Design Review		S	P	C	C	C	C	C	C	Checklist
7.	Pilot Run		C	P	C	C		C	C	C	B vs. C; Positrol; Process Certification; Pre-Control
8.	Design for Manufacturability		P	P	C						Boothroyd-Dewhurst Scoring
E.	**Design Cost Reduction**										
1.	Total Value engineering		C	P	C	C			C	C	V.E. Job Plan, "Fast" Diagram
2.	Part Number Reduction			P	C	C	C		C	C	SBU, Product, Model Reduction
3.	Early Supplier Involvement		C	P		C		C	P	C	Cost Targeting, Financial Incentive
4.	Patent Study		C	P						C	
F.	**Design Cycle Time Reduction**										

#	Task										Tools
13.	Outsourcing		C	P	C	C			P	C	"Black-box" Supplier Design
14.	"Human Inventory"		C	C					C	P	Integration of Cost-Time Curve
15.	Third Design Review	S	P	C	C	C	C	C	C		Checklist
G.	**Product Launch**										
16.	Customer/Field Tests		C	P	C	P		C			DOE at Customer Site
17.	Management Review	S	P	C	C	C	C	C	C	C	Authorization for Full Production
18.	Production Run		S	C	P	C		C	C		Mini-Meost, TDPU, COPQ
19.	Management Audit	P	P	C	C	C	C	C	C	C	Lessons Learnt
20.	Initial Customer Feedback		P	C	C	P	P	C	C	C	Survey Instrument

Responsibility codes:
S = Sign-off authority
P = Prime responsibility
C = Contributing responsibility

❖ Sign-off authority (S) for the program manager during three vital design reviews and at the start of production, to modify the scope or direction of the project.

❖ The major tools (see Chapter 11 for details) that should be used at each step of the design. Without the proper tools and disciplines, the design is likely to be weak, delayed, costly, and unresponsive to the customer's true needs.

Management Guidelines

There are several important management guidelines that should govern a design and that could be incorporated into the contract book:

1. The maximum 25 percent redesign rule
2. A stream of new products rapidly and frequently introduced
3. Supply management directives
4. Parts philosophy
5. Lessons learned log
6. Reverse engineering

The Maximum 25 Percent Redesign Rule

The maximum 25 percent redesign rule means that a minimum of 75 percent of the new design should be common to the previous design (by volume) unless there is a major new platform—which may occur only once in a few years. In the West, there is a tendency for a 100 percent new design based on the engineers' ego and their desire to have their names etched into the product in perpetuity. The result is a delay in design cycle time by a factor of ten as compared to the 25 percent rule.

A Stream of New Products Rapidly and Frequently Introduced

Even more important than the attendant cycle-time reduction, the 25 percent rule allows a company to introduce a new product every few months—say, every three months—as opposed to once per year or once in eighteen months. This constant stream of new products introducing new features and innovations moves a company four to six generations ahead of competition, leaving the latter in the dust. The classical example is Nike, Inc., where small changes are constantly, but so frequently introduced so that the company has surged ahead of its competitors in every measure of business performance.

Supply Management Directives

Early in the design cycle, management must stipulate to the team a few key practices in its dealings with suppliers:

- ❖ Select key suppliers based on ethics, trust, and mutual help to promote a win-win partnership. Verify a congruence of corporate values.
- ❖ Outsource not just piece-parts but higher-level assemblies and "black box" designs, when these are not core competencies of the company.
- ❖ Encourage, even mandate, the team's enthusiastic acceptance of early supplier involvement in the prototype stage of design as one of the best ways to reduce costs and improve quality.
- ❖ Prescribe the price paid to the supplier (so-called cost targeting) rather than the other way around (the old three-quote syndrome). The company does cost targeting based on computer-aided estimates of material and labor costs.
- ❖ Outlaw the totally destructive practice of sending out bids on the Internet on key parts with previously established partnership suppliers.

Parts Philosophy

To assure high standards of field reliability:

- ❖ Ninety percent of the parts must have a proven track record of reliability. There is a need for using state-of-the-art new parts, but such experimentation should be kept to a minimum.

❖ There should be an ironclad directive to the team that all critical parts must have a built-in factor of safety or derating. This subject is explained fully in the section on Design Quality/Reliability later in this chapter.

Lessons Learned Log

There is a predilection in many companies to sweep design mistakes under the rug. The objective of the lessons learned log—either a paper or computer trail—is not to embarrass the design team or punish it, but to assure that the sad lessons of historical errors are not condemned to repeat themselves, either for the next product or the next generation of engineers.

Reverse Engineering

Known earlier as competitive analysis, reverse engineering should be standard practice in any progressive company's design cycle. The competitor's product is stripped down and evaluated, part by part, for its design, reliability, materials, and cost (through cost targeting).

Reverse engineering is also supported by quality function deployment (QFD) and benchmarking studies to compare the customer's perception of the design vis-à-vis the company's best competitors.

Designing for the "Voice of the Customer"

Chapter 6 highlighted the several elements that combine to result in customer "wow," both for products (Figure 6-5) and for services (Figure 6-6). While no single element of customer "wow" is necessarily more important than any other, Bhote's Law states that it is that element of "wow" missing from a company's product or service (especially if it is truly important to the customer) that requires maximum top management attention.

The Elements of Customer "Wow"

The design team should ascertain, at the very start—the concept stage—the core customers' priorities associated with these elements of "wow." This is done with mass customization (Chapter 6) and with quality function deployment (Chapter 11). QFD, however, tends to concentrate on technical performance only at the expense of other elements that the customer may deem more important. The Japanese have shifted their customer attention from *atarimar hinshitsu*—taken for granted quality—to *miryo kuteki hinshitsu*—quality that fascinates, bewitches, and delights. The design team should therefore research features that core customer have not expected but that, when introduced, would thrill them.

Inaccurate Specifications: Ninety Percent of Them Are Wrong

In translating customer requirements into engineering specifications, techniques such as QFD are supposed to be employed. Yet, specifications are still pulled out of the air. Euphemistically, the practice is called "atmospheric analysis." The result is that 90 percent of specifications are wrong. Why are specifications so vague and arbitrary? There are several reasons:

❖ Specifications are lifted from older designs and drawings.

❖ Engineering relies on boilerplate requirements or supplier-published specifications.

❖ The computer is relied on to determine tolerances. This can only be done if the formula governing the relationship between the output (i.e., a dependent variable) and the independent variables is known. This is rarely the case in complex designs.

❖ Worst-case analysis is performed. Because of the very low probability of such occurrences, there is an appreciable addition to cost, with no value added.

❖ Statistical tolerancing is employed. This formula approach is inaccurate in actual practice.

❖ Engineering conservatism (otherwise known as "cover your hide") is in force. Engineers know that tight specifications and tight tolerances may be costly, but loose specifications and tolerances, resulting in product defects/failures, may cost them their jobs.

❖ Reliability is not a specification. Even if it is, it is not accurately predicted and tested in laboratory checks.

❖ Field environments/stresses are not measured, combined, or simulated in the laboratory.

❖ Ergonomics are not sufficiently considered. Engineers do not take into account the customer's lack of familiarity with the product or the misuses to which it could be subjected.

❖ Product safety is not pursued aggressively and product liability prevention is not enough of a discipline in design.

❖ Built-in diagnostics are not yet a part of design culture.

❖ Correlation studies, charting levels of customer preference versus levels of one or more product parameters, are not conducted.

CASE STUDY

The Importance of Realistic, Customer-Derived Specifications: Edge Defects in Contact Lenses

A manufacturer of contact lenses was convinced that edge defects—scratches, chips, and inclusions around the periphery of the lens—were objectionable to customers and had to be rooted out with four to five repeated inspections using high-powered microscopes. The company had spent millions of dollars with this brute-force approach and yet was losing market share to its competitor. In a reverse engineering study, it found, to its amazement, that the competitor's contact lenses had far more edge defects than its own. This initiated a correlation study comparing various levels of customer acceptance versus levels of edge defects. The study revealed little correlation. The edge defects were of minor concern to the lens wearers because they were only on the periphery of the lens, could not be seen with the naked eye, and did not affect vision. The competitor's superiority lay in 1) its closer adherence to the exact lens prescription (high c_{p_K}) and 2) its product's greater wearing comfort.

The company rapidly changed its concentration from edge defects to prescription accuracy (achieving $c_{p_K}s$ of 2.5 and higher) and to wearing comfort. In the process, it saved millions of dollars and restored its share within eighteen months.

A Comprehensive Specifications Checklist

Table 12-2 includes a series of questions that should be tackled by the design team at the concept stage and revisited at the prototype stage of design to assure that the various priorities of customer requirements are adequately translated into engineering specifications.

Design Quality/Reliability

Quality and reliability should be addressed at each stage of design:

1. Product feasibility study
2. Preliminary design
3. Prototype design
4. Engineering pilot run
5. Production pilot run
6. Production (covered in Chapter 14)
7. Field (covered in Chapter 15)

Product Feasibility Study (Concept Stage)

An invaluable starting point is to consult the "lessons learned" log from previous designs. The lessons learned log is a quick way for new engineers to avoid design pitfalls and shorten their learning time.

Table 12-2. A specifications checklist.

Customer Element	Checklist of Associated Requirements/Specifications
Quality	❖ Are acceptance/rejection criteria established, and are they based on customer expectations, not the company's? ❖ Are target COPQ, yields/cycle time, TDPU established? ❖ Are the Shainin/Bhote DOE requirements to prevent quality problems in production sufficiently well known and practiced?
Reliability	❖ Are targets (percent failure rate per year, MTBF etc.) "reach out" to meet future reliability expectations, such as: 1. Sharp reductions in the gap between failure rates of complex versus simple products? 2. Component failure rates below 1 ppm/year? 3. Use of incentive/penalty clauses for exceeding/not achieving reliability targets? ❖ Are maximum field environments/stresses measured? ❖ Are adequate test chambers in place to combine and accelerate stresses rapidly and to reliably predict and prevent failure rates, using MEOST? ❖ Is there a firm derating (factor of safety) protocol in place and followed?
Maintainability	❖ Are targets for mean time to diagnose (MTTD) and mean time to repair (MTTR) established?
Diagnostics	❖ Are MTTD targets helped by built-in diagnostics and MTTR targets by modular designs?
Uniformity	❖ Are target values and minimum c_{p_K}'s determined for key parameters?
Dependability	❖ Are warranty targets/times being extended toward lifetime warranties?
Availability	❖ Is there a targeted up-time as a percent of the total product use time?
Technical Performance	❖ Are target performance parameters established with the customer as the supreme court, and are they correlated with customer expectations? ❖ How do target performance parameters compare with competition?
Safety	❖ Is a formal, written process for product liability prevention in place?
Ergonomics	❖ Is user-friendliness tested with employee/focus groups, and how does it compare with competition?

Future Expectations	❖ Are future expectations continually solicited from customers through mass customization?
Service Before Sales	❖ Are sales/service parameters formulated to measure the effectiveness of service to the customer?
Services After Sales	❖ Are there targets for repair service accuracy and timeliness? ❖ Is there a systematic and continuous follow-up of customers after sale?
Price/Cost	❖ Have targeted prices, targeted costs, and targeted product life been determined? ❖ Has price elasticity been determined? ❖ Have incremental specifications versus incremental costs been tested with key customers?
Delivery	❖ Is there a targeted design cycle time that is at least half (or less) of a previous design of comparable complexity? ❖ Is the design inventory (which consumes design manpower) expressed in terms of a negative cash flow equivalent of months of sales?
Delight Features	❖ Are there a targeted number of delight features, unexpected by the customer, to generate customer "wow"?

Computer-Aided Design (CAD)

There are a number of software programs, ranging from Monte Carlo simulation, E-chip, MiniTab, and other modeling techniques that can be used at the product feasibility stage. Given the fascination with the computer, these methods are becoming ever more popular. Worst-case analysis, fail/safe conditions, and self-test architecture can all be examined.

However, an important prerequisite for CAD is a proven mathematical formula (i.e., equation) that governs the relationship between a number of independent variables and their dependent variable. Sometimes, such formulas can be developed based on experience—a sort of rearview mirror approach. Often, however, the computer cannot be programmed for unknown interaction effects and can lead to poor designs. One inviolate rule is reflected in President Reagan's pithy advice about the Russians: "Trust but verify!" Computer simulations must be verified at the prototype stage with actual hardware using powerful design of experiment tools.

Software Architecture

A review of the software architecture is also necessary at this product feasibility stage.

Reliability Block Diagram and Budgeting

The targeted reliability of the product as a whole must be subdivided into targets for each module and each subassembly comprising the total design. This is given the pompous name of reliability budgeting. Often, incompatibilities between the total reliability target and the feasibility of achieving a reliability target for a particular module show up at this early stage of design.

Preliminary Design (Breadboard Stage)

The elements of quality/reliability in preliminary design include:

- ❖ Thermal plots
- ❖ Derating and structural analysis
- ❖ Reliability prediction studies
- ❖ Failure mode effects analysis (FMEA)
- ❖ Fault tree analysis (FTA)
- ❖ Product liability analysis (PLA)
- ❖ Design reviews

Thermal Plots

In many products, especially electronics, heat is the enemy of reliability. A useful discipline, therefore, is to scan the preliminary product with infrared scanners to detect hot spots in the design that are likely to accelerate failure rates and can be corrected "a priori." Failure to conduct thermal plots is one of the reasons for early field failures.

Derating and Structural Analysis

When civil engineers design a bridge or building, they design it to bear a load six times greater than the load expected in actual use. In other words, they design in a factor of safety of six. But this discipline is seldom carried over into other branches of engineering, especially in electronics engineering. As stated earlier in this chapter, engineers are reluctant to use even a minimum 20 percent safety margin—or derating—because of fear of added material costs in the design.

The history of reliability disasters is replete with such design myopia. A multinational company used a one-amp diode in a diode trio subassembly that formed part of an alternator in a passenger car. The failure rates, low in early production, soon escalated into a major calamity costing the company millions of dollars. The engineers had been reluctant to derate the diode because a higher-rated part would cost a nickel more! Finally, the threat of the total loss of the alternator business forced them to design a three-amp diode—that is, a derating of 67 percent or a factor of safety of three. The result: On 800,000 new alternators with the larger diode, the total failure was one—a failure rate of 1.25 parts per million (ppm)—as compared to the previous failure rate of 15 percent or 150,000 ppm.

In the world of electronics, for instance, every important part should have a minimum derating of 40–60 percent in terms of voltage, current, power, and temperature. The parts should be able to "loaf" with the much-reduced stresses imposed on them. In mechanical products, the factory of safety should be at least 2:1.

Reliability Prediction Studies

The United States, Britain, France, and Germany issued cookbook reliability studies with failure rates on each part, based on applications. Most of them are almost worthless. The U.S. Defense Department uses a reliability study called Mil-Handbook 217E. Like its international counterparts, the predicted failure rates miss the actual field failure rates by 2:1 or even 10:1. They are either too low or too high, but invariably wrong. Computer programs for reliability, such as CB Predictor (by Decisioneering), are gaining currency. But they are no better than the cookbook approaches. The best practice is for a company to generate its own library of field failures based on the school of hard knocks—the actual field history, with superimposed stresses.

Failure Mode Effects Analysis (FMEA)

This thirty-year-old discipline is much touted as a major reliability tool. But it is a relatively weak technique, not worth the hours and hours spent on constructing it. Its objectives are good:

❖ To identify the weak links of design
❖ To predict the top failure modes and their effects on customers; prioritize and quantify the risks of using such parts; and reduce the risk through redesign, or redundancy, or tests, or supplier changes

Its weaknesses, however, overshadow its strengths. FMEA is only a paper study. It attempts to quantify the engineer's opinions and concerns. It is almost an opaque crystal ball in predicting true field failures. An FMEA is recommended only as a preliminary study. Not much time should be wasted in its development.

Fault Tree Analysis (FTA)

This is a mirror image of a FMEA. Whereas FMEA starts with potential causes of failure and determines their effect on the customer, an FTA starts with failures that could be observed by customers and traces their root causes in an attempt to prevent such root causes before the fact. It is somewhat more useful as a preliminary tool for product liability analysis.

Product Liability Analysis (PLA)

Many people consider the United States not a nation of laws but a nation of lawyers! Product liability lawsuits have become a bonanza for unscrupulous lawyers pressing for huge settlements in punitive damage. To protect against a plethora of frivolous lawsuits, companies must document their actions to prevent failures that can cause personal injuries. A product liability analysis, which is derivative of an FTA, can provide courts with evidence that preventive measures have been designed into the product. Table 12-3 is a product liability guideline for engineers if they are eager to keep out of jail.

Design Reviews

Design reviews are key ingredients in the launch of a new product. They are conducted at the concept stage, the prototype stage, and the production pilot run stage. These reviews should not be passed over lightly as a mere formality. The design team should spend considerable time in preparation to answer a volley of questions and likely challenges. Sometimes, senior engineers from different design groups perform the role of "devil's advocate" to unearth underlying weaknesses. At the end of each design review, the program manager decides to either continue with the project or terminate it if the design objectives simply cannot be met. Table 12-4 is a comprehensive checklist used during each of the three design reviews to assure that all bases have been touched and that weaknesses are identified and corrected before production begins.

Prototype Design Stage

Shainin/Bhote Design of Experiments and Variables Search

Chapter 11 introduced the simple but statistically powerful and cost-effective design of experiment (DOE) tools that make up the Shainin/Bhote methodology. One of these tools is especially vital at the prototype design stage—variables search. Because of its importance, it is explained in detail along with a case study.

The DOE tool of variables search[1] can be used to solve chronic quality problems in production, following the generation of clues from prior multi-vari, components search, paired comparisons, or product/process search experiments. Its greatest use, however, is at the design stage of a product or process, to prevent problems going into production in the first place. The objectives of the variables search are to:

Table 12-3. Product liability guidelines for engineers.

Definitions

1. Hazard: A condition with potential injury.
2. Risk: The percent probability of injury from hazard.
3. Danger: The combination of hazard and risk. To reduce danger, either hazard or risk must be reduced.

❖ Assume that the design will go to trial.
❖ Assume that every design decision will be scrutinized in a court.

Duties Imposed on Engineers for Product Liability

1. The duty to design for all foreseeable hazards—including guarding against foreseeable misuses of product and minimizing risks to users (and nonusers) if accidents are unavoidable.
2. The duty to investigate and test to discover risks—including investigation and correction of field problems.
3. The duty to minimize risk by alternate design, safety device, or warning. Courts do not insist on completely fail-safe designs; "reasonably safe" is adequate.

❖ The designer may reject a safer alternate design if it is not feasible, unduly expensive, or involves other dangers.
❖ Because warnings are inexpensive and practical, courts have even greater expectations of the efficiency of warnings than they do of alternate designs and safety devices.
❖ Warnings must be prominent enough to attract user attention and clear enough for user to understand the nature of the risk.

4. The duty to meet government/industry safety standards.
❖ If a government standard is not met, the manufacturer is liable per se, without an opportunity to defend the design.
❖ If an industry standard is not met, or is less safe than competitive products, the manufacturer is not liable per se, but will be found liable by a jury.

5. The duty to report field defects.

❖ Under the Consumer Product Safety Act, a manufacturer of consumer products must report to the commission if a defect can create substantial risk of injury to the public. Penalties for nonreporting are severe.

Successive Lines of Defense

1. Avoid the accident.
2. If hazard cannot be eliminated for functional reasons, protect against it.
3. Make the accident "safe"—design for minimum damage.
4. Predict or warn of impending accident.
5. Warn of possible (rather than impending) accident.
6. Protect the user in case of accident.

(continues)

Table 12-3. (Continued).

Trade-offs Between Safety Versus Cost; Safety Versus Function or Utility

❖ The objective should not be safety over other parameters but an optimization and elevation of all.
❖ Safety must be a specification, along with cost, performance, and reliability.
❖ Start with a product liability prevention analysis.

1. Use a method similar to a fault tree analysis, but confine yourself to safety hazards.
2. Assign probabilities and severities of failure to each branch and down to root cause using the following formula:

$$\text{Probability} \times \text{Severity} = \text{Risk Priority}$$

3. Determine critical path (highest risk priorities).

❖ Separate the important variables from the unimportant ones.

❖ Tightly control the important variables with a $^c p_K$ of 2.0 or more.

❖ Open up the tolerances of the unimportant variables to reduce costs.

The variables search procedure, outlined here, is ideal if there are five or more variables to investigate.

Variables Search Procedure

1. Define the problem (called the Green Y) and assure its measurement accuracy.
2. List the most important variables or factors (e.g., A, B, C, D, E, F, and so on), in descending order of perceived importance, that can contribute to the Green Y variation.
3. Assign two levels to each factor—a best (B) and a marginal (M) likely to give best and marginal results.

Stage 1: Selecting the Right Variable

4. Run three trials, each with all factors at their best (B) levels and three more trials, each with all factors at their marginal (M) levels. (The six trials should be run in random order.)
5. Test of significance in selecting the right variables.

❖ Rule A: The three B levels should outrank the three M levels.
❖ Rule B: The $D/\overline{d}$ ratio should be a minimum of 1.25, where D is the difference *between* the medians of the three B levels and the three M levels and $\overline{d}$ is the average of the ranges seen *within* the three B and three M levels.

Table 12-4. A comprehensive design review checklist.

No.	Checklist	Concept Stage	Prototype Stage	Pilot Run/ Production Stage
	Management Guidelines			
1.	Strategic business interface with customer done ahead of project start?	X		
2.	Specifications checklist (Table 12-2) conducted?	X		
3.	"Lessons learned" file consulted?	X		
4.	Maximum 25 percent redesign rule followed?	X	X	X
5.	Stream of new products introduced rapidly and frequently?	X	X	X
	Quality/Reliability			
6.	Computer simulation studies made and verified with DOE?	X	X	
7.	Software architecture reviewed and software standardized?	X	X	X
8.	Thermal scanning conducted?		X	
9.	FMEA and FTA studies made and risks reduced?	X	X	
10.	Product Liability Analysis (PLA) made and documented?	X	X	X
11.	Important parts sufficiently derated?		X	
12.	Reverse engineering (competitive analysis) made?		X	X
13.	DOE product characterized using variables search?		X	X
14.	DOE product improvements verified using B versus C?		X	X
15.	DOE product optimized using scatter plots or response surface methodology (RSM)?		X	X
16.	Process gains frozen, using positrol?			X
17.	DOE process certification conducted for best signal-to-noise ratio?		X	X

(continues)

Table 12-4. (Continued).

No.	Checklist	Concept Stage	Prototype Stage	Pilot Run/ Production Stage
18.	Design to target values? c_p, c_{p_k} greater than 2.0 on all important parameters?			X
19.	MEOST: Field failures predicted and older failures replicated?		X	X
20.	Metrology: Instrument: Product accuracy ratio of 5:1 achieved?		X	X
21.	Data retrieval timely, transparent, and visible to all employees?			X
22.	Engineering change control: accurate, coordinated, rapid?			X
23.	Poka-yoke for reducing operator-controllable errors reduced?			X
24.	Field escape control: Early warning of potential field reliability problems in place?			X
25.	Failure analysis capability: Staff and equipment adequate?	X	X	X
26.	Quality system audits effective?	X	X	X
27.	Patent studies made and patent applications submitted?	X	X	X
28.	Hard tooling adequate, rapid?		X	X
29.	Software "walk through" successful?		X	X
30.	Traceability requirements (e.g., date codes, bar codes) adequate?			X
31.	UL (or equivalent) approvals granted?			X
32.	Packaging and transportation control adequate?			X
33.	Precontrol used to maintain gains in DOE.			X
34.	Cost of poor quality (COPQ) gathered, analyzed, and reduced?			X
	Cost Reduction			
35.	SBU, product, model, and parts reduction measures in place?		X	X

36.	Modular design for mass customization and diagnostics achieved?			
37.	Total value engineering to maximize customer "wow" and minimum cost done?		X	X
38.	Design for manufacturability (DFM): Score?		X	X
39.	Total productive maintenance (TPM): Targets? Results?			X
40.	Partnership with key suppliers (in fact, not just in name) in place?	X	X	X
41.	Financial incentives/penalties with partnership suppliers in effect?			
42.	Cost targeting employed?		X	
	Cycle Time Reduction			
43.	Outsourcing of modules, "black boxes" for key suppliers to design?		X	X
44.	Early supplier involvement (ESI): Results?		X	X
45.	Parallel development and daily team interchanges?	X	X	X
46.	Focus factories instituted?			X
47.	Product flow in place of process flow?			X
48.	Pull systems; MRP II obsolete?			X
49.	Small lots and drastic reductions in setup times?			X
50.	Human inventory reductions calculated?	X	X	X

6. If the $D/\bar{d}$ ratio is greater than 1.25, stage 1 is successful. It means that the right variables have been captured.

7. If the $D/\bar{d}$ ratio is less than 1.25, either:

> The right variables were not selected.
> The right B and M levels were not selected.
> There may be an interaction effect between an even number (but not an odd number) of variables.

In which case, stage 1 should be repeated with other levels of the variables selected or with different variables.

Stage 2: Separation of the Important from the Unimportant Variables

Follow these steps to separate the important from the unimportant variables:

1. Run a pair of tests with the marginal level of the most important factor (labeled A_M) along with the best levels of all the remaining factors (labeled R_B). Repeat with the best level of A (A_B) along with the marginal levels of all remaining factors (R_M). Four outcomes are possible:
   (a) $A_M R_B$ gives the best result of stage 1, and $A_B R_M$ gives the worst result of stage 1—in which case A is not powerful enough to change the outcome of stage 1 and is therefore unimportant.
   (b) $A_M R_B$ gives the worst result of stage 1, and $A_B R_M$ gives the best result of stage 1—in which case A reverses the results of stage 1 and is very important (the Red X).
   (c) $A_M R_B$ give a result *worse* than the center of the best and worst results of stage 1, and $A_B R_M$ give a result *better* than the center of the best and worst results of stage 1, in which case A can be declared partially important along with another factor—possibly an interaction effect.
   (d) $A_B R_M$ give results *better* than the center of the best and worst results of stage 1, and $A_B R_M$ give a result *worse* than the center of the best and worst results of stage 1—in which case A can be considered unimportant.

2. If the results are in outcome (c) above, repeat stage 2, step 1, with factor B and use the rules similar to 1(a), (b), (c), (d). If the results are (a) or (d) above, B is unimportant; if the results are (b), B is most important; if the results are (c), B is partially important, along with another factor.

3. Repeat stage 2, step 2 with other factors (e.g., C, D, E, F, and so on).

Stage 3: Capping Run—Verification Test

The purpose of the capping run is to verify that the important factors are, indeed, important and that the unimportant factors are, indeed, unimportant. Let us say, in stage 2 of the process previously outlined, that factors A and B are partially important and that the others are unimportant. In this case:

1. Run a test with $A_M B_M R_B$. If this gives the worst results of stage 1, A and B are declared important, along with an AB interaction.

2. If this gives the best results of stage 1 or a reading between the

worst and best of stage 1, factors A and B cannot be verified as truly important.

CASE STUDY

Variables Search: The Engine Control Module

A microprocessor-based engine control module with 600 electronic components was being developed by the engineering department of a large company for an important automotive customer. The module monitors twenty to twenty-five engine parameters in a passenger car and optimizes the engine for maximum gas mileage and minimum pollution. The customer considered idle speed current as a critical parameter. It had to be between 650 ma (milliamperes) and 800 ma. If the idle speed current went below 650 ma the car could stall or stop. If it went above 800 ma, some of the components could burn up.

Only one physical module was available—a sample size of one! The engineer selected seven factors in descending order of importance and determined the best level of each factor (the target value) and the marginal level of each factor (at one end or the other of the tolerance limit). The variables search design, as well as the stage 1 and 2 experiments, are shown in Table 12-5.

Stage 1 was successful (meaning that the list contained the important variables) because the $D/\overline{d}$ ratio of 312/23, or 13.6, was much greater than 1.25.

Stage 2 indicated that none of the variables A through F were important, contrary to the engineers' judgment. Only variable G, the last on their list, was truly important—a solid Red X. This meant that the tolerances of the first six variables could be opened up to reduce costs, while variable G's tolerances had to be tightened to achieve a c_{PK} of 2.0. As a result, the company saved $450,000 at the end of its first year production run.

Multiple Environment Over Stress Testing (MEOST)

Chapter 11 has amply described the need, objectives, and benefits of MEOST, along with its principles and its powerful methodology. MEOST

Table 12-5. Variables search case study: engine control module.

Factor	Factor Target Value	Factor Tolerances	Chosen Factor Levels Best (B)	Chosen Factor Levels Marginal (M)
A. Resistor R85	0.68 ohms	±5%	0.68 ohms	0.65 ohms
B. Power Supply Voltage	5.0 volts	±5%	5.0 volts	4.75 volts
C. Resistor R77	100 ohms	±1%	100 ohms	99 ohms
D. Resistor R75	787 ohms	±1%	787 ohms	779 ohms
E. Transistor (Q8) Voltage	75 mV	150 mV max	75 mV	105 mV
F. Resistor R79	43 ohms	±5%	43 ohms	40.85 ohms
G. IC4: Off Set Voltage	0 mV	±8 mV	0 mV	−8 mV

Stage 1	All Factors at Best Levels	All Factors at Marginal Levels
	742 ma	1,053 ma
	738 ma	1,050 ma
	725 ma	1,024 ma

$D = 1,050 - 738 = 312 \text{ ma } \overline{d} = (17+29)/2 = 23 \text{ } D/\overline{d} = 312/23, >1.25.$

Stage 2	Test No.	Combination	Results	Center Line	Conclusion
	1	$A_M R_B$	768	894	A not important
	2	$A_B R_M$	1,020	894	
	3	$B_M R_B$	704	894	B not important
	4	$B_B R_M$	1,051	894	
	5	$C_M R_B$	733	894	C not important
	6	$C_B R_M$	1,028	894	
	7	$D_M R_B$	745	894	D not important
	8	$D_B R_M$	1,018	894	
	9	$E_M R_B$	726	894	E not important
	10	$E_B R_M$	1,022	894	
	11	$F_M R_B$	733	894	F not important
	12	$F_B R_M$	1,020	894	
	13	$G_M R_B$	1,031	894	G very important
	14	$G_B R_M$	718	894	

is the tool, par excellence, for reliability just as DOE is the tool, par excellence, for quality and problem solving/prevention. Yet, of all the tools described in Chapter 11, MEOST is almost totally foreign to industry, with only one percent of leading companies even aware of it and 0.05 percent using them. For the average company, MEOST is a totally unknown black hole.

MEOST is most effective at the prototype stage of design, where potential failures that take months to occur in the field can be replicated in design, typically in less than twenty-four hours with extremely small prototype samples of three or less. It should be repeated in the pilot run stage to make sure that engineering changes and tooling have not adversely affected reliability. It should also be repeated in production before shipment to the customer to assure that production workmanship/processes and suppliers have not degraded the design intent.

My experience with MEOST has resulted in product field failure rates of 4 ppm to 10 ppm per year; and in one product line with hostile environments, we achieved zero failures in over seven years of field use, with each engine traveling over 500,000 miles.

c_{p_K} of 2.0 or More for All Important Parameters

Following variables search where all the important parameters of a product have been defined, their target values (i.e., design centers—center of a given parameter) should be aimed for in design and their c_{p_K}s should be specified at 2.0 or higher. This also applies to the important parameters for suppliers and should be part of their specifications and drawings as well.

Instrument Accuracy

There is an inviolate rule in the quality discipline that the accuracy of the instrument measuring the product should be at least five times the accuracy of the product. Instrument accuracy is made up of the variations within-instrument (V_{WI}); instrument-to-instrument (V_{I-I}); and operator-to-operator (V_{O-O}) as expressed by the formula:

Total Instrument Variation (V_{IT}) $= \sqrt{V_{WI}^2 + V_{I-I}^2 + V_{O-O}}$
Product variation (^VP) = Specification Width
This means that V_P/V_{IT} should be equal to or greater than 5:1.

Many a problem in design and in production goes unsolved because this 5:1 instrument accuracy rule is not heeded.

Modular Designs

Modular designs should also be a "must" for the design team. They facilitate diagnostics in the field, ease of service and, above all, the postponement of options in mass customization as far down the assembly line as possible.

Quality Systems Audit

It is at the prototype stage that an existing quality system audit that supposedly is in place should be reexamined and updated for effectiveness. The self-assessment/audit contained at the end of each chapter of this book goes well beyond just quality. Nevertheless, an audit is useful in covering weaknesses in all areas, starting with design. Under no circumstances should weak quality systems, such as the Malcolm Baldrige National Quality Award or the European Quality Award, or the even weaker ISO-9000 and QS-9000, be given a second thought!

Failure Analysis Capability

A major weakness in many companies is the inability to conduct a comprehensive failure analysis on products and components. Dependence on suppliers is risky, both in terms of time and accuracy. Dependence on outside failure analysis laboratories is, at the least, expensive. There is no substitute for a professional failure analysis staff within the company, along with costly but necessary failure analysis equipment, such as scanning electron microscopes (SEMs) and spectral analysis, to get to the root cause (Red X) of a problem. This is especially true today, as more and more products use a larger share of electronics, such as integrated circuits and microprocessors. (It is estimated that more than 50 percent of a passenger car, in dollar value, will be composed of electronic content in a few years.)

Pilot Run Stage

B Versus C Validation Tests[2]

B versus C is an integral part of design of experiments (DOE). C stands for a current product, B for a better or improved product. The purpose of a B versus C test is to validate or verify that the product improvement is effective and permanent. Often, designers think that they have achieved a product improvement, only to discover a week or two later that the improvement has vanished. They find themselves thrown back to square one. To avoid these premature judgments, it is necessary to go back to the old design to see if the problem has reappeared; then return to the new design to see if the improvement is reconfirmed. This back-and-forth process must be repeated twice. It is the equivalent of turning a light switch "on" and turning it "off."

Then, the three "B" readings and the three "C" readings must be arranged in descending rank order. Only if the three Bs outrank the three Cs can there be a 95 percent confidence that the B product is significantly better than the C product. A mixed rank order would indicate that a significant improvement could not be validated. (Note: The sequence of testing three Bs and three Cs must be randomized.)

The B versus C test is also useful in determining a reliability improvement very quickly. Usually, a design change for reliability takes months of exposure in the field to verify. No one can afford that length of time. Consequently, half-baked improvements are prematurely made, only to be shown to be ineffective later.

MEOST offers a fast and reliable way to determine the effectiveness of design changes "a priori" instead of "a posteriori." Three Bs and three Cs are subjected to MEOST tests. The times-to-failure or the stresses-to-failure are then arranged in rank order. Only if the three Bs outrank the three Cs can there be 95 percent confidence that a design improvement has been effected.

Product Optimization

Earlier in this chapter, there was a blunt statement that 90 percent of engineering specifications are wrong and the reasons for it. How, then, can realistic specifications and realistic tolerances be established by the design team? One simple, graphic, and effective way is with scatter plots. (Scatter plots can be used if there are negligible interaction effects between two or more input variables. If the interactions are suspected to be significant, a DOE technique called response surface methodology, or RSM, is used. A detailed treatment of scatter plots and RSM is available in Chapters 16 and 17 in my book *World Class Quality—Using Design of Experiments to Make It Happen*.[3]

Positrol: Freezing The Gains

One of the weaknesses of industry is that engineers try to control a process by checking the product it produces. That is too late. An analogy would be to steer a boat by looking at its wake! Once important process variables have been identified through variables search (and optimized through scatter plots and response surface methodology, with realistic specifications and realistic tolerances), the purpose of positrol[4] is to freeze the process gains by continuously monitoring the process. In this way, important variables or parameters—the "what"—never exceed their carefully crafted tolerances. This is done by a regimen in which each parameter is monitored with a:

Who (either an operator, maintenance person, or automatic controls)

How (accurate instrumentation)

Where (at the most critical point in the process)

When (that is, how frequently—either once a day, once an hour, or continuously)

Positrol is a discipline that 90 percent of industry does not even know, much less follow. Yet, it is vital if the gains through careful DOE studies are not to be frittered away by well-meaning but "diddle artist" technicians. I have seen so many good DOE studies set back because process technicians take the law into their own hands. They must be persuaded to believe that positrol is not a spy system on their behavior but a police system on the behavior of the process.

Process Certification[5]

If St. Patrick is the patron saint of Ireland, St. Murphy is, undoubtedly, the patron saint of industry. In fact, the humorous but very real foundation of process certification is Murphy's Law, which states: "If something can go wrong, it will." Murphy's Law is omnipresent in industry. The task of process certification is to round up these little Murphies and incarcerate them.

There are a number of peripherals that can contribute to poor quality. They can be divided into five broad categories:

- ❖ Management/supervision inadequacies
- ❖ Violation of good manufacturing practices
- ❖ Plant/equipment inattention
- ❖ Environmental neglect
- ❖ Human shortcomings

Some misguided companies feel that the discipline of ISO-9000 documentation could cure these ills. But ISO-9000 standards are so contractually determined that if the contract allows scrap, ISO-9000 will provide it consistently! Process certification goes way beyond ISO-9000 or QS-9000 or any other quality standard. Its purpose is to so reduce these uncontrollable factors that their collective noise is much less than the purity of the signal required for a DOE study. In other words, it assures a good signal-to-noise ratio and gives DOE a chance to really succeed. It is one of the best ways to make a process robust. As such, process certification should both precede as well as follow a DOE study.

Unfortunately, as in the case of positrol, 99 percent of companies do not understand process certification. Yet, its methodology is simple:

- ❖ It is best conducted by an interdisciplinary team.
- ❖ The team prepares a list of quality peripherals (based on Murphy's Law, where every procedure, process, practice, housekeeping measure, environmental control, and piece of test equipment is suspect).

❖ The team then audits the process to highlight potential quality problems and performs a "process scrub" before certification is granted for the process to start manufacturing the product.

❖ Periodically, the process is recertified so that the old, bad practices do not sneak back in and new ones are not introduced.

Poka-Yoke

This important tool to prevent operator-controllable errors is explained in Chapter 11, along with references that give many examples of poka-yoke sensors. The Japanese use poka-yoke extensively in place of control charts. It is a 100 percent check instead of the sampling that is used in control charts. Yet, because the 100 percent check is automatic at an operator's position, it costs less than superimposed control charts and is a decided improvement on product quality.

Design for Manufacturability (DFM)

Two disciplines that have been introduced in the pilot run stage of design in the last fifteen years are design for manufacturability and design for automated assembly. They have been perfected by two university professors, Jeffrey Boothroyd and Peter Dewherst.[6] Any assembly can be quantified in terms of a numerical score to measure the ease of manufacturability. They range from one (worst) to 100 (perfect). Most companies that have not been exposed to DFM have scores below twenty. Progressive companies do not allow a design to go into production with any score less than eighty.

There are ten basic rules for achieving best results in DFM:

1. Minimize the number of parts.
2. Minimize assembly surfaces.
3. Design for Z-axis assembly.
4. Minimize part variation (through DOE).
5. Design parts for multiuse and ease of fabrication.
6. Maximize part symmetry.
7. Optimize parts handling.
8. Avoid separate fasteners, where possible.
9. Provide parts with integral "self-locking" features.
10. Drive toward modular design.

Cost Reduction in New Product Design

Total Value Engineering

The distinction between traditional value engineering (VE), which concentrates on cost reduction while not deteriorating quality, and my total value engineering (TVE),[7] which concentrates on maximizing customer "wow," loyalty, and retention, has been covered in Chapter 11 and highlighted in Table 11-10. Both VE and TVE are one of the most powerful cost reduction tools available to design engineers for two chief reasons:

- ❖ An average reduction of 25 percent can be achieved, with a minimum of a 10 percent reduction. A 75 percent reduction is not uncommon.
- ❖ A minimum return of 10:1 in investment is possible, with 100:1 returns frequently registered. Furthermore, the amount of investment required in design is modest.

Total value engineering's gains are even more spectacular because it concentrates on customer loyalty. A 5 percent increase in customer retention can achieve a 35–120 percent increase in profit.

SBU, Product, Model, and Part Number Reduction

Part number proliferation is a design and manufacturing disease. Every engineer wants his own part number to meet the exact requirements of design. Line managers in manufacturing want their own unique part number on a common part so that other lines would not raid their hoarded stock. The result is part number pollution! One defense contractor, with a base of 30,000 parts, had ballooned the total to more than 300,000 part numbers.

The starting point for reduction is not at the part level but at the strategic business unit (SBU) level. Figure 12-2 shows the well-known Boston Consulting Group (BCG) portfolio analysis that separates SBUs into four categories on the basis of their relative competency versus their industrial attractiveness. In Figure 12-2:

1. The dog SBUs, with low competency and low industry attractiveness, should be divested.
2. Cash cow SBUs, with high competency but low industry attractiveness, should be "harvested" (i.e., milked for cash, with minimal resources poured into them, to feed the question mark and star SBUs).
3. Question mark SBUs, with low competency but high industry attractiveness, should be "nurtured" with cash inflows to move them into the star category.
4. Star SBUs, with high competency and high industry attractiveness, should be a major focus for the company.

Unprofitable "dog" SBUs should be divested first. As the great Peter F. Drucker stresses, a company must make a concentration decision and

Figure 12-2. Boston Consulting Group (BCG) portfolio analysis on value of business units in a corporation (competency focus).

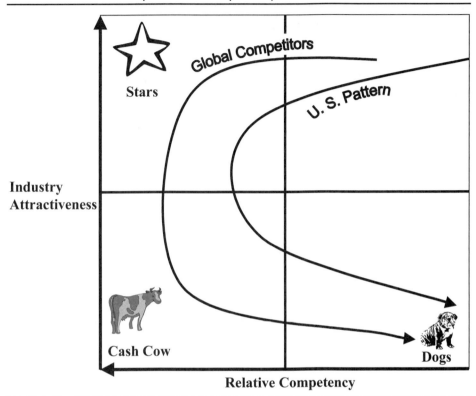

shed its unprofitable businesses from time to time. This could reduce a large volume of part numbers in one stroke. As an example, a medium-size company, in deciding to exit its consumer business, reduced its total part number base from 64,000 to 37,000.

Next in line should be the reduction of unprofitable product lines that are also relatively unimportant to customers. A third step should be to reduce the vast array of model numbers within a product line. Only then can a systematic attack on part number reduction be attempted through standardization or group technology.

Group Technology (GT)

The principle of group technology[8] is that parts that are similar in shape, tolerance, size, and other design characteristics can be consolidated into fewer parts, and that parts using similar manufacturing processes can be grouped together in smaller, dedicated processes called group technology cells. GT's main use is in parts design and design retrieval coding.

Before the use of computers, a seven-digit part number could not describe a part, other than its broad category. With computers, a code of twenty to thirty digits can be assigned and completely characterize the part for shape, dimensions, tolerances, chemistry, surface finish, manufacturing processes, supplier, and price, among other descriptions.

Group technology is the needed medicine for an astonishing parts proliferation malady. At General Dynamics, a virtually identical nut and coupling unit had been designed on five separate occasions by five different design engineers and draftsmen.[9] To compound the irony, the parts were purchased from five different suppliers at prices ranging from 22 cents to $7.50 each.

Proponents of the GT methodology claim the following cost savings potential:[10]

52 percent in new part design

10 percent in the number of drawings

30 percent in new shop drawings

60 percent in industrial engineering time

20 percent in production floor space

42 percent in raw material stocks

69 percent in setup time

70 percent in throughput time

62 percent in work in process (WIP) inventory

82 percent in overdue orders

A *Harvard Business Review* article on GT goes even further, stating that the cost of introducing a new part into a company ranges from $1,300 to $12,000, including expenses for design; planning and control; procurement; inspection; storing; tools; and fixtures. Reducing no more than 500 part numbers would save a company from $650,000 up to $6 million.

Early Supplier Involvement (ESI)

Early supplier involvement represents a symbolic continental divide between the archaic, old-line procurement practices of soliciting three bids and modern supply management principles involving the preselection of a partnership supplier (see Chapter 13). Several benefits accrue both to the company and the partnership supplier through ESI:

❖ Substantial cost reductions through value engineering ideas and cost targeting

- Shorter design cycle times
- Optimal designs and materials
- Mutually determined specifications and tolerances
- Classification of characteristics (highlighting important parameters)
- Utilization of supplier's technology
- Optimal match between part design and supplier's process technology
- Improved first piece-part quality
- Improved delivery
- Shorter tooling time and costs
- Improved customer-supplier communications and trust

The most important benefit of ESI is that it permits parallel development of a design in an iterative interaction with the partnership supplier rather than the "series" handover of the design to the latter supplier, with the design cast in concrete and incapable of any significant improvement. The methodology for ESI is outlined as follows:

- *Concept Phase.* Two or three of partnership suppliers capable of producing the product are chosen.
- *Feasibility Phase.* ESI is started with all suppliers to generate a maximum number of ideas for quality, cost, and cycle-time improvement.
- *Prototype Phase.*
    - Selected ideas are incorporated into the design.
    - Financial incentives are awarded to each supplier in proportion to the projected savings contributed by each of them. (Chapter 13 elaborates on incentives/penalties.)
    - One supplier is then selected as the partnership supplier for the design.
    - A determination is made whether the supplier should be confined to making the piece-part or the subassembly or the total product.
    - Broad outline specifications are developed for the supplier.
    - Close linkages are started between commodity teams (see Chapter 13) and the supplier.
    - A second round of value engineering ideas are generated in ESI meetings, then purified, distilled, and implemented.

❖ Cost targeting is established by the design team.

❖ Joint DOE studies are started with the supplier to achieve c_{p_K}s of 2.0 or more on important parameters and to open tolerances on the unimportant ones to reduce costs.

❖ Derating (factors of safety) is established on each part.

❖ Reliability targets are determined and MEOST planning started.

❖ *Pilot Run Phase.*

❖ c_{p_K}s of ≥ 2.0 are achieved on important parameters through DOE.

❖ Optimization studies (scatter plots) are run to establish realistic specifications and realistic tolerances.

❖ Positrol is established to freeze process gains.

❖ Process certification is conducted to assure design robustness.

❖ MEOST experiments are conducted to validate supplier reliability targets.

Cost Targeting

For decades, supplier costs were determined by soliciting bids from three suppliers and either selecting the lowest cost supplier or whipsawing them into negotiated cost reductions. Cost targeting, introduced twenty-odd years ago, is based on working with just one supplier—the best—who has been preselected as the partnership supplier. The part or product cost, however, is not determined by the supplier, but by the customer's design team, based on a computerized determination of the expected material cost and the estimated labor and overhead costs. This is then given to the supplier, as a target, to meet or to beat especially vis-à-vis competition. Then, using ESI and value engineering, a mutual agreement is reached to lower even the targeted cost as well as how to attain the reduction.

Cycle-Time Reduction in New Product Design

Cycle time can be defined as the total clock time that it takes from the beginning of any activity to its completion. Generally, this cycle time, or clock time, is ten to 100 multiples of the actual labor time spent in advancing the product or service. Direct labor is the only value-added portion of the total cycle time. As examples:

❖ In manufacturing, the nonvalue-added portions are waiting (queue) time; transport time; setup time; inspection time; test time; machine/equipment breakdown time; and quality rejection time.

❖ In businesses processes, the nonvalue-added portions are waiting time; transport time; setup time; inspection time; test time; quality rejection time; approval time; and rework time.

❖ In design, the nonvalue-added portions are lost time due to customers' changing requirements; disjointed customer-marketing-engineering interfaces; development in series rather than in parallel; suppliers not given any design responsibility; lack of commonality/standardization; parts proliferation; supplier proliferation; vertical integration; piece-part rather than "black box" buying; poor quality evaluations; and poor and lengthy reliability evaluations.

Cycle-Time Goals

Cycle time, as the other side of the quality coin, is as important as quality itself. Progressive companies have established reach-out goals for cycle time almost as aggressive as quality goals. At Motorola, we have 10:1 cycle-time reduction goals every five years—in manufacturing, in business processes, and even in design. Of these, design cycle-time reduction is the most important because the company is pitted against its predatory competitors to be the first with a product in the marketplace.

Measures to reduce design cycle time are listed in the following sections. If not a 10:1 reduction, at least 4:1 reductions can and have been registered.

Customer-Marketing-Engineering Interface

Keeping a finger on the customer's pulse should long precede product development, well ahead of its concept stage. The task starts even before the end of the launch of the previous design. The changing needs, requirements, and future expectations of the customer must be carefully monitored through quality function deployment, focus groups, clinics, panels, and—above all—one-on-one mass customization sessions. Preparing the groundwork reduces the numerous delays that can be caused by customers changing their minds in the middle or at the end of a current design cycle. Furthermore, this should be a team effort where marketing, sales, and engineering work together in what is called the strategic business process.

Parallel Development

Older design cycles were characterized by "series" practices where one activity was not started until the previous one was finished. This added considerably to cycle time. It also added to higher cost and poorer quality as various steps in the design cycle had to be abandoned and a fresh start made all over again. Parallel development allows different portions of the design to move forward simultaneously and significantly reducing design cycle time.

Daily Team Interchanges

You might ask whether parallel development leads to confusion and a situation where the left hand doesn't know what the right hand is doing. This problem can be eliminated by daily interchanges between the different teams. The interchanges are facilitated by Internet conferencing, where product, process, material, and die designs are exchanged between the teams.

Outsourcing

One of the common weaknesses in Western design activities is a preference for vertical integration—that is, 100 percent of the design is done in-house. No company—as even General Motors belatedly discovered—is an expert on everything. The movement today is outsourcing—not just for the procurement of piece-parts, but for the wholesale transfer to partnership suppliers of design responsibility on those models, subassemblies or "black boxes" where the company does not possess core competencies. The company specifies its requirements in broad terms to the supplier, who then delivers the finished design for the company's approval. This is the ultimate in parallel development, and the very significant reduction in cycle time is worth any incremental cost of outsourcing.

Early Supplier Involvement (ESI)

The power of ESI in reducing design cost has already been detailed in this chapter in the section on Cost Reduction in New Product Design. By preselecting partnership suppliers and working with them from just an outline drawing to fill in detail specifications and tolerances, the design team can also reduce valuable design cycle time.

Part Number Reduction, Preferred Parts, and Parts Qualification

Part number reduction through SBU, product, and model reductions as well as through the use of group technology has also already been covered in the section on Cost Reduction in New Product Design. For each of these measures, there is a reflected cycle time reduction as well.

The use of preferred parts not only helps with reliability but eliminates the need to go through lengthy parts qualifications. If the latter is necessary, it need not be undertaken separately by a component-engineering group, but in parallel with the partnership supplier, at the supplier's facility.

Design of Experiments (DOE)

Described in Chapter 11 and earlier in this chapter, DOE not only achieves outstanding quality and cost improvements, but cycle time reductions as well. When exposed to DOE, engineers may initially object to the time it takes to conduct these experiments. These counter arguments are convincing:

❖ DOE does not take more than a couple of days.

❖ Running the experiment can be done by technicians.

❖ The greatest time savings is not having to redesign because of unsolved quality problems. It is amazing how designers do not have time for good designs the first time, but have all the time in the world to redo bad designs three and four times!

Multiple Environment Over Stress Tests (MEOST)

The tremendous advantage of MEOST is that it can predict failures months before they can occur in the field and be corrected. It reduces the number of units on test from hundreds in conventional reliability testing to three and ten. Above all, it reduces reliability test time in the laboratory from hundreds of hours and days to twenty-four hours (for the most part). No self-respecting company can afford to do without MEOST.

Reduction of Human Inventory in Design

Figure 12-3 shows how cost and cycle time accumulate in a factory. The company purchases material from suppliers, adding cost but no cycle time.

Figure 12-3. Cost/cycle-time profile in manufacturing.

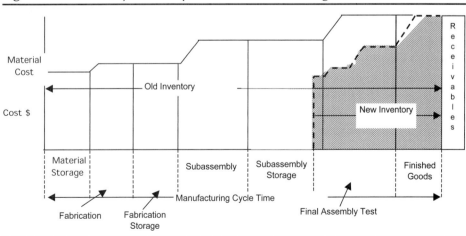

In storage, the material adds no cost, but consumes cycle time. The next stage—fabrication—consumes both cost and time, and so on, through the subassembly, subassembly storage, and final assembly/test to finished goods. Most of the cycle time is wasted in setup, transport, and waiting. The integral of this cost-time line (that is, the total area under the curve in Figure 12-3) is inventory. None of this large inventory cost can be recovered until the product is sold and payments collected through accounts receivable.

If the cycle time is reduced by factors of 10:1 and more, the inventory costs are dramatically reduced (shown by the shaded area in Figure 12-3).

An even more dramatic portrayal of cost/cycle time is Figure 12-4, which shows the design cost/cycle time curve. Costs and time are spent in developing the strategic business, designing the product, and procuring materials. Finally, there is the manufacturing cycle time of Figure 12-3, shown as a "pimple on the log" of the design/cost cycle time of Figure 12-4. The area under the curve of Figure 12-4 also represents inventory costs of a different type—the invisible human inventory "costs" of design people. Not one penny of these invisible inventory costs is recovered until manufacturing ships the first product and receivables are collected. This makes for an astounding cash flow problem. It has been estimated that a typical company has a negative cash flow equivalent to one year of sales! If design cycle time can be cut by 50–80 percent, not only would the human inventory costs be greatly reduced, but the negative cash flow can be appreciably reduced as well.

Figures 12-3 and 12-4 are just as valid for a company's suppliers as well as for these suppliers' suppliers. If each supplier in the chain reduces its manufacturing inventory (and also the more important invisible design people inventory), the savings in reduced overall cycle times and material

Figure 12-4. Cost/cycle-time profile in design in relation to manufacturing.

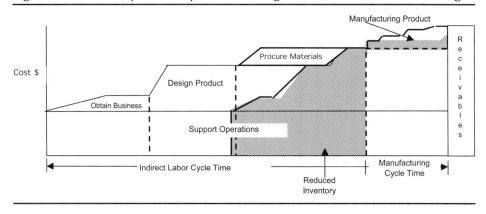

costs to a company can snowball. Extrapolated to an entire country, this practice could make it cost-competitive with any other nation in the world.

CASE STUDY

Automotives in Japan—A Benchmark Industry in the Area of Design

Instead of choosing a particular company as a benchmark, I have selected a country and an industry within the country as a model for new product development. The country is Japan and the industry is its famed automotive industry (although one company—Toyota—dominates the field of car companies in Japan and worldwide, it is unquestionably the benchmark par excellence).

Two comparisons are made:

1. *Features.* Table 12-6 compares features associated with new product introduction in Japan versus the West and helps explain the radical differences in approaches between Japan automakers and their Western counterparts.

2. *Performance Results.* Table 12-7 shows how performance results differ between Japan, the United States, and Europe. It is true that general financial and economic fluctuations on the world scene may have changed some of these figures in absolute terms, but the relative differentials remain. They should be a wake-up call to the West.

SELF-ASSESSMENT/AUDIT ON DESIGN ROBUSTNESS

Table 12-8 is a self-assessment/audit on the design discipline, as it pertains to new product development, that a company or external Ultimate Six Sigma auditors can use to gauge its progress. It lists six key characteristics and fifteen success factors each worth five points, for a total score of seventy-five points.

Table 12-6. Case study: new auto introduction in Japan versus the West.

Feature	Japan	West
Project Manager	Leader, total authority, "shusa;" great power, most coveted position.	Coordinator must beg for resources, team member time.
Team	Stays on project full-time; team members evaluated by shusa.	Team members part-time; functional managers demand priority and do evaluations.
Engineer Rotation	Three months on assembly line; three months in marketing, one year in various engineering departments, then routine work on NPI, then more fundamental work, then more training, then more advanced work, then other disciplines.	Little or no rotation; start on the bench; one or two disciplines in a lifetime.
Communication	Conflicts resolved early— "ringi" and "neemawashi."	Conflicts persist.
Development Method	Parallel development; die production and body design start at same time.	Series (i.e., sequential) development. Die making delayed until designers complete detailed part specs.
Range of Models	Wider range, short life, frequent replacement. Result: Innovations are 3 or 4 product generations ahead of the West.	Narrower range, longer life for each model.
Outsourcing	Only core competencies are retained in-house.	Vertical integration still favored in design.

Source: James P. Womack, Daniel T. Jones, and Daniel Roos, *The Machine That Changed the World* (New York: HarperCollins Publishers, 1997).

Table 12-7. Case study: automotive performance results in Japan versus the West.

Parameter	Japan	U.S.	Europe
Average engineering hours/new car (millions)	1.7	3.1	2.9
Average development time/new car (months)	46.2	60.4	57.3
Number of employees in project team	485	903	904
Number of body type/new car	2.3	1.7	2.7
Average ratio of shared parts	18%	38%	28%
Supplier share of engineering	51%	14%	37%
Ratio of delayed products	1 in 6	1 in 2	1 in 3
Die development time (months)	13.8	25.0	28.0
Time from production start to first sale (months)	1	4	2
Return to normal productivity after new model (months)	4	5	12
Return to normal quality after new model (months)	1–4	11	12
Reliability (complaints/100 cars)	66	140	128
Quality (delivered defects/100 cars)	30	61	61
Productivity cars/employee/year	47.3	31.0	24
Cost: Japanese landed cost (into U.S.) advantage			
Small Cars	$2,600	—	?
Intermediate Cars	$3,000	—	?
Large Cars	$4,200	—	?

Source: James P. Womack, Daniel T. Jones, and Daniel Roos, *The Machine That Changed the World* (New York: HarperCollins Publishers, 1997).

Table 12-8. Design: key characteristics and success factors.

Key Characteristic	Success Factors	Rating				
		1	2	3	4	5
7.1 Organization for New Product Introduction (NPI)	1. Concurrent engineering—a cross-function team drawn from engineering, manufacturing, quality, sales/marketing, sourcing, service. and finance—is the full time organizational building block for NPI.					
7.2 Management Guidelines	1. Management guidelines are firmly established for: ❖ A maximum 25 percent redesign rule ❖ A stream of new products frequently and rapidly introduced to keep competition off-balance ❖ A "lessons learned" file to prevent mistakes in the next generation of design ❖ Reverse engineering to evaluate strengths/weaknesses of competition					
7.3 "Voice of the Customer"	1. All elements of customer "wow" are considered in the concept phase of design, with special attention to those features unexpected by the customer that would cause customer delight. 2. The checklist of all relevant specifications has been considered.					
7.4 Design Quality/ Reliability	1. A thorough Product Liability Analysis (PLA) has been conducted and documented. 2. Design of experiments (DOE) studies are conducted to prevent quality problems entering production. 3. Important parts are derated (factors of safety built-in) to reduce stresses and enhance field reliability. 4. Multiple Environment Over Stress Testing (MEOST) is conducted to predict and correct potential field failures with very small sample sizes and short test (24 hours) duration. 5. Process certification is conducted to make processes robust against noise; and positrol is put in place to freeze gains made in DOE.					

7.5 Design Cost Reduction	1. Total value engineering is employed to simultaneously increase value to the customer and reduce costs. 2. Concentration decisions are made to reduce the number of unimportant/unprofitable SBUs, products, and models. Group technology is used to reduce part numbers substantially. 3. Early supplier involvement (ESI) is promoted with partnership suppliers to outsource designs, establish meaningful specifications, and meet target costs set by the design team.					
7.6 Design Cycle-Time Reduction	1. Reach-out design cycle-time reduction goals are formulated. 2. "Parallel" rather than "series" development is conducted between design teams and daily team exchanges facilitated. 3. The cost/time profiles for development are made and reduced to improve cash flow in design significantly.					

Thirteen

FROM A CUSTOMER-SUPPLIER WIN-LOSE CONTEST TO A WIN-WIN PARTNERSHIP FOR THE ENTIRE SUPPLIER CHAIN

❖   ❖   ❖

A supplier's main objective for its customer: To create and nurture satisfied, repetitive, and loyal customers who have received added value from the supplier. . . . A customer's main objective for its supplier: To establish a firm partnership with key suppliers, at least at two

levels in the supply chain,

rendering active, concrete

help to them to improve its

own cost, lead time, and

quality, while increasing

their profitability.

—KEKI R. BHOTE

The March to Supply Chain Management

There has been a steady evolution in procurement practices in the last fifty years:

❖ In the first half of the twentieth century, it was purchasing—a clerical appendage.

❖ In the 1960s and 1970s, purchasing was expanded to materials management, adding inventory control, and logistics to the buying activity.

❖ In the 1980s, supply management evolved to add key suppliers as partners in procurement.

❖ In the 1990s, supply chain management extended the frontiers of this partnership in both directions—to the whole chain of suppliers, from raw materials to first-tier suppliers and from the customer company to its distributors, dealers, retail chains, servicers, and, as always, to its end-customers. The main reason for this radical approach is that high material costs, long lead times, and poor quality do not start with first-tier suppliers but are caused throughout the supply chain, going all the way back to raw material suppliers. (This subject is elaborated in the next section.)

The Escalating Importance of Supply Chain Management

There are six factors that underline the ever-increasing importance of supply chain management vis-à-vis other disciplines within a company:

1. The enormous leverage for profit improvement
2. The price-material cost squeeze

3. The natural advantages of outsourcing
4. The trend toward suppliers as design partners
5. Gains in supplier quality, cost, and lead-time improvement
6. Unbelievable economic gains in strengthening all links in the supply value chain

The Large and Growing Leverage of Supply Management

Figure 13-1 shows that direct labor as a percentage of sales accounts for not more than 5 percent (on an average) of the sales dollar, whereas purchased materials account for at least 50 percent. Supply management has over ten times the leverage of direct labor. Yet, where does management allocate its time? Fifty percent to production and a grudging 5 percent to purchased materials! Is it any wonder that industry's weakest link is management? Furthermore, in the factory of tomorrow, direct labor will shrink to 2 percent of the sales dollars and materials will increase to 60 percent, raising supply management's leverage to thirty times direct labor. In the "lights out" factory of the near future—a peopleless, paperless environment where computers will control computers that run machines— direct labor will evaporate to almost zero while purchased materials will soar to a staggering 70 percent of the sales dollar. What a profit potential!

The Price-Material Cost Squeeze

Another dangerous phenomenon is shown in Figure 13-2. Gone are the days when companies could blissfully raise prices. Customers are not only resistant to higher prices, but they push down prices as their buying options from global competitors increase, as they shop for value, and as customer loyalty can no longer be taken for granted. This price erosion is a minimum of 2–5 percent a year in stable markets and as much as 25 percent a year in high-technology markets. The drastic price erosions in the computer and cellular phone businesses are but two examples of the power of technology to slash prices.

While price erosion is a fact of life, the cost of purchased materials from suppliers can continue to escalate because of raw material cost increases, inflation, and other factors. Caught in the middle of this "scissors action" of customers and suppliers is the squeezed manufacturer. If the trend should continue, the two lines in Figure 13-2 can intersect. Long before that happens, a company will have gone out of business. Supply chain management is the best way to actually reverse the trend. Material costs must decrease at a steeper rate than customer prices.

The Natural Advantages of Outsourcing

The old penchant for vertical integration in companies has been giving way to outsourcing for the last twenty years. Gone are the battles of make

Figure 13-1. The leverage of purchased materials.

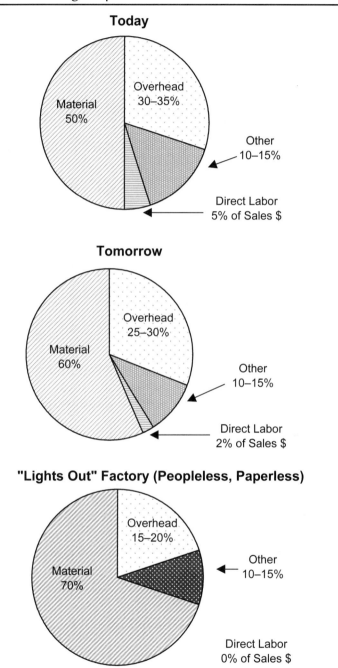

Figure 13-2. The scissors effect of customer price erosion and material cost increases.

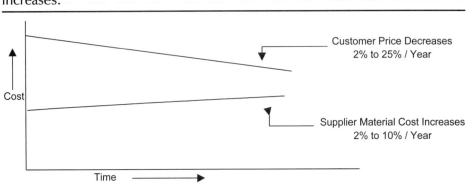

versus buy decisions, where internal advocates generally used to win this tug of war. Today, almost any activity within a company—manufacturing, accounting, management information systems, or payroll—is a candidate for outsourcing. The famous Peter F. Drucker concentration decision now extends beyond product. Any activity that is viewed as neither critical nor where the company does not perceive itself to have a core competence can be economically justified to be handed over to partnership suppliers. Witness the large number of electronic companies that are turning over the stuffing, soldering, and testing of printed circuit boards—once considered a manufacturing prerogative—to hundreds of contracting suppliers.

The Trend Toward Suppliers as Design Partners

The core competency reasoning now extends to the design function, which can no longer afford to be an expert in all modules of a product design. It should retain those portions of a design that are strategically important and where it has a decided competency and outsource all the others to "black box" partnership suppliers. This decision saves valuable design time, design manpower, design costs, and design quality. This is one of the reasons the Japanese are more productive in new product launches. Japanese car manufacturers retain engine development as a core competency and outsource almost all other major assemblies to their roster of loyal suppliers. (The supplier share of engineering in Japan is 51 percent as compared with 14 percent in the United States.)

Gains in Supplier Quality Improvement

Supplier quality has been the second most important category of poor quality—next only to poor design. Despite the quality revolution, most suppliers still hide behind acceptable quality levels (AQLs) of one percent defective. World-class companies expect defect levels to be no more than

10–50 parts per million (ppm) and their world-class suppliers deliver such levels at reduced cost. Effective supply management is the key.

Gains in Supplier Lead-Time Improvement

Of the various elements that contribute to long manufacturing cycle time—customer order time, factory order time, setup time, build time, and ship time—supplier lead time is the longest and most costly. It is generally five to twenty times as long as manufacturing cycle time. A hard-hitting supply management thrust can reduce supplier lead time by factors of 10:1, even 100:1. This is achieved not only at the supplier's facility, but also by extending supply management to subsuppliers and sub-subsuppliers.

Gains in Supplier Cost Reductions

Typically, customer companies attempt supplier cost reductions by pounding the table, changing suppliers, and even by virtual blackmail, as the notorious Ignatio Lopez at General Motors did with his supplier base while he was vice president of purchasing. The result: a grudging 2–5 percent cost reduction, with trust and goodwill becoming casualties. (There is an amusing story about Lopez when he was at General Motors. His unpopularity with his partnership suppliers had reached fever intensity because of the brutal treatment accorded to them. One of his associates told Lopez: "I have good news and bad news." Lopez relied: "Give me the bad news first." The answer: "There is a contract out on your life by your suppliers." "And what is the good news?" asked Lopez. The answer: "Why, it's been given to the lowest bidder!") True supply management renders active, concrete help to its partnership suppliers as the best way to help itself. In the process of receiving significant cost reductions year after year, (typically 5–15 percent per year), the customer company makes sure that the profits of its partnership suppliers have a corresponding increase.

Avoiding the Siren Song of Internet Bidding

There is an absolutely pernicious practice percolating through industry of soliciting bids from a whole host of suppliers through the Internet and selecting the lowest bidder. This can be a virus that can destroy a company in several ways:

❖ It can sour relations between the company and its close partnership suppliers who, after years of loyal service, find themselves betrayed and left twisting slowing in the wind.

❖ It trades continuous quality, cost, and cycle-time improvement associated with partnership suppliers for the vagaries, disruptions. and deceptions from suppliers of unknown, unproven reputation and trust worthiness.

❖ It produces short-term gain for a certain, guaranteed long-term loss.

Unbelievable Economic Gains From Strengthening All Links in the Supply Chain

Supply chain management, for all its rhetoric, has, in fact, been restricted to the main link—namely, the customer–first-tier supplier interface. Very little has been done to push the frontiers of partnership to either upstream supplier links (those second-tier suppliers, let alone third-tier and raw material suppliers) or downstream links (the distributor, dealer, retail chains, and servicers).

Upstream Supplier Links

Historically, relations between firms all along a value stream—from raw material suppliers to original equipment manufacturers (OEMs)—are either nonexistent or characterized by the minimum cooperation needed to get the product made. Each company wants to maximize its profits regardless of others in the supply chain. There is little recognition that this is a zero-sum game and even less recognition that "a rising tide lifts all boats"—that the envelope of total profits in the value chain can be much larger for all and for each supplier link if there is mutual trust, full transparency, and full cooperation to:

❖ Remove or reduce all nonvalue-added operations along the entire supply chain

❖ Concentrate at each supply link on reducing cost, defects, and cycle time

❖ Learn from one another and help one another, on either a one-on-one basis or through a network of supplier associations

As an example, Toyota—a model of excellence in manufacturing and lean production—extends its supply management reach to four levels of supplier-subsupplier links. As a result, its raw material inventory turns of 248 is three times the best U.S. figure of 69.[1] But even Toyota has not optimized the cost of its total supply chain. Its manufacturing costs, incurred along its value stream, is characterized as follows:

Supply Chain	Costs
Toyota itself	22 percent
First-tier suppliers	22 percent
Second-tier suppliers	10 percent
Third- and fourth-tier suppliers	3 percent
Raw material suppliers	43 percent

Obviously, even Toyota needs to work with or influence its raw material suppliers more effectively. The principles, type of help, infrastructure, and development of second- and third-tier suppliers are similar to the ones for first-tier suppliers detailed later in this chapter.

Downstream Distributor/Wholesaler/Franchiser/Dealer/Servicer Links

Partnerships with distributors and dealers are even rarer and more tenuous than those with first-tier suppliers. They look upon one another with suspicion. Yet distributors and dealers—or retail chains—play a key role, because it is almost impossible for a company to keep its finger on the pulse of hundreds and thousands of individual consumers and their requirements and their loyalty. Furthermore, dealer, retail chain, and service personnel can influence, for better or for worse, sales to consumers with their recommendations.

While the principles, types of help, infrastructure, and development of these downstream links are similar to those for first-tier suppliers, a few general points need to be stressed:

❖ Just as it is important to reduce the customer base and the supplier base, it is equally important to reduce the distributor/dealer/servicer base. Not all of them are worth keeping. Concentrating on those with true partnership potential conserves a company's limited resources and allows it to service this important constituency.

❖ Spending time with them is almost as important as spending time with core customers and with customer-contact frontline troops within your company. Listening to them is an attribute of leadership.

❖ Supporting them with training, especially technical training, is essential. This training includes seminars, newsletters, videotapes, and field bulletins.

❖ Sharing equitably in profits with the distributor/dealer/servicer base assures profit enhancement for both sides.

World-class companies state that dealer loyalty is a major reason for their success. Here are some examples:

❖ Caterpillar's dealer organization is placed almost on a par with its customer base by its CEO. In fact, the dealer is the last production step in the complex assembly of its giant products.

❖ Deere & Company attributes its 98 percent customer retention rate per year to a well-trained dealer network that serves no other company.

❖ Lexus (the benchmark case study in Chapter 6) states categorically that its "key to customer loyalty lies in a loyal partnership with our dealers—the single most important element in our success."

The Tangible Benefits of Supply Chain Management

Few disciplines offer as wide a range of benefits as can supply chain management (SCM). Table 13-1 lists several of these tangible benefits.

Table 13-1. Primary benefits of supply chain management.

Benefit	Level
1. Profitability	Up to double current levels; up to 4 times if the entire value chain is exercised
2. Return on Investment	Up to triple current levels; up to 6 times if the entire value chain is exercised
3. Customers	Lower prices; higher retention; greater longevity
4. Quality	Up to 1,000 times current levels—toward zero defects
5. Reliability	Up to 100 times current levels—toward zero field failures
6. Material Cost	5–15 percent reductions across the board each year
7. Cycle Time/Lead Time	A 10:1 to 100:1 reduction in supplier lead time
8. Inventory Turns	Up to 100 and more
9. Suppliers, Distributors, Dealers	Larger volumes; greater longevity; higher profits

Why Supplier Partnerships Fail

Given these tangible benefits, why do most partnerships either fail or chug along in a loveless marriage? There are several reasons, some where customers are responsible and others where suppliers are responsible:

Reason for Failure	Responsible for Failure
Principles of partnership not known or not followed	Customer and supplier
Supply management not a key corporate strategy	Customer
Dictatorial, remote control management	Customer
Supplier base not reduced sufficiently	Customer
Part number base not reduced sufficiently	Customer
Unequal economic leverage: small customer, large supplier	Supplier
Insufficient business to generate customer loyalty	Supplier
Top management steering committee not established	Customer
Commodity teams not established	Customer
Measurements based on purchase price, not on total costs	Customer
Poor knowledge of quality, cost, and cycle-time improvement tools	Customer
Perceived threat of supplier's cozy relations with the company's competitor or the supplier turning competitor	Supplier

To a large extent, the same reasons are valid for explaining why downstream partnerships with distributors, dealers, and services fail.

Principles of Supply Chain Partnership

The principles of partnership in supply chain management are similar to the principles governing a company and its customers (described in Chapter 6). After all, the end-customer is the last link in the supply chain. Yet there are significant differences necessitated by a different set of relationships. There must be:

1. Commitment to partnership, by both the customer and supplier, for there to be a win-win bond rather than the traditional win-loss contest—a zero-sum game.

2. Partnership based on a foundation of ethics and uncompromising integrity (discussed in Chapter 7).

3. Mutual trust, where each side earns the other's trust and strives to live up to the other's trust.

4. Agreements reached with a handshake, not hidden behind the fig leaf of legal contracts that spell distrust in capital letters.

5. Active, concrete help on both sides, but especially on the part of the company as senior partner, as the best—no, the only—way to achieve its supply management goals.

6. *Supplier cost a ceiling, not a floor.* Instead of the old base price, from which the supplier raises costs (e.g., because of inflation, raw material increases, and union pressures); there is a firm understanding that the supplier will lower prices continuously in return for the customer's concrete help.

7. *Supplier profits a floor, not a ceiling.* The objective is not to squeeze the supplier dry like a lemon, but to help improve the supplier's profits in return for cost, lead time, and quality improvements to the customer.

8. An "open kimono" policy, where both sides share their strategies, costs, and technology in the spirit of true partnership. Transparency is the key.

9. Commitment to the partnership as a long-term marriage, not a short-term weekend fling.

10. Belief in the supplier as an extension of the customer company—except for separate finances.

Even though the words *customer* and *supplier* have been used in the abovementioned context, these principles apply with equal validity to the relationships between:

❖ A supplier—as customer—and its subsuppliers along the supply chain

❖ A company and its distributors, dealers, and retail chains

Types of Mutual Help

Help is at the very foundation of supply chain management. Yet, companies have shied away from rendering it. Why? Companies have historically felt that their role is to define the requirements and specifications to the supplier, sign the contract, pay the price—and from there on, it is the supplier's responsibility to deliver the product. This hand-off syndrome has characterized purchasing for a century. Companies further believe that helping a supplier is an expenditure of its time and resources, with no return on such investment. In addition, helping a supplier, they feel, may end up helping their competition.

Suppliers, for their part, are also suspicious of such help:

- ❖ They do not want to share proprietary information or to expose their weaknesses.
- ❖ Their help, especially in the design stage, is likely to be benefiting only their customer, leaving them nothing for their helpfulness.
- ❖ Worse, despite their design help to the customer's engineers, the purchasing department can, and often does, place the order with a competitor.

All of these concerns violate the very principles of true partnership. The customer's help to the supplier and vice versa is the best way to improve the bottom line for both. Table 13-2 indicates the specific types of help each partner should render to the other. Most of these are tools that are detailed in Chapters 11, 12, and 14. Unfortunately, most companies do not know many of them. Even if known, they have no experience in using them. Even if experienced, it takes time to go from novice to expert. As a result, companies remain limited in the help they can render suppliers and in the accruing financial benefits.

Criteria for Selecting Partnership Suppliers

Although several factors need to be weighed in selecting a partnership supplier for a company, a few among them are most important:

- ❖ *Values.* There should be compatibility of values of the supplier with those of the customer. A strong ethical foundation must govern their relationship.
- ❖ *Location, Location, Location.* Preferably, it should be in close physical proximity to the customer.
- ❖ *Supplier Size.* Small is super; big is bad. The chances for reform are much greater with a small supplier than a large, bureaucratic entity that may be muscle-bound and inflexible.
- ❖ *Supplier Attitude.* Select a supplier that is humble and willing to learn. Often, survival (that is, a threat of extinction) can provide powerful inducement.

Supplier Location

There is a traditional view that the best supplier should be chosen for a particular commodity, regardless of whether the supplier is in the United

Table 13-2. Specific types of help in supplier partnerships.

A. From Customer to Supplier	B. From Supplier to Customer
1. Quality ❖ Design of experiments (DOE) ❖ Multiple Environment Over Stress Tests (MEOST) ❖ Cost of poor quality (COPQ) ❖ Meaningful specifications, classification of characteristics ❖ Positrol, process certification, precontrol ❖ Poka-yoke	1. Quality ❖ $c_{p_K} \geq 2.0$ on important parameters ❖ Certification through process control ❖ Failure mode effects analysis (FMEA) ❖ Virtual elimination of incoming inspection ❖ Virtual elimination of line defects ❖ Virtual elimination of field failures
2. Cost ❖ Cost targeting ❖ Total value engineering ❖ Group technology ❖ Audits ❖ Financial incentives/penalties	2. Cost ❖ Continuous cost reduction (5–15 percent per year) ❖ Early supplier involvement (ESI) ❖ Value engineering ideas ❖ Standardization assistance
3. Cycle Time ❖ Flowcharting ❖ Cycle time audits ❖ Focused factories ❖ Product versus process flow ❖ Total productive maintenance (TPM) ❖ Pull Systems ❖ Less necessity for forecasting and master schedules	3. Cycle Time ❖ "Black box" design help ❖ Reduced lead times ❖ Inventory reduction ❖ Lead time reductions in the whole supply chain
4. Help in General ❖ Frequent visits (once per week) ❖ Training and coaching ❖ Larger volumes, longer contracts ❖ Drastically reduced competition ❖ Higher profits and return on investment (ROI) ❖ Long-term partnerships	4. Help in General ❖ Commitment to terminate or reduce work with customer's competition ❖ Electronic data interchange (EDI) ❖ Help with next tiers of suppliers ❖ Networking with other suppliers in the "family"

States, Europe, Asia, or elsewhere. Others argue that labor costs dictate a country—China, for instance—where cheap labor would be an advantage.

Both these approaches have serious flaws. On the first point, the best suppliers may be separated from the customer company by thousands of miles, breaking a fundamental rule that the customer should be able to "kick the tires," as it were. The second point makes even less sense. Cheap labor has shifted in the last twenty years—to Japan, Korea, Singapore,

Hong Kong, Malaysia, Thailand, and now to China. It is not impossible, in the next twenty years, that labor costs may be the lowest in the United States—if not in wages, in terms of productivity. There is a firm principle governing supplier location:

- ❖ A company should manufacture its products in the same country or region where it sells.
- ❖ A company should buy its parts or subsystems in the same country or region where it manufactures.

Table 13-3 marshals the cogent arguments against the much-too-common practice of manufacturing in one country and buying products, supposedly "cheaply," in another. Not only are there hidden cost penalties of unstable governments in many countries and currency fluctuations, among other issues, but there are also strategic/tactical drawbacks. Worst of all, there are tangible costs associated with long cycle time, poor quality, and other factors.

The conventional wisdom is that an offshore purchase should cost at least 25 percent less than one onshore. But if the tangible factors alone, listed in Table 13-3, are estimated, then an offshore purchase should cost 75 percent less—not 25 percent—than an onshore purchase to break even.

Table 13-3. The hidden cost penalties of offshore purchases.

Governmental/Political	Strategic/Tactical	Tangible
❖ Political instability	❖ Customer bonds weakened	❖ Inventory/cycle time increased
❖ Currency exchange rate fluctuations	❖ Customer-marketing-engineering links poor	❖ Poor tooling
❖ U.S. trade barriers	❖ Early supplier involvement and value engineering much more difficult	❖ Quality often a casualty
❖ Foreign/U.S. government regulations		❖ Extra source inspection costs
❖ Public Law 98-39	❖ Visits/tech help/training to suppliers ruled out	❖ Freight costs higher
❖ EPA restrictions	❖ Longer product cycles	❖ U.S. tariffs
❖ Product liability exposure	❖ Design changes more difficult	❖ Customs delays
❖ Patent infringements	❖ Longer lead times	❖ Brokerage fees
❖ Foreign content laws	❖ Schedules must be frozen	❖ Port entry fees
❖ Foreign tax structure	❖ Perils of long-range forecasting	❖ Labor cost escalations
❖ Foreign work rules	❖ Danger of forward integration	❖ Cash flow problems
❖ Standards differences (e.g., EIA versus JIS)	❖ Technology transfer drain	❖ Cost of frozen cash
	❖ Travel/communication costs increased	❖ Premature payments with no guarantee of good, usable products
	❖ Cultural/language barriers	

Infrastructure of Supply Chain Management

There are companies that operate their supply management processes without a systematic structure and get by. However, a well-crafted infrastructure can assure better and more sustained results. Its elements include:

1. A steering committee
2. Supplier and distributor councils
3. Commodity teams and distributor/dealer teams

The Steering Committee

With the CEO of a company (or the division manager in a business) as the chairperson, the steering committee consists of members of the senior staff. Their responsibilities include:

- ❖ Drafting a vision/mission statement, committing the company to the entire chain of supply management
- ❖ Communicating the importance of supply chain management to the whole organization
- ❖ Establishing goals for supply management (e.g., quality; cost; lead time; reduction of the supply/distributor/dealer base, and reduction of the part number base)
- ❖ Selecting commodities for prioritization
- ❖ Selecting commodity teams
- ❖ Monitoring partnership progress; and acting as a "supreme court" to resolve conflicts internally, with suppliers or with distributors/dealers
- ❖ Providing recognition and rewards to commodity teams and to outstanding partnership suppliers, distributors, and dealers

Supplier/Distributor/Dealer/Servicer Councils and Conferences

The objectives of these councils and conferences are to:

- ❖ Advance partnership for mutual economic benefit
- ❖ Provide a forum for policy, expectations, plans, and consultations
- ❖ Resolve conflicts and evaluate recommendations
- ❖ Measure partnership progress

The structure of the councils consists of an equal number of company and supplier representatives (and distributor/dealer senior management).

They meet two to four times a year and plan an annual conference for sharing state-of-the-company messages, technological developments, and other issues. The most productive aspect of these council meetings is the ability to air mutual concerns and frustrations and see them quickly and effectively resolved. No company can afford to do without them.

The Commodity Team

This is the workhorse of supply chain management. No company can achieve breakthrough results without it. The commodity team's main purpose is advancing supplier development, along technical and business lines, so that a supplier can grow into becoming a viable partner of the company. (The formulation of the downstream company—distributor/dealer teams and their main objective, which is the development of the distributor/dealer network as a viable partner—is the same as that of the commodity team.)

The commodity team consists of a team leader along with members drawn from the old department structures of engineering, purchasing, quality, and the partnership supplier as a core group. It is reinforced, as needed, by specialists in the technology of the commodity, tooling, and other disciplines. It should be a full-time activity. Companies can evaluate full-time commodity team effectiveness by estimating team costs (1) of payroll and travel, for example, and projecting potential savings, and (2) in terms of cost, quality, and cycle time improvements. A formula that any reasonable management would find hard to deny is to show that $\frac{1}{2}$ B $\geq$ 2A—in short, a 4:1 return on investment (ROI). This is because traditionally costs are underestimated and savings are overestimated. Our experience has shown that $\frac{1}{3}$ B $\geq$ 3A—a 9:1 ROI.

Just as in the case of a concurrent engineering team, a part-time team spells part-time, mediocre results. To be most effective, the team should be at the plants of the partnership suppliers most of the time. Sitting chained to their desks, because of headcount and travel freezes, is the surest way for the supply chain to atrophy. Commodity teams are part of the order fulfillment process in a horizontal process-centered organization (see Figure 8-2) as opposed to a vertical departmentalized organization. The commodity team's initial responsibilities include:

1. Establishing goals and timetables for the commodity: quality, cost, cycle time, maximum number of partnership suppliers (typically two or three) per a major commodity, and reduced part numbers

2. Reviewing the technology of the commodity for growth, obsolescence, or breakthroughs

3. Designing guidelines for engineers that would be helpful to partnership suppliers

4. Running DOE studies with design teams to translate product specifications into component specifications

5. Promoting outsourcing the entire "black box" design to capable partnership suppliers

6. Benchmarking "best in class" companies for superior supplier chain practices

7. Benchmarking new suppliers, including transplant suppliers

8. Training commodity team members in a number of supplier development techniques especially quality, cost, and cycle-time improvement (see Table 13-4)

9. Thinning out the supplier base down to a finalist list

10. Selecting partnership suppliers (typically two) based on actual visits and detailed audits

Partnership Supplier Development

It is often argued that the partnership supplier is more a master of the component's or product's technology than is the customer company. If that's the case, what need is there to help the supplier's development? In many ways, however, the supplier is far less knowledgeable than you think about all nontechnological disciplines, including cost, quality, cycle time, and leadership. Table 13-5 is a list of development techniques where coaching becomes essential. (Of course, the commodity teams must become professional in their own companies in these techniques before they can presume to help the supplier. Otherwise, they could be accused of the Ford slogan coined by skeptical suppliers: "Do as I say, don't do as I do"—a preacher command, rather than a practitioner's helpfulness.) The list is a veritable miniversion of the Ultimate Six Sigma contents of this book. To be effective, the commodity team should visit the partnership supplier once a week and follow-up assignments and "to do's" of the previous week. The process continues until the supplier's development no longer requires nursing, and there is a momentum for business excellence, as determined by the customer.

Supplier/Distributor Evaluation of the Company as Partner

Most of this chapter has been to evaluate, measure, and improve upstream supplier performance or downstream distributor/dealer performance. It is equally important to know how well the partnership is working from the supplier/distributor perspective. Progressive companies like to be in the

Table 13-4. Supplier training—a comprehensive curriculum.

Course/Topic	Content Objective	Supplier Participants	Duration
1. Design of Experiments (DOE) Overview	Shainin/Bhote DOE (not classical or Taguchi DOE): Impact on problem solving, profitability, customer, and design	Senior Management	½ day
2. DOE Problem Solving	Shainin/Bhote DOE: Ten powerful tools to help engineers, direct labor, and suppliers solve chronic quality problems	Management and all technical staff	1 day
3. DOE Workshop	Road map to solve chronic supplier quality problems	Management and all technical staff	1 to 2 days
4. Multiple Environment Over Stress Testing (MEOST)	Drastic reduction of field failures and application to supplier field failures	Management and all technical staff	Course: ½ day Applications: 1 to 2 days
5. Cost of Poor Quality	Tool for 50–100 percent profit improvement	Senior Management and Accounting	½ day
6. Lean Manufacturing	Improvement in inventory turns from single digits to 20 to 80	Senior and Manufacturing Management	1 day
7. Total Productive Management	Improvements in yield, uptime, and changeover time (FOE from less than 40 percent to greater than 85 percent)	Manufacturing Management	½ day Applications: 1 day
8. Poka-yoke	Prevention of operator-controllable errors with visual, electrical, or mechanical sensors	Manufacturing	½ day Applications: 1 day
9. Next Operation as Customer (NOAC)	Improvement in white-collar quality, cost and cycle time	Management and all support services	Course: ½ day Applications: 1 day
10. Total Value Engineering (VE)	Increase in customer enthusiasm with decrease in product costs	Engineering, Manufacturing, Purchasing	1 day Applications: 1 day

Table 13-5. Supplier development: issues to address, by area.

Area	Development Techniques
Leadership/ Management	❖ Encourage a shift from management to leadership. ❖ Encourage supplier's leaders to free the creative genie in every employee. ❖ Encourage driving out fear among the employees. ❖ Encourage leaders to walk the talk. ❖ Conduct comprehensive supplier training/workshops.
Organization	❖ Encourage horizontal cross-functional teams rather than a vertical department organization. ❖ Encourage the reduction in the number of layers of management. ❖ Encourage the reduction of paperwork, e-mail, policy manuals, and endless meetings to reduce bureaucracy.
Employees	❖ Encourage job redesign to tap the fountain of inner motivation in all employees. ❖ Encourage employee study of financial statements and the formation of self-directed minicompanies.
Quality/Cycle Time/Cost	❖ Conduct detailed quality and cycle-time audits. ❖ Review audit weaknesses and work with supplier to overcome them. ❖ Conduct a plant tour to pluck the "low lying fruit" in quality, cost, and cycle time. ❖ Draw joint plans and timetables for longer-range improvements. ❖ Gather, analyze, and reduce the cost of poor quality (COPQ). ❖ Flowchart the supplier cycle time—from purchase order to customer delivery. ❖ Eliminate or drastically reduce all nonvalue-added steps. ❖ Encourage forming focus factories and manufacturing cells. ❖ Encourage plant rearrangement from process flow to product flow. ❖ Institute total productive maintenance (TPM) and measure overall equipment effectiveness (OEE) progress. ❖ Introduce "pull" system and lessen dependence on forecasts and master schedules. ❖ Significantly reduce setup time/changeover time. ❖ Institutionalize Shainin/Bhote DOE, MEOST, Total VE, and NOAC with demonstration projects.
Design	❖ Encourage formation of concurrent engineering team. ❖ Encourage design for manufacturing (DFM) principles and practices. ❖ Act as liaison between customer's design team and supplier's technical personnel to make early supplier involvement (ESI) a way of life.

Second- and Third-Tier Suppliers	❖ Encourage partnership relationships with second- and third-tier suppliers.
	❖ Strengthen the purchasing power of suppliers by banding them together for joint purchases..
	❖ Encourage helping these subsuppliers with the same type of development help being rendered by the company.
	❖ Institutionalize commodity teams.
Manufacturing	❖ Introduce the disciplines of process certification, postirol, and precontrol.
	❖ Equip operators with poka-yoke sensors to prevent operator-controllable errors.
	❖ Encourage monitoring effectiveness of analyzing quality defects.
	❖ Introduce "field escape" control.
	❖ Ensure that the accuracy of measuring instruments (product accuracy is 5:1).
Field	❖ Ensure adequacy of packaging and transportation.
	❖ Encourage professional failure analysis capability.
	❖ Ensure completeness of manuals and installation instructions.
Support Services	❖ Encourage the concept, implementation, and measurement of the Next Operation as Customer (NOAC).
	❖ Encourage making the internal customer the scorekeeper and primary evaluator of the internal supplier.
	❖ Encourage "out of the box" thinking to revolutionize business processes.
Results	❖ Establish metrics in all areas (see Chapter 10).
	❖ Set up a "customer effectiveness index" to monitor customer feedback.

position of being perceived by the partnership supplier or distributor as their best customer. Table 13-6 is a typical scorecard by which suppliers or distributors rate their customer company vis-à-vis their best customer. Feedback of this nature can have a sobering effect on a customer company's ego.

CASE STUDY

Toyota—A Benchmark Company in the Area of Supply Chain Management[2]

Toyota Motor Corp. has been recognized not only as the most effective car company in the world, but also as the preeminent

Table 13-6. Supplier/distributor evaluation of the customer company.

Item**	Company Rating*					Best Customer Rating				
	1	2	3	4	5	1	2	3	4	5
1. Effectiveness of overall partnership										
2. Equitable sharing of profit increases										
3. Increases in business volume										
4. Longer-term agreements										
5. Ethics, trust (in action, not words)										
6. Transparency (sharing of costs, strategy, technology)										
7. Concrete development help (in quality, cost, cycle time, leadership)										
8. Development help to second- and third-tier suppliers, dealers, and servicers										
9. Training (seminars, workshops, demonstration projects)										
10. Early supplier/distributor involvement										
11. Granting "black box" design responsibility to supplier										
12. Clear, meaningful, mutually determined specifications										
13. Financial incentives for quality, delivery, performance										
14. Royalties for ideas (even without an order)										
15. Receptivity to suggestions										
16. Forecast accuracy, dependability										
17. Schedule sharing										
18. Electronic data interchange										
19. Volume variable pricing										
20. Networking initiatives										

*Rating: 1 = lowest; 5 = highest
**It is also possible to rate the importance of each item on a scale of 1 to 5 and arrive at an overall score, similar to that developed in Table 10-4.

manufacturing company anywhere on the globe (Chapter 14 tells more of the Toyota story). What is not generally recognized is the fact that it is also a giant in the area of supply chain management, has been in that leadership role since shortly after World War II, and has completely revolutionized relationships between a company and its partnership suppliers.

❖ As early as 1949, Toyota made a dramatic departure from vertical integration to create three famous companies as partnership suppliers—Nippondenso, Aisin Seiki, and Toyoda Gosei. Each was also given major design responsibilities.

❖ By the late 1950s, Toyota reduced its own added value to the average vehicle from 75 percent down to 25 percent. Even 50 percent of its final assembly was farmed out to partnership suppliers.

❖ Toyota's supply base is only 190 as compared with over 2,000 at General Motors and over 1,500 at Ford. It manages this base with a supply management staff of 185, compared to the nearly 4,000 employees at GM and more than 2,000 at Ford.

❖ Toyota pioneered the concept of continuous cost reductions from its suppliers when the common practice was to accept constantly increasing costs in an era of rampant inflation.

❖ When the "oil shock" of the Arab embargo hit Japan in 1973, Toyota extended the concept to its second-tier suppliers, not by squeezing them, but by teaching them its world-famous Toyota production system (TPS) to go from batch production and the "push" system to single-piece flow and the "pull" system.

❖ Toyota has further extended supply chain management to third-tier suppliers. (Although it has yet to capitalize on the last extension to its fourth-tier raw material suppliers that still control more than 40 percent of its manufacturing costs.)

Other supply chain management innovations at Toyota include:

❖ *Establishing Mutual Help Groups.* These help groups are established among forty-two of the car maker's largest and most important suppliers, divided into six groups of seven suppliers and requested, with Toyota's help, to conduct one major improvement activity each month. These suppliers were then persuaded to set up similar networks with their second-tier suppliers with the objective of lean production as a way of life. The result: A continuous cost reduction on every part, every year, from every supplier. Furthermore, the suppliers were delighted because of the extra profits they could realize with their other customers.

❖ *Early Supplier Involvement.* ESI became an integral part of Toyota's new product development (see Chapter 12) with its hard-driving shusa—or team leader.

❖ *Cost Targeting.* This practice made obsolete the three-bid syndrome and put the customer-company in the driver's seat for determining the price

SELF-ASSESSMENT/AUDIT ON SUPPLY CHAIN MANAGEMENT

Table 13-7 is a self-assessment/audit on supply chain management that a company or external Ultimate Six Sigma auditors can use to gauge its progress. It lists six key characteristics and fifteen success factors each worth five points, for a total of seventy-five points.

Table 13-7. Management: key characteristics and success factors.

Key Characteristic	Success Factors	Rating				
		1	2	3	4	5
8.1 Importance of Supply Chain Management	1. The company maintains an increasing differential between its price erosions in the marketplace versus the cost decreases it seeks from its suppliers. 2. The company retains key elements of its product portfolio that it considers its core competency, but outsources all other elements. 3. The company strengthens its upstream links to its first- and second-tier suppliers and its downstream links to its key distributors/dealers/servicers.					
8.2 Supply Partnership Principles	1. The principles governing supplier partnerships include: ❖ Commitment to a win-win bond, not a win-lose contest ❖ A foundation of ethics; mutual trust; active, concrete help; and transparency 2. Supplier costs are negotiated as a "ceiling" (rather than as a "floor") beyond which costs are not allowed to rise. But supplier profits are a floor, not a ceiling, to allow the supplier an equitable share of the overall savings.					
8.3 Types of Mutual Help	1. The customer company renders active, concrete help in the areas of quality, cost, and cycle time; the supplier renders corresponding help (see Table 13-2).					
8.4 Selection of Partnership Suppliers*	1. Partnership preference is to give suppliers who are small, humble, willing to learn, and whose values are similar to those of the company. 2. It is preferable to coach and improve an existing but deficient supplier and develop that supplier into a worthy partner rather than select another thousands of miles away who may not be as responsive. 3. Offshore suppliers are not chosen unless the price is 75 percent (or more) under the price of an onshore supplier.					

(continues)

Table 13-7. (Continued).

Key Characteristic	Success Factors	Rating				
		1	2	3	4	5
8.5 Infrastructure	1. A top management steering committee is in place to set supply management goals, select commodity teams, and monitor progress. 2. Supplier and distributor/dealers councils are in place to advance partnerships for mutual benefit and to resolve conflicts. 3. The commodity team, as the workhorse of supply management, consists of a core membership drawn from the old engineering, purchasing, and quality departments, along with the selected 2 or 3 partnership suppliers.					
8.6 Supplier Development	1. The commodity team's main objective is to encourage and coach the supplier to a level of proficiency similar to that of the company, using the development techniques listed in Table 13-5. 2. The commodity team trains the partnership supplier's personnel in a number of disciplines (see Table 13-4) needed for development. 3. The company encourages and receives periodic evaluations by its partnership suppliers on the effectiveness of partnership from their perspective.					

*The term *supplier* is used in a generic sense in this table, referring both to upstream suppliers as well as downstream suppliers (i.e., distributors, dealers, and servicers)—in other words, the entire supply chain.

Fourteen

From Second-Class Citizen to Manufacturing as a Major Contributor to Business Excellence

❖ ❖ ❖

A growing number of companies in electronics, toys, and consumer goods are coming home from abroad. . . . [C]ompanies that sought cost reductions and manufacturing efficiencies through foreign

*operations have learned the
hard way that it is easier to
control your destiny
through better management
techniques, local
manufacturing
effectiveness, just in time,
subcontracting, and
training. The higher-quality
product may again be your
U.S. competitor that uses
these techniques. . . .*

—DAVID M.
RICHARDSON,
The Wall Street Journal

Why Has Manufacturing Been Sidelined?

For the last twenty years and more, manufacturing has been looked upon as a "sunset" discipline. It has lost the aura it used to have in the 1950s and 1960s. It has lost the respect of top management; and its salaries are well below those of finance, marketing, and design. There are a number of reasons for this disenchantment:

❖ The surge of the service industry, which accounts for 70 percent of jobs as compared to manufacturing, which has declined to 25 percent. (There is a parallel in the decline of agriculture, which has gone from providing 90 percent of jobs in the early-nineteenth century to 10 percent half a century ago to 3 percent today.)

❖ The third economic revolution—the digital age (after the agricultural revolution around 9000 B.C. and the Industrial Revolution a century and a half ago). E-commerce and the ubiquitous Internet now account for more than 75 percent of engineering jobs, while hardware is reduced to 25 percent.

❖ The flight offshore to low-wage countries, with China leading the parade. The illogical rationale is lower cost, not recognizing the hidden costs of offshore procurement (see Table 13-3).

❖ The influence of the business schools churning out MBAs who have no clue on what it takes to design or manufacture a product, but who aspire to be CEOs in five years after graduation.

❖ The attitudes of top management with primarily a single background—finance or law*—and a hazy concept of manufacturing, at best. What insights can these two fields provide for manufacturing or running a successful business for that matter?

The Key Role of Manufacturing

Yet manufacturing need not be a stepchild, an appendage of other disciplines that have shunted it aside. Scores of progressive companies have brought forth the renaissance of manufacturing in the past twenty years.[1] These companies have run circles around their best offshore competitors with a much lower unit cost, higher productivity, better quality, faster delivery, higher inventory turns, and higher profitability, despite higher pay to their line workers.

Why can manufacturing contribute so much to a company's welfare?

❖ It makes a direct contribution to the customer. It is the "final answer"—to use the popular phrase from the *Who Wants to Be a Millionaire?* television show—to what the customer wants.

❖ It integrates the functions of sales, marketing, engineering, and other disciplines. It is the only place where they come together.

❖ It can be a marketing weapon, as Tom Peters, in his book *Thriving on Chaos,* advocates. "It is from hands-on interaction—among foremen, line operators, customers, suppliers, researchers, and distributors—that day-to-day advances in innovation and responsiveness flow."[2]

❖ It is a breeding ground for small but continuous improvements, as opposed to the large but rare breakthroughs generated in design. An analogy is that the engineering function is the Babe Ruth who would hit a home run but had many strikeouts in his career,

---

*William Shakespeare, "First, let's kill all the lawyers."

whereas manufacturing is the slugger Wee Willy who could be counted on for a hit almost every time at bat.

Restoring Manufacturing to a Place of Honor I: What Not to Do

Before examining several key initiatives that can truly restore manufacturing to its rightful importance in a company, it is useful to list practices that should be given a wide berth:

1. *Mass Production.* Except in the rare instances where the market can sell millions of units of a standard product, mass production is as dead as the dodo bird. Today, more and more, we must consider a customer as one who wants what he wants, how he wants, where he wants, when he wants, and at what price he wants! Ours is the age of mass customization (see Chapter 6).

2. *Automation.* In the 1970s and 1980s, the plaintive cry of the entrepreneur was "emigrate, automate, or evaporate." Emigrating industry to offshore locations is proving to be less and less of a bargain each year. But there is still an unseemly fascination with automation. For example, an automated firm processes molding, cutting, and painting parts for its product.[3] Robots stack the parts from each automated fabrication step on pallets taken by automated guided vehicles to an automated storage. From there, the parts are taken automatically to an automated final assembly line that can adjust its fixtures to hold any one of 100 models of the product and assemble it with pick-and-place robots. The direct labor headcount: zero. The manufacturing support count: 3,600. Is the factory in the United States? Wrong— it is in Japan, the land of Kanban! Automation can become a narcotic.

Automation is good if it reduces human slavery—fatigue, boredom, and drudgery. And it should only be used if it meets an inviolate rule: The cost of indirect technical support and high-tech tools, added together, should be less than the savings in direct labor.

Ross Perot bemoaned General Motors' expenditure of $40 billion for robotics-equipped plants and capital equipment. It still lost market share and went from being the low-cost producer to the high-cost producer among the Big Three automakers.

Even Toyota, which had religiously followed the precepts and practice of its great founder of just-in-time (JIT) manufacturing—Taichi Ohno— was tempted to go the route of automation in its Tahara plant near Toyota City. It was a disaster. Toyota has never repeated the mistake again.

3. *MRP and MRPII: Systems Whose Time Has Long Gone.* Materials requirements planning (MRP), a computerized system used to determine the quantity and timing requirements for materials, and manufacturing resource planning (MRPII), used for capacity planning and master sched-

ules, were staple tools of the 1970s and 1980s. They are still used in those industries that have not switched to JIT. MRPII's main weakness is that it uses the old "push" system, where each workstation is gated by a preordained master schedule. As a result, inventories can pile up at each station that are not required by the next station.

Restoring Manufacturing to a Place of Honor II: What Must Be Done

An Infrastructure of Support for Manufacturing Excellence

In a larger sense, the previous chapters in Part II of this book have prepared the groundwork for manufacturing excellence. Let's recap the highlights of their contributions:

Leadership
- ❖ Releases the genie currently locked up in the direct labor worker
- ❖ Increases trust in the line worker
- ❖ Gives workers freedom—even to make mistakes
- ❖ Provides training for every worker
- ❖ Improves the quality of work life—creating joy in the workplace

Organization
- ❖ Eliminates mind-numbing bureaucracy
- ❖ Converts companies from a department structure to a team-based process structure
- ❖ Changes the ways workers are hired, trained, evaluated, compensated, and promoted.
- ❖ Designs meaningful and egalitarian gain sharing for all workers

Employees
- ❖ Redesign the dull, boring jobs of the line worker
- ❖ Participate in the power and joy in team competitions
- ❖ Adopt one of three approaches—open book management, self-direct work teams and minicompanies—to industrial democracy
- ❖ Move up in the ten stages of empowerment

Measurement
- ❖ Provides simple, fair, and meaningful metrics for the line worker
- ❖ Involves workers in the formulation of parameters by which they are to be measured

❖ Provides measurements transparent and visible to all workers

❖ Provides quality, cost, and cycle-time measurements for manufacturing (see Table 10-3)

Tools for the Twenty-First Century

Of the ten tools, seven apply directly to manufacturing and should be used as a way of life. These are:

❖ Design of experiments (DOE) for chronic problem solving

❖ Multiple Environment Over Stress Testing (a miniversion) to prevent production processes and workmanship from degrading design intent

❖ Total productive maintenance (TPM) to optimize machine/process effectiveness

❖ Benchmarking to simulate and improve best-in-class company practices

❖ Poka-yoke to prevent operator-controllable errors

❖ Total value engineering to reduce manufacturing costs

❖ Lean manufacturing to reduce inventory and cycle times

Design

❖ Design for minimum variability: c_{p_K} 2.0

❖ Design for manufacturability (DFM)

❖ Minimizing model numbers and part numbers

❖ Postponing mass customization to almost the end of a production line

Supply Chain Management

❖ Sends perfect parts directly from supplier to the line

❖ Provides drastic reductions in supplier lead time yet assures there are no line stoppages or stockouts

The Twin Engines of Quality and Cycle Time in Manufacturing

For manufacturing excellence, two disciplines are vital—quality and cycle time. Motorola calls them the twin engines that give a company its powerful thrust for world-class achievement. It's 10:1, 100:1, and 1,000:1 quality improvement drive has already been discussed. Motorola has a similar drive for a 10:1 and 100:1 improvement in cycle-time reduction. Each of these engines of change, as it pertains to manufacturing, is described in detail in the next sections.

Engine One: Quality

Quality starts with defining the customer's requirements, designing for zero defects and zero field failures, and getting materials from suppliers with zero defects. In an ideal world, that would make production a breeze. In the real world, though, quality disciplines need to be extended to manufacturing—a backstop to prevent items falling through the cracks.

Making the Product Robust

In production and in the field, there are a number of uncontrollable factors, sometimes called noise factors, that can adversely affect the quality of the product. Ambient temperature, humidity, static electricity, line voltage fluctuations and transients, lack of preventive maintenance, and some degree of customer misuse can degrade product quality and reliability. Product robustness means making the product impervious to these noise factors, This cannot always be done, but robustness should be attempted in production before the product is launched into the field.

The necessary discipline is again DOE. Using a second round of a variables search experiment (see Chapter 12), the uncontrollable (i.e., noise) factors are deliberately introduced to assess their individual or collective impact upon the response (or Green Y). If the noise factors prove to be unimportant, they can be ignored. If they are important, some of the other product parameters must be modified so that we can live with the noise factors present.

As an example, an appliance manufacturer was experiencing an unacceptably high failure rate of its dryer in consumer homes. Two of the principal reasons were 1) poor venting, with outlets clogged up, and 2) line voltage fluctuations. Both were beyond the manufacturer's control. Consumers could not be expected to worry about the venting, nor did they have control over line voltage fluctuations caused by the power utilities. A variables search experiment was run in production. It indicated that airflow was a parameter that could be increased with minimal cost increases, making the dryer impervious to venting constrictions and line voltage fluctuations. The company now has a "robust" dryer.

Process Characterization and Optimization

One of the frequent reasons for poor quality in manufacturing is that while much attention is paid to the design of the product, the process that makes the product is often treated as a stepchild. It is often selected as an afterthought to the product, with little compatibility between the two. Development engineers do not feel responsible for the process. They relegate that to the process engineer—who leans on the supplier of the process equipment. The result: arbitrary specifications, antiquated procedures, and arbitrary process parameters.

World-class quality companies start their process research one to two years ahead of product development, exactly the opposite of practices in the average company. When the product and process come together, a team consisting of the process engineer, the development engineer, and the equipment supplier conduct a DOE study—typically a variables search experiment—to define and characterize the process (i.e., to separate the important process variables from the unimportant ones, open up the tolerances of the latter to reduce costs, and optimize the levels of the former to achieve ^cp_Ks of 2.0 or more, with a subsequent scatter plot technique).

Positrol: Freezing Process Quality Improvements

The purpose of positrol, following optimization of the process, is to make sure that process gains and improvements are "frozen." Each important parameter (the what) is monitored by a "who, how, where, and when." It is described in somewhat greater detail in Chapter 12.

Process Certification: Improving the Signal-to-Noise Ratio

The purpose of process certification is to identify as many quality weaknesses of a peripheral nature and nip them in the bud. It makes the process robust by removing noise factors and thus improving the signal-to-noise ratio. It is described in somewhat greater detail in Chapter 12.

Control Charts: A Technique Whose Time Has Gone

Control charts, under the banner of statistical process control (SPC), have been widely used in Western industry ever since the 1980 NBC television documentary *If Japan Can, Why Can't We?* The documentary purported to show that Japan was way ahead in quality because of control charts, so the West must do the same—even though, for more than twenty-five years, Japan has abandoned control charts as being of marginal use. (Though occasionally they would roll out the old control charts to impress visiting Western "firemen.")

As a result, control charts that had gone into limbo after World War II were recalled from exile and given a coronation in the West. It has been a tyrannical reign, with several original equipment manufacturers (OEMs)—especially some of the automotive companies—demanding the use of control charts as a passport to doing business with them. They force control charts down the throats of unwilling suppliers, and they bludgeon into submission those knowledgeable suppliers who dare to point out that the control chart emperor wears no clothes! As often happens, the royal court is filled with hangers-on and charlatans who exploit the desperation of companies to gain a foothold on the control chart bandwagon by offering courses, tutorials, and ubiquitous computer software programs on their methodology. It is a sad fact that control charts have done very little for the quality revolution along with their equally bumbling cousins, ISO-9000 and QS-9000.[4]

Precontrol: A Technique Whose Time Has Come

Precontrol is much simpler, more cost-effective, and more statistically powerful than control charts. Another huge advantage is that it can be taught (the mechanics, not the theory) to line operators in five minutes. In fact, direct labor people and suppliers never, ever want to go back to control charts once they have sampled the ease of precontrol.

Its rules are simple and shown in Figure 14-1. It can be charted in the same manner as control charts, but is much easier for the operator to perform and for its interpretation of quality progress.[5]

Figure 14-1. Precontrol: a technique whose time has come.

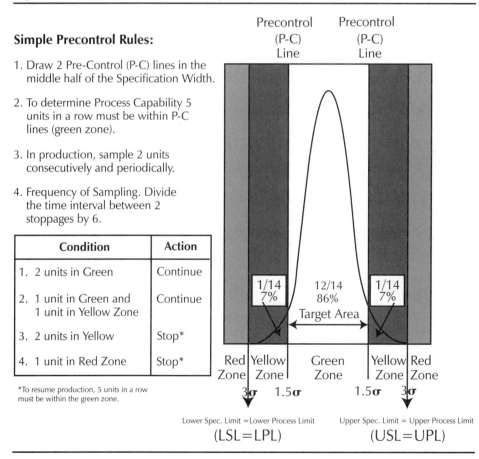

Simple Precontrol Rules:

1. Draw 2 Pre-Control (P-C) lines in the middle half of the Specification Width.

2. To determine Process Capability 5 units in a row must be within P-C lines (green zone).

3. In production, sample 2 units consecutively and periodically.

4. Frequency of Sampling. Divide the time interval between 2 stoppages by 6.

Condition	Action
1. 2 units in Green	Continue
2. 1 unit in Green and 1 unit in Yellow Zone	Continue
3. 2 units in Yellow	Stop*
4. 1 unit in Red Zone	Stop*

*To resume production, 5 units in a row must be within the green zone.

Precontrol (P-C) Line   Precontrol (P-C) Line

1/14 7%   12/14 86%   1/14 7%

Target Area

Red Zone | Yellow Zone | Green Zone | Yellow Zone | Red Zone

3σ   1.5σ   1.5σ   3σ

Lower Spec. Limit =Lower Process Limit
(LSL=LPL)

Upper Spec. Limit = Upper Process Limit
(USL=UPL)

Despite its simplicity and statistical power, precontrol—as well as the other quality disciplines of product robustness, process characterization/ optimization, positrol, and process certification discussed thus far—are no better understood, let alone practiced, than are the ten powerful tools of the twenty-first century (covered in Chapter 11). Table 14-1 is my estimate of the awareness and implementation of these quality disciplines in manufacturing. No wonder manufacturing is so pathetically behind and the quality chase within it a will-o'-the-wisp.

Problem-Solving Tools: "Talking to the Parts"

One of the brilliant features of the Shainin/Bhote DOE (see Table 11-1) is the ability to generate clues in production problem solving by "talking to the parts" as compared with relying on the guesses, hunches, theories, and opinions of engineers, which form the basis of the poorer classical and Taguchi DOE.

There are four simple but powerful clue generation tools—all of which talk to the parts. They are multi-vari, components search, paired comparisons, and product/process search. Each is described briefly here with a summary of its objective, applicability, sample size, and methodology. A practical case study of the relevance of each tool in manufacturing is also provided.[6]

Multi-Vari

Objective. To reduce a large number of unmanageable variables to a smaller family of related variables containing the root cause (the Red X). The major families are time-to-time, unit-to-unit, and within-unit, with subfamilies in each.

Where Applicable. In process-oriented manufacturing. The multi-vari is

Table 14-1. Awareness and implementation of five quality disciplines in manufacturing.

	Leading Companies		Average Companies	
Discipline	Percent (%) Aware	Percent (%) Implementing	Percent (%) Aware	Percent (%) Implementing
1. Product Robustness	5	0.1	—	—
2. Process Characterization/ Optimization	2	0.05	—	—
3. Positrol	5	0.1	0.1	—
4. Process Certification	2	0.1	0.1	—
5. Precontrol	40	15	10	2

a quick snapshot of product variations going through the process, without massive historical data that is of very little value.

Sample Size. Minimum nine to fifteen units, or until 80 percent of the historic variation is captured.

Methodology. Take periodic product samples of three to five units at a time drawn from the process and plot the variations in time-to-time, unit-to-unit, and within-unit families. The Red X is always in the family with the largest variation.

Case Study of the Application of Multi-Vari. A printed circuit board with 1,000 solder connections was running at a defect rate of 1,500 parts per million (ppm) for several months at Motorola's Melbourne plant in Australia. I ran a multi-vari study there at three intervals: 9:00 A.M., 9:30 A.M., and 10:00 A.M.—and with five panels, each with five boards, for each of the three times. The locations of the solder defects were noted on each of the seventy-five boards.

The multi-vari results indicated that there was little variation in the three time trials; little variation from panel to panel; and little variation from board to board. The Red X family was within each board. Concentration charts of the exact location of these solder defects indicated 1) an accumulation of pinhole rejects in the top-middle area of the board and 2) two intermediate frequency (I.F.) cans displaying poor solder. The remaining 980 connections had perfect solder.

The manufacturing team working with me eliminated the pinhole defects by removing the slight tilt of the panel fixtures as they progressed through the wave-solder machine. That reduced the defect rate to around 500 ppm—a 3:1 improvement in half a day. We then tackled the I.F. can solder problem by changing the ratio of the hole size in the board to the lead diameter in the I.F. cans. That took one day of trial-and-error, but we got the defect rate down to zero parts per million—not bad for one and one half day's work.

Components Search

Objective. To home in on the Red X from among hundreds of components in a product.

Where Applicable. In assembly-oriented manufacturing, where the units are capable of disassembly and reassembly.

Sample Size. Two units—one very good, the other very bad, with as wide a separation between the two as possible.

Methodology. Swap a suspected part or subassembly from the good unit to the bad and vice versa. If the good unit remains good, the part or subassembly is not the problem. If the good unit becomes completely bad, and the bad unit completely good, the part or subassembly is the Red X. If the good unit becomes partially bad, and the bad unit becomes partially good, the part or subassembly is important along with another untried part (indicative of an interaction effect).

Case Study of the Application of Components Search. A tape deck being supplied from Motorola's Seguin plant in Texas to General Motors was experiencing an azimuth (stereo sound unbalance) problem for over two months. A team of direct labor people decided to tackle the problem. They selected one tape deck that had a low unbalance of 2 decibels (db) and another an unacceptable unbalance of 5 db. Figure 14-2 shows a simple components search experiment. When the team swapped the diaphragm from the good unit to the bad and vice versa, the good unit stayed good, the bad stayed bad. Therefore, the diaphragm was not the problem. There were similar results with the capstan and flywheel and with the pressure rollers. However, when the head arm and lever assembly were swapped, a complete reversal of good and bad took place. It was the culprit Red X.

Paired Comparisons

Objective. To determine, among many quality characteristics or parameters common to both good and bad units, which of them are important in explaining the difference between good and bad units and which parameters are unimportant.

Where Applicable. In assembly-oriented manufacturing, where the units cannot be disassembled and reassembled without damaging or destroying them.

Sample Size. Six or eight good units and six or eight bad units, with as large a difference between them as possible.

Methodology. Select six or eight of the best and six or eight of the worst units. List as many parameters as possible that might explain the difference. Measure each parameter. Rank each parameter on all sixteen units

Figure 14-2. Components search case study: tape deck azimuth.

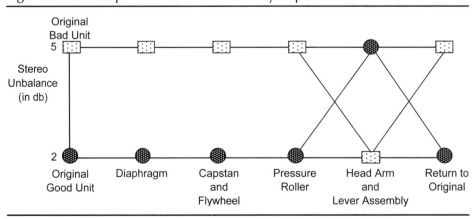

from the smallest reading to the largest (or vice versa). This rank test is called the Tukey test, named after the renowned John Tukey of Princeton University, and it is one of the most powerful tools in DOE work. (Dr. Tukey also gave expression to two terms widely used today in the Internet age—digital and binary.) Divide this ranking into three zones. The top zone (called the top end count) should have readings only from the good units (or vice versa). The bottom zone should have readings opposite to the top zone—that is, from bad units if the top zone has only good units (or vice versa). The middle zone (called overlap) contains both good and bad units. Add the numbers in the top and bottom zones. This is the total end count. Only if the total end count is six or more is there 90 percent or higher confidence that the particular parameter is important in explaining the difference between good and bad units.

Case Study of the Application of Paired Comparisons. In an investigation of the optical accuracy of a contact lens, six of the best lenses and six of the worst lenses were selected. Five lens parameters—cylinder (i.e., the curvature in the front side of the lens), cylinder BP side (i.e., the curvature in the back side of the lens), ultraviolet absorbance, polarizer, and mold fabrication—were measured on each of the twelve lenses. Table 14-2 shows two of these parameters arranged in ascending rank.

The results show that UV absorbance, with a total end count of five has less than 90 percent confidence, and therefore is not important as a parameter in explaining the difference between good and bad lenses; whereas the cylinder with a total end count of twelve is very important— hence a huge 99.6 percent confidence—in explaining the difference between good and bad lenses. It is an extremely important clue to pursue to get the culprit (Red X) root cause.

Product/Process Search

Objective. To determine, among many process parameters, which are important and which are unimportant in explaining the difference between good and bad product produced by the process.

Where Applicable. For all processes and machines in process-oriented or assembly-oriented manufacturing.

Sample Size. Six or eight good units (product) and six or eight bad units (product), with as large a difference as possible.

Methodology. Allow a sufficient number of units (product) to go through the process.

Measure each of the important process parameters (actual readings, not settings) that are associated with each product as it goes through the process. At the end of the process, select six or eight of the best units (product) and six or eight of the worst units—with as large a difference between the good and the bad as possible. Then rank each process parame-

Table 14-2. Paired comparisons ranking and Tukey test.

UV Absorbance Percentage (%)		Cylinder (Curvature: mm)	
↑ Top End Count	7.4 B 7.8 B 8.7 B 8.8 B	↑ Top End Count	0.016 G 0.018 G 0.020 G 0.026 G 0.030 G 0.030 G 0.048 B
✗ ↑ Overlap ↓	8.9 G 9.1 G 9.3 B 9.8 B 9.9 G 10.9 G 11.1 B 11.2 G	✗ Bottom End Count ↓	0.051 B 0.051 B 0.053 B 0.055 B 0.056 B
Bottom End Count			

B = Bad lens; G = Good lens

Parameter	Top End Count	Bottom End Count	Total End Count	Confidence
UV Absorbance	4	1	5	Not enough confidence: <90 percent
Cylinder	6	6	12	>90 percent (actually 99.6 percent)

ter associated with the twelve or sixteen products in ascending or descending order of magnitude. Apply the Tukey test. As in paired comparisons, if the total end count is six or more, that parameter is important in the process and should be carefully monitored. If the total end count is less than six, that process parameter is not important and its tolerance can be opened up and its cost reduced.

Case Study of the Application of Product/Product Search. In a plastic injection molding machine that was producing up to a 40 percent defect rate for short shots in a toy product, the process parameters in Table 14-3 were considered important.

A product/process search was conducted and the eight best and eight worst products were ranked in terms of each process parameter and their total end counts determined as shown in Table 14-3. Only material temperature, back pressure, and injection speed with end counts of six or more were the important parameters that needed tight control. The tolerances of the others were opened up to reduce costs.

Table 14-3. Product/process search case study: plastic injection molding machine—important vs. unimportant parameters.

Parameter	Total End Count	Confidence (Percentage)	Important
Mold Temperature	4	<90	No
Material Temperature	8	>90	Yes
Pressure	2	<90	No
Back Pressure	8	>90	Yes
Injection Speed	6	90	Yes
Screw Speed	2	<90	No
Mold Vents	2	<90	No
Injection Time	2	<90	No

A Commentary on the DOE Clue Generation Technique

At a first exposure to these clue generation techniques, engineers tend to dismiss them as too simple. After all, they say, the parts did not go to college! The higher the engineers are on the technical ladder, the more skeptical they are. Yet, as often happens, manufacturing engineers, manufacturing technicians, and manufacturing line operators reach for these techniques as a drowning man reaches for a floating log. They've been stuck with problems they had not been able to solve. They've gotten little real help from cloud nine engineers. These clue generation techniques have solved problem after problem for them. One of the distinct pleasures in the DOE business is to see how line operators solve problems with these tools, to see their eyes light up with joy at their accomplishments. And all that the engineering leaders can say is, "There go my people, I must follow them, for I am their leader!"

Additional Quality Bases to Be Touched in Manufacturing

1. *Mini-MEOST.* The use of Multiple Environment Over Stress Testing (MEOST) for reducing and virtually eliminating field failures has been covered in Chapter 11. It is primarily used at the design stage to replicate and correct potential field failures before the product goes into production. But it is also a necessary discipline in manufacturing to assure that workmanship, production processes, and supplier materials have not degraded design reliability.

However, a full-blown MEOST is not necessary here. Mini-MEOST is a truncated version with fewer stresses and somewhat lower levels of stresses. Again, the sample size for mini-MEOST testing remains unbelievably small—a maximum of ten units if the product is repairable or a maximum of thirty units if it is not repairable.

2. *The Elimination of Burn-in as a Reliability Tool.* Many companies subject their electronic products to a 100 percent high temperature test, often

with power cycling, to assure that early-life failures (known as infant mortality) are removed before the product gets to the customer. The rationale is that high temperature—the enemy of reliability—can weed out the weaker components and thus improve reliability.

However, burn-in has proven to be an ineffective technique. If a company does not know what to do for reliability, it falls back on burn-in as a stopgap, brute-force measure. The reasons for its ineffectiveness are:

❖ It does not stress the product high enough in temperature.

❖ It does not accelerate the rate of temperature change.

❖ It does not thermal cycle the product (i.e., go from high temperature to cold and back to high temperature over several cycles of hot-cold swings).

❖ It does not combine other stresses such as humidity, vibration, or voltage transience.

❖ It adds cost and lengthens throughput time, without adding value.

As a consequence, burn-in results are poor. It rarely, if ever, catches field failures. It is much better to substitute mini-MEOST in manufacturing, along with fast corrective action on two or more failures with the same failure mode.

3. *Field Escape Control.* Companies do not recognize that their earliest and largest service station is not in the field but in their own factories. During production testing, tabs should be kept on defects of a reliability nature (i.e., catastrophic failures). The discipline is called field escape control, which I introduced at Motorola sixteen years ago. During the course of a day or two, a single failure can be ignored. But if there are two or more reliability-type failures, with the same failure mode and the same failure mechanism, the panic button should be pushed. The failures should immediately be analyzed and corrections instituted. If action is not taken on these factory failures, the probability of the same failures appearing in the field, in the hands of the customer, is unacceptably high.

4. *Analyzing Effectiveness.* Analyzing product fallout in the factory by human analyzers is not always 100 percent effective. Lack of time, lack of knowledge, lack of accurate instrumentation, and intermittent failures are the main causes. To some extent, computer testing of the product can overcome some of these weaknesses. A spot-check should also be conducted on the products the analyzers troubleshoot to determine their effectiveness. A practice of "seeded defects," where a known defect is deliberately created to see if it would be detected by the analyzer, is encouraged. This practice can also be used to detect the effectiveness of mini-MEOST checks.

5. *Metrology.* Several product quality problems can be solved by mak-

Reducing Instrument Variation

This aspect of metrology—reducing instrument variation—should precede reducing product variation if the discrimination ratio is less than 5:1. The total instrument variation or tolerance, T_T, is made up of three subvariations—within-instrument tolerance (T_{W-I}), instrument-to-instrument tolerance (T_{I-I}), and operator-to-operator tolerance (T_{O-O}). The total instrument tolerance is based on the root-mean-squared law as follows:

$$T_T = \sqrt{T_{W-I}{}^2 + T_{I-I}{}^2 + T_{O-O}{}^2}$$

The 5:1 rule means that the specification tolerance, T_P (P stands for product), should be at least 5 times T_T. The methodology calls for scatter plots where:

1. Twenty to thirty units of product, covering the full range of the specification width, are measured twice with the same instrument and same operator and the results plotted on the X axis the first time and on the Y axis the second time. The width of the resultant tilted scatter plot determines the within-instrument variation, or tolerance T_{W-I}.

2. The same twenty or thirty units are measured in the second round twice with the same operator, the first time with instrument one plotted on the X axis and the second time with instrument two on the Y axis. The width of this resultant tilted scatter plot determines the instrument-instrument variation, or tolerance T_{I-I}.

3. The same twenty or thirty units are measured in the third round, with the same instrument as round 1, twice, with operator 1 results plotted on the X axis and with operator 2 results plotted on the Y axis. The width of the resultant tilted scatter plot determines the operator-to-operator variation, or tolerance T_{O-O}.

4. The total instrument variation or tolerance is then calculated using the root-mean-square formula stated previously.

5. Then, the ratio of the product tolerance (T_P) to the instrument tolerance (T_T) should be equal to or greater than 5:1.

ing sure that the measuring instruments are accurate. There are four parameters in metrology— precision, bias, accuracy, and discrimination— that must be defined:

> *Precision* is defined as the spread or range of a parameter and can be measured as a range that can be called ^{c}p. *Bias* is defined as the deviation of the average of a parameter from the target value. It is the noncentering ($\overline{X}$-D), where $\overline{X}$ is the average and D is the design center.
>
> *Accuracy* combines precision and bias. It can be called ^{c}p$_K$ if tied to a specification width. *Discrimination* is the ratio of product spread to measurement spread, with a minimum ratio of 5:1.

Figure 14-3 depicts these metrology terms pictorially. There are three panels in Figure 14-3:

❖ Panel A shows four frequency distributions with constant precision—that is, constant range and constant ^{c}p—but with biases ranging from high at the top to zero at the bottom. ^{c}p$_K$ also improves.

❖ Panel B shows four frequency distributions, all with zero biases, along with precision ^{c}p and ^{c}p$_K$, all improving from poor at the top to very good at the bottom.

❖ Panel C shows four frequency distributions with varying accuracies—poor at the top, excellent at the bottom—because precision, range, bias, and ^{c}p, ^{c}p$_K$ are all improving.

Other aspects of metrology include reducing instrument variation (discussed in the sidebar); correlation traceable to the Bureau of Standards or a national equivalent; calibration frequency; and calibration control (used to make sure that a due date for calibration is not missed).

6. *Routine Quality Disciplines in Manufacturing.* Several other routine quality disciplines are part of any good quality system. In the interest of brevity, they are:

❖ Plant safety guidelines
❖ Electrostatic discharge (ESD) protection
❖ Underwriters Laboratory (UL) or equivalent approval
❖ Quality walkthrough of software
❖ Storage, transport, packing, and shipping control
❖ Effective engineering change control system
❖ Traceability and bar-coding control
❖ Effective failure analysis down to root cause and a well-equipped, comprehensive failure analysis lab

Figure 14-3. Metrology.

Precision = Repeatability; spread; range, C_P (with specs)
Bias = Deviation from design center, D; X - D, where
  D = Design center or target
  X = Average; R = Range
Accuracy: combines precisionn and bias; C_{PK} (with specs)
Discrimination: Ratio of product spread: measurement spread

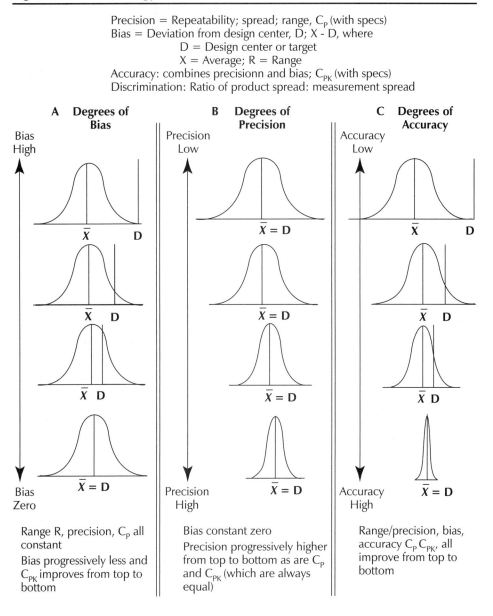

A Degrees of Bias	B Degrees of Precision	C Degrees of Accuracy
Bias High	Precision Low	Accuracy Low
Bias Zero	Precision High	Accuracy High

Range R, precision, C_P all constant

Bias progressively less and C_{PK} improves from top to bottom

Bias constant zero

Precision progressively higher from top to bottom as are C_P and C_{PK} (which are always equal)

Range/precision, bias, accuracy C_P C_{PK}, all improve from top to bottom

❖ Destructive physical analysis (DPA)/ tear-down audits

❖ Manufacturing flowchart with maximum defect targets at each workstation.

Engine Two: Cycle-Time Reductions in Manufacturing—The War on Waste

The Importance of Inventory Reduction

Inventory was widely recognized twenty years ago as a liability, not an asset, as nineteenth century accountants still label it in the twenty-first century. It plays an important role in substantially increasing a company's return on assets (ROA), which is calculated as follows:

$$\text{Return on Assets} = \text{Sales/Total Assets} = \text{Sales/(Inventory + Fixed Assets + Receivables)}$$

Of the total asset base, inventory often accounts for 40 percent. Therefore, reducing inventory to one-third could increase return on assets by 30 percent. Reducing it to one-tenth could increase return on assets by almost 60 percent—a figure that world-class companies have far exceeded.

Cycle Time—the Integrator

Yet inventory, like profit, is a lagging indicator. Like profit, you cannot work on inventory. You can, however, work on cycle time. In fact, cycle time is the great integrator of our times. In terms of calculus:

$$\text{Cycle Time} = \int \text{Quality, Cost, Delivery, and Effectiveness}$$

As such:

❖ If quality defects are reduced, cycle time is reduced.

❖ If costs are reduced, cycle time is reduced.

❖ If delivery is improved, cycle time is reduced.

❖ If effectiveness is improved, cycle time is reduced.

Cycle time, therefore, can be the *single metric* to assess improvements in quality, cost, delivery, and effectiveness. It is especially a vital metric in white-collar operations, which have derailed measurements of any kind in the past.

Nonvalue–Added Operations

In any industrial process, whether it is manufacturing or white-collar work, there are many nonvalue-added operations. Figure 14-4 is an example of typical nonvalue-added operations in production, each of which lengthens the cycle time from start to finish. Cycle-time management is, therefore, a war on waste. The elements of waste include poor quality, machine breakdowns, poor space utilization, long setup time, long transport time, and the killer—waiting time.

Figure 14-5 is a graphic portrayal of cycle time and waste in various operations within a company. The cycle time in production (labeled A) is made up of actual direct labor, setup time, transport time, and waiting time. Of these elements, only direct labor is value-added. Just as direct labor (D/L) constitutes only 5 percent or less of the sales dollar, it also constitutes 5 percent or less of the total production cycle time. If the nonvalue-added operations of setup, transport, and waiting that consume 95 percent of the total cycle time can be significantly reduced, production cycle time can be compressed to a value no more than twice direct labor

Figure 14-4. Nonvalue-added operations in production.

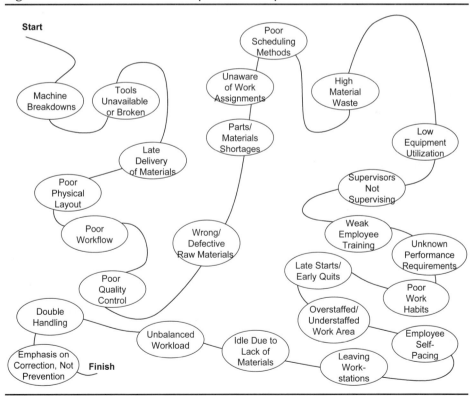

Figure 14-5. The elements of production cycle time and nonvalue-added cycle time.

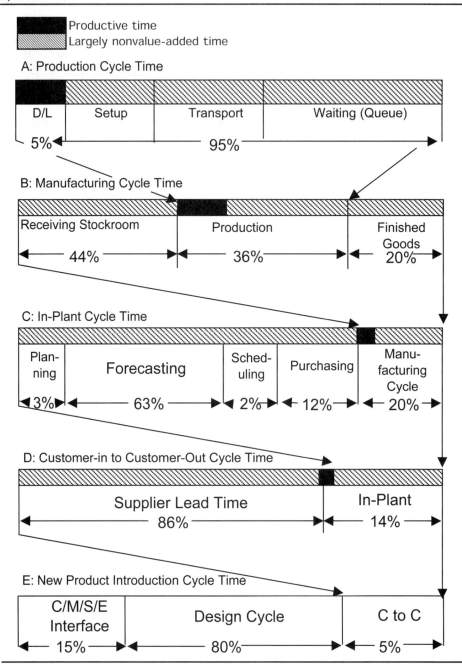

time. This is called theoretical cycle time, a target figure that can be and has been achieved.

Similarly, in the total manufacturing cycle-time loop (labeled B in Figure 14-5), receiving, incoming inspection, and the stockroom are at the front end of production and finished goods at the back end. Ninety percent of these support operations as well as 90 percent of the production operations (from A) are a waste. Therefore, the total manufacturing cycle time can be reduced by 90 percent.

The in-plant cycle-time loop, C, consists of planning, forecasting, scheduling, and purchasing—ahead of the manufacturing cycle. In an ideal just-in-time (JIT) plant, these preliminary activities can be eliminated or drastically reduced in time. As an example, with a pull system, small lots, and focused factories, forecasting is no longer necessary. Consider also the typical waste in the purchase cycle. Purchase order releases, acknowledgments, expediting, counting, inspecting, sorting, scrap, rework, repackaging, and invoices add almost no value. The other functions can all be reduced to achieve a total in-plant cycle-time reduction of 90 percent.

The last repetitive cycle-time loop, D in Figure 14-5, is from the time the customer's order is booked to the time it is shipped to the customer. The longest single element in this loop is supplier lead time, which can also be reduced by 90 percent and more.

Finally, there are similar cycle-time loops in all support services and white-collar work, of which the new product introduction cycle time, E, is the most important. The first cycle-time element is the customer/marketing/sales/engineering interface. The second and largest element is design cycle time, which is so important that it can make or break a company, as discussed in Chapter 12. In summary, the "water" of cycle time can be squeezed out of all operations (from A to E in Figure 14-5) in industry.

A Blueprint for Drastic Cycle-Time Reduction

Several major characteristics of cycle-time management must be present for a significant breakthrough in cycle-time reduction. These attributes of cycle-time management include:

- ❖ The focused factory
- ❖ Unified teams
- ❖ Product versus process flow
- ❖ The flat pyramid
- ❖ Continuous flow
- ❖ Push versus pull systems

❖ Small lot sizes

❖ Setup time reductions

❖ Linear output

❖ Material control

❖ People power

❖ Dependable supply and demand

❖ Tracking progress

Figure 14-6 shows how these attributes are linked together for a major breakthrough in manufacturing cycle time reduction.

The Focused Factory

As opposed to the behemoth factories of yesterday that produced a wide variety of products, the focused factory is a plant within a plant that manufactures only a family of closely related products and is managed by a dedicated, semiautonomous, interdisciplinary team.

To assure the transition from the all-purpose plant to the focused factory, the product designs should be scrubbed for best manufacturability, part number reduction, component standardization, modularization to accommodate some model variety, a simplified bill of material structure, and simplified routing.

Figure 14-6. Manufacturing cycle-time optimization.

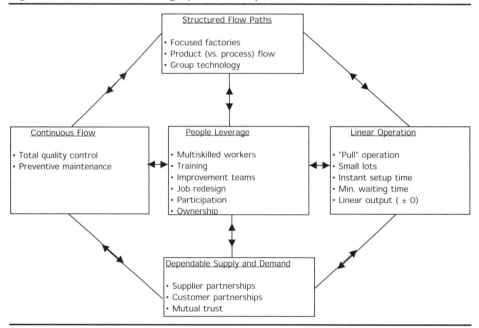

Unified Teams

In the old, unfocused factory, line support operations reported to bosses who were detached from manufacturing except at the very top rung—the plant manager. Departments such as manufacturing engineering, industrial engineering, process engineering, plant quality, plant purchasing, and maintenance protected their own turfs. Teamwork was difficult at best.

In the focused factory, these departmental walls come tumbling down. Its one focal point is a business manager or product manager to whom all support functions report (the only exception being a few specialists to serve as internal consultants covering several focused factories). In the ideal focused factory, this not only includes all manufacturing support functions but others such as purchasing, engineering, and sales as well.

Product Flow Versus Process Flow

Conventional thinking was premised on common processes being located in one area, generating "process islands" scattered throughout a plant. Backtracking and crossovers were common. As a result, the enemies of cycle-time reduction—transport time and waiting time—escalated. In one factory, a part actually traveled nine miles within a plant area of 30,000 square feet before shipment. One wag said that part "went round and round the plant until it developed enough kinetic energy to go out the door!"

Structured flow paths, by contrast, make for a smooth product flow that minimizes transport and waiting time. Preferably, the product flow is a U-shaped design that allows maximum operator flexibility and control. Although this creates some duplication of processes, the advantages of substantially improved inventory turns and other cycle-time benefits greatly outweigh the increased capital equipment costs. Furthermore, even these capital costs can be minimized by the use of smaller, simpler, and more flexible processes and equipment. The old order believed that people and equipment must be kept busy at all times, no matter how much excess inventory they produced. The new order believes that permitting workers and equipment to stand idle is not a crime, but that having idle material is a sin.

A simple way to assess product flow is to measure travel distances for all products and then to establish plans for their systematic reduction. Another method is to determine product flows between plants. There have been horror stories of product fabricated in the United States, assembled in Korea, tested in Hong Kong, quality audited in the States again, and distributed abroad again—all in the name of chasing direct labor or getting around tariff barriers and other intergovernmental regulations. The damage caused by these practices to cycle time and return on investment is almost too painful to measure. In one company, there was actually a case of a part that crossed the oceans seven times during the total manufacturing cycle! Any departure from a single plant, even a feeder plant, for product movement should be considered only as a last resort.

The ideal focused factory layout is line-of-sight manufacturing, where the production status can be visually ascertained by all at a glance. Production rates, quality levels, maintenance charts, and so forth should be posted for all to see. Storage racks, the handmaiden of work-in-process (WIP) inventory, should be banished.

The Flat Pyramid

The organization in a focused factory resembles a shallow pyramid, with authority and accountability permeating down to the lowest levels. Cycle-time training is given to all people in order to maximize understanding, commitment, and enthusiastic involvement in cycle-time management. Through cross-training, the number of job classifications are significantly reduced. The worker's flexibility and problem-solving ability now become the criteria for pay grade elevation and eventual promotion. In one of Motorola's divisions, the pay grades in one of its focused factories are based on the total number of machines and processes that an operator is certified to handle. Its focused factory, located in a high-wage area in the United States, has not only successfully challenged Japanese competition, it has knocked out several of its Japanese competitors.

Continuous Flow

Continuous flow as an attribute of cycle-time management represents quality at its best. The object is to assure that poor designs, unstable processes, defective materials, and marginal workmanship are not just corrected but prevented so that there can be a continuous, uninterrupted, and one-way flow of product with zero defects, 100 percent yield, minimal variation within specification limits, and no inspection and test. This, of course, is an ideal, but the techniques of design of experiments, poka-yoke, and total productive maintenance (TPM) offer an excellent blueprint for the eventual attainment of this ideal. TPM, discussed in Chapter 11, plays a specially important role in reducing cycle time, increasing throughput, and reducing costs up to 4–9 percent of sales.

Pull Versus Push Systems

The fame of the Japanese Kanban system has heightened interest in the "pull" system of product control as opposed to the old "push" system. In the push system, operators pile up product at a workstation in adherence to a master schedule or to keep machines and people busy, regardless of the pile-up of inventory or the lack of need for product further down the line. In the pull system, the last workstation in the line paces the entire line, with each previous workstation producing only the exact amount needed by the next station.

In the push system, problems are hidden. The cushion of large work-in-process inventories allows for a leisurely approach to problem solving. The best feature of the pull system is that it heightens the visibility of any problem. If there is a quality or delivery or other problem that shuts down a given workstation, the previous workstation, sensing that product is not needed at the next station, also shuts down. The ripple effect is fast, and soon the entire production line is down. There is nothing more visible than a whole line shut down. This accentuates the urgency of immediately solving the problem at the offending workstation. Workers, technicians, and engineers swarm over the station like bees to rapidly restore it to health so that the whole line can start up again.

Small Lot Sizes

A central feature of the pull system is that lot sizes are drastically reduced. In batch production, which is prevalent in most companies, large runs are the norm. If a process produces a 1,000-part run, with ten at the exact same time, 990 parts do nothing but wait and twiddle their thumbs. If the lot size is reduced to ten, however, waiting time is reduced to near zero. This is the crucial difference between batch production and short-cycle manufacturing using continuous flows.

Setup Time Reduction

Small lots, however, require significant reductions in setup time in order to facilitate a line changeover rapidly from one model to the next. It has been demonstrated that setup time can be reduced by factors of sixty to one and more, given the use of ingenious industrial engineering methods fueled by workers' ideas.[7] One such ingenious method is to videotape the changeover process to pinpoint where time and motion are wasted.

Another technique is readying the workplace so that all tools and materials will be instantly available and there will be no need for even first piece-part quality evaluations through external audits. A classic example of this would be changing a tire in record-breaking time. The operation takes the average motorist at least twenty minutes. At the Indianapolis 500, taking as much as fifteen seconds to change four tires could mean losing the race. Granted, the manpower and expense involved in effecting such a rapid changeover, though necessary to auto racing, is impractical in most industrial situations. Nonetheless, the sport's masterful organization of material and labor, honed to a fine science through practice and drill, can provide many useful tips to manufacturing. Another example is provided by Toyota Motor Corp. In the early 1970s, it took three hours to make a die change in its large stamping machines. Hearing that Volkswagen had perfected a method to make the change in ninety minutes, Toyota engineers went to study VW's techniques, installed these techniques at Toyota, and went on to shave another half-hour off setup time. Proudly, they announced to their general manager that they had beaten VW. The manager said: "Good. My congratulations to your team. Now change the dies in three minutes!" Unattainable as the new goal seemed, by the late 1970s the Toyota engineers had achieved this industrial equivalent of the four-minute mile by proceeding to shave setup time down to one and a half minutes.

Linear Output

The benefits of a pull system are limited if the total quantity required by master schedules is allowed to vary from day to day. Such schedules should have nearly constant rates—known as fidelity or linearity—over short periods of time, with quantities being ramped up or down slowly. There can be model mixes within this linearity, but the total output of the mix should be held to an ideal ± 0 deviation from the constant rate.

Material Control

The old "just in case" system was characterized by incomplete picks, partial builds, expediting, parts chasing, inventory auditing, and an antediluvian cost-accounting system—a bloated superstructure built on the vanishing foundations of a direct labor base.

In cycle-time management, with a steady and dependable stream of just-in-time parts from suppliers, incomplete picks and partial builds are no longer necessary. Neither are parts chasing and expediting by supervisors, who can now divert the 90 percent of time previously spent on these useless activities to helping and coaching their people. The burdensome chores of physical inventory audits conducted once a month now disappear. The line-of-sight layout, the pull system, and small lot sizes facilitate easy counting, at no extra cost, directly by the operators themselves on a daily, and even an hourly, basis. Cost accounting, too, is beginning to enter the twenty-first century—directly from the nineteenth—with cycle time rather than direct labor as the base for overhead cost allocations.

People Power

Granting workers a sense of "ownership" is probably the most important attribute of cycle-time management. The concepts and techniques of ownership are detailed in Chapter 9.

Dependable Supply and Demand

It is only when a company has put its own house in order by slashing its work-in-process inventory that it has a right to approach its suppliers and customers as partners in cycle-time reduction. With the aid of these techniques, suppliers can be encouraged, especially within the framework of partnership, to deliver a linear output of their products in smaller, frequent lots.

Customers may also become converts to cycle-time management when a company can deliver their order with shorter and shorter lead times. They can then begin, with greater confidence, to order smaller quantities more frequently. In time, they can also be persuaded to give longer-term contracts, blanket orders, and more stable forecasts—in short, to be partnership customers.

Tracking Progress

Cycle-time management in manufacturing requires only two macroscopic measurements to track progress: yield and cycle time. Figure 14-7 shows how a product line in a computer company's focused factory was tracked for yield and cycle-time improvements. It also indicates the major techniques used to achieve these improvements both in quality and in cycle time.

Figure 14-7. Manufacturing yield/cycle time: a graphical plot.

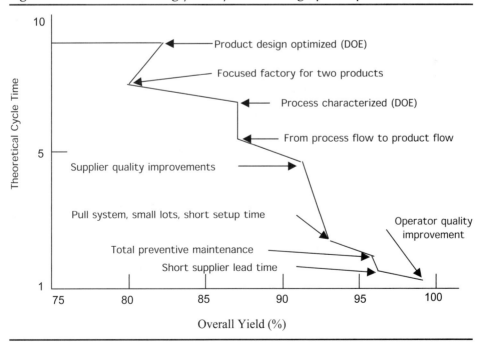

Cycle-Time Management: From Theory to Action

The most frequently asked question about cycle-time management is: "Where do we begin implementation?" As with other processes (as opposed to programs) of importance, the start should be with top management exposure, education, and commitment. From that point forward, the following is a suggested sequence. It is not necessary, however, that these steps be taken in order. Parallel and iterative steps can be accommodated.

1. *Setting Performance Measurement Parameters.* The focus must shift from profit to return on investment (ROI) for each strategic business unit (SBU); from direct labor to manufacturing cycle time; from single workstation defect analysis to overall yield improvement; from manufacturing requirements planning (MRP II) to just-in-time (JIT) manufacturing; and from the vagaries of forecasting to build-and-ship-to-order with lightning speed.

2. *Launching a Pilot-Focused Factory.* Rather than immediately breaking up a large plant into several focused factories, a single focused factory should be formed to act as a pilot to manufacture a product family based on similar processing requirements and using similar parts.

3. *Creating a Focused Factory Management Team.* Forming a focused factory management team may be the hardest step of all from the point of view of human relations because it disrupts traditional functional organizations and breaks up "empires." A first step is to assign a task force role to a selected focused factory management team. As the task force begins to pull together and succeed, it can become a permanent focused factory organization, with an autonomous team consisting of a product manager and members from every function within manufacturing, followed soon after by all supporting functions, such as sales, engineering, supply management, and quality assurance.

4. *Generating People Power.* A focused factory, with its small size, natural work units, client relationships, quick feedback on performance, and sense of ownership, has a far greater chance than a traditional factory of welding its people into a family and unleashing full people power. The Hawthorne effect (named after the famous Hawthorne plant in Chicago where, in the 1920s, the noted Elton Mayo found that workers responded to the attention and support from management to achieve productivity gains) may also come into play as its success becomes more widely recognized. Training of the workforce in cycle-time principles, multiple skills, and problem solving should begin with the creation of the focused factory.

5. *Poor Quality: The First "Rock to Blast."* In the "rocks in the river" analogy, there is no rock more formidable than poor quality to blast, so that the ship of production does not flounder. Tackling all quality problems must be the highest priority in the focused factory—from product

yields to total productive maintenance, from incoming inspection, to supplier process control.

6. *Adopting a Pull System.* To convert from a push to a pull system, lot sizes must be reduced gradually but systematically, with some inventory banks between stations at the start leading to an eventual elimination of such banks. Setup time reduction should be pursued and an attempt made at stabilizing the total model mix at a nearly constant rate.

7. *Demanding Dependable Supply and Demand.* The last steps involve influencing suppliers and customers. A demonstration of achievement in one's own plant is worth a thousand exhortations to suppliers and pleas to customers. Such achievement can then become the best sales tool for requesting blanket contracts and smaller, more frequent, deliveries from suppliers and to customers.

CASE STUDY

Hewlett-Packard—A Benchmark Company in the Area of Manufacturing

While Taichi Ohno started his famous Kanban system at Toyota and launched modern JIT, several companies have successfully emulated the Toyota production system. (Incidentally, JIT did not start in Japan. Its originator was Henry Ford in the 1910s. From the time iron ore landed on the docks at Dearborn, was processed into engine blocks and machined, engine assembly completed, and the car readied for shipment to dealers, the total lapsed clock time was forty-eight hours! Today, Ford Motors has not been able to capture old Henry's recipe.) Intel Corp., Motorola, Dell Computer, Omark Industries, Lincoln Electric Company, and scores of others have rejuvenated their manufacturing operations. One company, in particular, Hewlett-Packard (HP), is a model of quality, JIT, and manufacturing excellence. Table 14-4 documents the typical changes HP has wrought on a number of criteria.

SELF-ASSESSMENT/AUDIT ON MANUFACTURING EXCELLENCE

Table 14-5 is a self-assessment/audit that a company can conduct to score its manufacturing prowess. It has three key characteristics and fifteen success factors. Each success factor has a maximum of five points for a total score of seventy-five points.

Table 14-4. Breakthrough improvements in manufacturing at Hewlett-Packard.

Criteria	Before	After
Solder Quality	5,000 ppm	1 ppm
Repair/Scrap (percentage of direct labor)	35 percent	2 percent
Overall Equipment Effectiveness	40 percent	99 percent
Number of Suggestions/Employee/Year	0.15	6.5
Plant Space (typical)	30,000 sq. ft.	7,500 sq. ft.
WIP Inventory	30–40 days	3 days
Number of Part Numbers	20,000	450
Number of Suppliers	2,000	200
Average Lot Size	500	5
Number of Skills/Operator	1	5
Actual Time: Direct Labor Time	3,000–14,000:1	2:1
Process Speed: Sales Rate	100–1,000:1	3:1

Table 14-5. Manufacturing: key characteristics and success factors (75 points).

Key Characteristic	Success Factors	Rating				
		1	2	3	4	5
9.1 Manufacturing Resurgence	1. The company has an enlightened policy to manufacture in the country or region where it sells and to buy materials in the country or region where it manufactures. 2. The company has moved away from mass production, complex automation, and MRP II as being counterproductive.					
9.2 Quality Improvement in Manufacturing	1. The product is made robust by examining and compensating for uncontrollable (noise) factors present in production and in the field. 2. Important processes have been characterized and optimized with variables search and scatter plots, respectively. 3. Important parameters on each process are monitored in production with a "who," "how," "where," and "when" positrol chart to "freeze" process gains. 4. Process certification is conducted to ensure that peripheral quality weaknesses (e.g., poor preventive maintenance, violations of good manufacturing practices, equipment inaccuracies, and noncalibration, environmental changes, operators not following critical procedures, etc.) are corrected ahead of full production and ahead of formal DOE studies. 5. Precontrol is used to monitor products and processes which have been optimized with DOE to levels of c_{pk} greater than 2.0. 6. Clue generation techniques such as multi-vari, components search, paired comparisons, and product/process search are used for chronic quality problems instead of engineering guesswork, hunches, and opinions. 7. Mini-MEOST is used to monitor design integrity that may have degraded in production because of workmanship and materials. 8. Metrology is used to ensure instrument precision, nonbias, accuracy, and discrimination.					

(continues)

Table 14-5. (Continued).

Key Characteristic	Success Factors	Rating				
		1	2	3	4	5
9.3 Cycle-Time Reduction in Manufacturing	1. Structured flow paths are employed to create: ❖ Focused factories of dedicated product with dedicated people ❖ Product flow—not process flow—to minimize transport time ❖ Group technology (GT) cells to conserve space ❖ Multiskilled operators to staff GT cells effectively 2. Continuous flows are achieved through: ❖ Elimination of defects and rework loops ❖ Overall equipment effectiveness (OEE): maximizing yields, uptime, and machine efficiency 3. Pull systems and Kanban are used to: ❖ Minimize the need for master schedules and forecasts 4. Drastic reductions in lot sizes and setup time 5. Customers and suppliers are encouraged to order and ship, respectively, small quantities frequently					

FIELD OPERATIONS: FROM AN APPENDAGE TO A MAXIMUM SERVICE TO DOWNSTREAM STAKEHOLDERS

❖ ❖ ❖

Every day, there are 50 million moments of truth, 50 million contacts between customers and the frontline employees of a company. These 50 million moments of truth integrate in the customers' minds, forming an indelible impression that can make or break a

company.

—JAN CARLZON,

FORMER PRESIDENT,

SCANDINAVIAN

AIRLINES

The Service Paradox

The two most underrated areas of a company are manufacturing and field service. Yet, excellence in these two disciplines is vital for business excellence. Chapter 14 was devoted to manufacturing excellence. This chapter concentrates on field service excellence. The paradox is that the most underpaid, understaffed, undertrained, and undermotivated people are precisely the frontline troops that come into daily contact with customers and can affect a company's fortunes for better or for worse. Their influence on the customer is second only to the top management of a company and on a near par with sales. A special section of this chapter is devoted to the care and feeding of these frontline troops.

The Dimensions of Field Service

Field service encompasses:

1. Product, including reliability, maintainability, availability, uniformity, dependability, safety, diagnostics, product liability, user-friendliness, resale value, and price
2. Predelivery service, including handling and transport, storage, installation, instruction manuals, training, human error, billing, and accounts receivables
3. Service to downstream supply chains, including distributors, dealers, retail chains and retailers, installers, servicers, and parts support
4. Service to users, including frontline troops, repair service centers, field maintenance, spare parts, and field complaints

Field Services: Product Highlights

Several aspects of customer loyalty have been described in Chapter 6 (Table 6-4). Table 15-1 is a capsule summary of each aspect of a product that is important to customers/field operations and the technique most appropriate for its success. Chapters 11, 12, and 14 also outlined the tools to assure product excellence in quality and reliability.

Failure Analysis

Failures can occur anywhere in the life cycle of a product—at the design/ development stage, in pilot runs, in production, and in the field. The following priorities, in descending order of urgency, are recommended:

1. Multiple Environment Over Stress Testing (MEOST) failures in design (see Chapter 12).
2. Mini-MEOST failures in production (see Chapter 14).
3. Field escape failures in production repair (see Chapter 14).
4. "Zero time" failures. These are dead-on-arrival (DOA) failures at the customer's site that have escaped reliability safety nets (items 1, 2, and 3 in Table 15-1). Each DOA failure should be analyzed to determine how it escaped the safety nets.
5. Field failures—after one month, three months, six months, and one year in service. Hopefully, the earlier "infant mortality" failures

Table 15-1. Product characteristics and associated techniques for field success.

Product Characteristics	Technique Appropriate for Success
1. Reliability	Multiple Environment Over Stress Testing (MEOST) and failure analysis
2. Maintainability	Mean time to diagnose (MTTD) and mean time to repair (MTTR)
3. Availability	Toward 100 percent uptime through items 1 and 2 and total productive maintenance (TPM)
4. Uniformity	Design of experiments (DOE); c_p, c_{pk}
5. Dependability	Lifetime guarantees with MEOST and item 2
6. Safety	Safety analysis and fault tree analysis (FTA)
7. Product Liability	Liability prevention analysis
8. Diagnosis	Built-in diagnostics in design
9. User-Friendliness/ Ergonomics	Evaluations by focus groups, panels, and employees
10. Delivery	Cycle-time reduction; lean manufacturing
11. Price	Total value engineering
12. Resale Value	Earned reputation; customer loyalty

have been caught and a low constant failure rate (of the reliability bathtub curve) can be analyzed.

6. Out-of-warranty and lifetime guaranteed failures, needed in order to be several steps ahead of competition.

Failure analysis is so important as a reliability tool that every company should develop its own independent failure analysis capability, manned by state-of-the art equipment and highly competent failure analysis professionals. Failure analysis should be pursued until the root cause of the failure is isolated. Often, the root cause is buried under two or more layers of effect-cause pairs. When this occurs, a few special techniques should be used:

❖ Components search

❖ Paired comparisons

❖ Failure simulation, where the presumed failure cause is deliberately introduced and removed to see whether the observed failure mode can be artificially reproduced (This is known as "switching the cause on and off.")

❖ B versus C to validate the effectiveness failure correction[1]

Predelivery Services

While many companies have concentrated on product quality in the hands of customers, nonproduct factors associated with predelivery services have suffered from benign neglect. One large multinational company found that less than 30 percent of customer dissatisfactions were associated with its products whereas over 70 percent were caused by nonproduct factors. An Ultimate Six Sigma system covers all the bases where customer dissatisfaction can arise and prevents them, rather than waiting for customer complaints and, worse, customer defections. Quality audits should be made on all of the following predelivery services:

❖ *Handling and Transport.* It is so easy to lay the finger of blame on sloppy workers in assembly and shipping, on truckers, and on installers for handling and transport damage. Even assuming that Murphy's Law reigns supreme in handling, as it does in other parts of a company's operation, prevention actually starts with adequate packaging design and testing for product protection against temperature and humidity fluctuations, vibration, shock, carelessness, and even sabotage. For some of these perils, the product is in greater danger from handling and transport than from use. One noteworthy appliance manufacturer cited damage as its number-one warranty problem even before customer installation. Damage claims amount to more than $6 billion per year in the United States alone.

Packaging generally stays with the product through its life. Packing, by contrast, is a throwaway after installation or use. Both are important. Both have to be tested—separate from MEOST in design and mini-MEOST in production—for simulated transport stresses such as shock, vibration, drop, and rough handling. One enterprising company regularly sends its product out with its specified packing (using the specified carrier to one of its remote plants) and then has it shipped back with the same carrier to assess the reliability of the packing ahead of production. "Before" and "after" readings can then measure the degree of degradation.

❖ *Storage.* Nothing can be further from the truth than to assume that there is no deterioration or degradation of products in storage. Even nonuse time has an impact on reliability as environmental stresses, oxidation, humidity, dust, radiation, electromagnetic exposure, and other factors take their toll in storage. In one study of four classes of complex systems, based on one million hours of total storage time, two of the four systems were nonoperational after five years. The lessons can be particularly painful for military systems whose products are long kept in storage until crises anywhere in the world require their immediate use.

Storage can also be simulated and accelerated with modified MEOST studies. The stresses would be similar to MEOST for product reliability, with the exception that no power would be applied to the product. In addition, periodic audits should be conducted on stored product to assure no degradation or, in the case of a product with limited shelf life, that the shelf life can be extended or the product purged before time limits are reached.

❖ *Installation.* Before actual use, a product goes through additional processing at distributors and/or dealers, along with assembly, installation, and turnkey checks at customer sites. Complex products require distributor or dealer "preps" before turning over the products to the customer. In fact, dealers today perform the role of the last stage of final assembly and test for a company. This is especially true in mass customization, which is better postponed until the last moment before the customer receives his individually customized product. The key to success, again, is simplicity. Just as design for manufacturability (DFM) is gaining importance and is being quantified, design for dealer prep must also be considered by the concurrent engineering team for ease and mistake-proofing at the dealerships.

Installation is either performed by skilled specialists or by the consumer. In the first case, special tools, instruments, and instructions for error proofing must be provided to the specialists. Often, these aids are overlooked. Skilled installers play an even more important part. For complex products, installer certification based on tests and periodic recertifications should be carried out.

For installation by the consumer (i.e., user), the record is dismal. Parents assembling toys for their children at Christmastime can attest to the frustration felt in following unclear, complicated, or downright wrong instructions. For lay people, comprehension is best enhanced with a liberal use of illustrations, pictures, audiotapes, and even free videotapes. Some enterprising companies even have an 800 number or web address printed on the product that accesses a company's service center. Others test the instructions with their own relatively unskilled workers "acting as guinea pigs."

❖ *Use.* Several studies have been made to quantify the extent and effect of human error on product reliability and maintainability. Up to 50 percent of failures in major systems can be assigned to human errors. There are several types:

1. *Failure to Allow Sufficient Time for Training and Learning the Operation of New Products.* Computer illiteracy is a classical drawback.

2. *Failure to Use Available Information.* It's "macho" to ignore the owner's manual, except as a last resort.

3. *Use of the Product in Applications or Stresses Never Intended.* An extreme example was a customer using a lawn mower as a hedge trimmer!

4. *Failure to Maintain.* Consumers are notoriously lax in following a prescribed schedule for lubrication, cleaning, and replacement of expendables.

In many cases, prevention of human errors in use can and should take place in product design. This includes ergonomics, built-in diagnostics, and fail-safe features. The use of service contracts for proper maintenance cannot only reduce warranty, but also add to business profitability. One manufacturer of dryers, where the clogging of vents was a perpetual problem in users' homes, offered a periodic vent inspection and correction to the homeowner for a small fee. In another case, where dairy farmers blamed the milking equipment for a reduction of milk price based on bacteria counts, the equipment company offered a free audit of the entire installation and captured 60 percent of the business away from its competition.

❖ *Billing and Accounts Receivables.* Often, the accounting department of a company can dissipate the positive aspects of good product quality. Wrong prices, wrong counts, and wrong addresses are bad in themselves. But dunning the customer for payments and the general insensitivity of account personnel can often result in customer defections—the worst sin of all.

❖ *Configuration Control and Product Traceability.* Once a product is completely defined in terms of bill of materials and the issue or revision number of each part as well as software sequences, the design configuration is supposedly frozen. Yet, design changes do occur in production and in the field. Questions arise as to when to make these changes and whether only some or all units need the changes, parts availability, field retrofit, and so on. All of these issues are addressed in configuration management or configuration control.

Traceability of the exact issue or revision number of each critical part becomes important when human or product safety is involved. It necessitates the identification and purging of all the "brothers and sisters" of these parts having the same vintage. There is an amusing story of an Air Force contractor that was forced to identify parts still in its factory with the same vintage as the failed parts on a critical mission. The contractor's production people searched high and low for the stored parts but could not find them. In desperation, they turned to a psychic, who divined correctly that the parts were in a remote warehouse behind a pillar! Good configuration control and traceability are essential disciplines, especially when product liability, recalls, and lawsuits are involved. Then, there would be no need for psychics!

Services to Downstream Supply Chain

Many companies pay great attention to their customers (i.e., users), but they do not give the time of day to the downstream members of the supply chain—their distributors, dealers, retail chains and retailers, installers, servicers, and parts support operations. They do not regard them as their immediate "customers." The same principles of supply chain partnerships, highlighted in Chapter 13, apply with equal force to a company's downstream partners. These principles are a commitment to partnership; ethics and trust; handshake agreements; active, concrete help; mutual escalation of profits; an "open kimono" policy; a commitment to a long-term marriage; and a belief in downstream partners as an extension of the company.

The company must also emphasize the following points in its relations with distributors, dealers, retail chains, and servicers:

❖ Reducing the size of the distributor/dealer/servicer base (similar to the reduction of the customer base and the supplier base)

❖ Spending valuable time with this base in order to build important bridges of personal relationships

❖ Providing regional service representatives

❖ Distributing pamphlets for diagnosing user complaints

❖ Conducting seminars/clinics (live or videotaped) for distributor, dealer, and servicer personnel

❖ Establishing exchange stations and exchange units, where defective units can be quickly exchanged for good units with a one-day turn-around time

❖ Distributing newsletters and field bulletins on service tips

❖ Feed-forwarding production plant experiences (e.g., MEOST data, field escape analysis, zero-time failures, and so on)

❖ Providing assistance in design of service tools and equipment

❖ Networking through distributor councils, dealer councils, and servicer councils

❖ Establishing parts standardization to facilitate parts interchangeability, not only on the company's parts, but also for parts from its competitors

❖ Providing parts inventory reduction assistance

❖ Providing starter kits of unique and critical parts made available to service stations ahead of product introduction in the field

❖ Assuring that distributorships/dealerships are not shared with competition

Services to Users

In the final analysis, the end-user is the ultimate evaluator, judge, and economic arbiter of the effectiveness of a company's field service.

The Frontline Troops

The customer game is ultimately won on the frontline—where the customer comes into contact with any member of the company. The frontline team *is* the company in the customer's eyes. It is easier to visualize these frontline troops in the service arena than in the manufacturing sector. Consider the following:

❖ In airlines, it is the flight attendant, the ticket and gate agents, and the baggage claims office who provide Jan Carlzon's "50 million moments of truth."

- ❖ In hotels, it is the front desk, the concierge, the porter, housekeeping, and room service that are the most visible to the customer. It is a fact that the chairman of the well-known Marriott chain plays the role of a porter to get unfiltered information about performance at his hotels.

- ❖ In manufacturing, frontline troops start with the apex—the chief customer officer (CCO) and the quality function that serves as the customer's advocate down to the ombudsmen (or "fixers" of customer concerns). The troops include the sales force, the marketers, customer service operations, order entry personnel, and—last but not least—telephone operators.

- ❖ In the field, it is the retail salespeople, the installers, and the servicers that most influence customer perceptions.

A Strategy for Conversion of Frontline Troops From Deadbeats to Heroes

The objective is to turn these frontline troops from being neglected and despised as second-class citizens into the heroes they genuinely should be. The recommended strategy demands:

1. *Selective Hiring.* Instead of brining in bodies that are barely warm off the streets, there must be careful screening to determine the degree of customer empathy and team player potential in prospective candidates. A sunshine personality is always a joy to customers.

2. *Adequate Staffing.* Numbers matter. Instead of skimping on headcounts, there should be an adequate number of personnel to prevent long customer lines and consequent irritations. Supervisors and members from other departments should be roped into service during peak periods.

3. *Training.* Training should cover issues such as customer sensitivity, product knowledge, the skill and art of selling, and customer anger dissipation.

4. *Walking the Talk.* Management by walk around (MBWA) is a must. Motorola's vice chairman urged all of the company executives to set aside at least one hour a week to spend time with their people. It is especially important to mingle with frontline troops. Listening to them for their ideas and suggestions is one of the best ways to get customer feedback. It represents the greening of the executives on customer issues.

5. *Compensation.* Instead of being the lowest on the payroll totem pole, this group should receive compensation commensurate with their unique position in the inverted pyramid—second in importance only to the customer at the apex.

6. *Employee Empowerment.* If empowerment is desirable for all employees, it is vital for the frontline troops. No manual can ever spell out the unique and unexpected situations they encounter with customers. Noth-

ing frustrates a dissatisfied customer more than to be told that the employee is "merely following rules." Most of these rules are for internal control, not for customer care. One of the best ways to empower frontline troops is to grant them financial discretion—albeit up to established limits—to settle genuine customer complaints and borderline out-of-warranty claims. As an example, a customer at a hotel chain complained to the doorman about minor damage to his car in the hotel parking lot. To his utter surprise, the doorman pulled out a hotel checkbook and handed the customer a $50 check—no questions asked—to assuage the customer's anger. One can only imagine how this actual story did the rounds with the customer's friends and neighbors and converted an annoyed customer to a loyal one.

7. *Celebration.* Heroic acts by frontline personnel need to be widely publicized within companies and the heroes feted. A supervisor in a Frito-Lay warehouse found he could not deliver product to the distributor because of a raging snowstorm that had all the roads blocked for miles around. Instead of accepting defeat, the supervisor rented a plane to deliver the merchandise to the distributor. He feared he would be fired for this "irrational exuberance"—to use the immortal Alan Greenspan's phrase. Instead, the company not only turned the incident into an extravaganza, but it received national publicity that millions of dollars of advertising could never achieve.

Repair Service Centers

Service shops have been severely and justly criticized for late responses, incompetent and indifferent mechanics, high prices, and even downright fleecing. Two of the main reasons for this shoddy state of affairs are that many of these service shops are too small to be efficient and too independent, in terms of financial ownership, to be supported by the manufacturer. A study conducted on service shops several years ago indicated that 76 percent were one-person operations. The parallel is the small grocery store, which has been outmoded by food store chains. Today, manufacturers are reducing the number of small, independent repair shops as well as replacing them with company-owned professional shops.

For example, at Motorola's automotive and industrial electronics group, we used to have more than 7,000 service stations, with many serving our competition as well. When I inherited the field service operations there, in addition to my quality assurance responsibilities, I set up our own service stations with two main objectives:

❖ To get an early and accurate estimate of our real field reliability problems

❖ To give our customers the most accurate and reliable repair service

Among the larger and more professional independents we created "listening posts," where any failure trend (we defined two failures of the same type as a trend) was immediately reported to our service headquarters while the product was flown back for analysis and correction. As a result, we reduced our field failure rates by an average of 7:1. We also created exchange service stations, where a customer was given an exchange unit of the same model, with an option to keep it or return it when the original unit was repaired. We further developed a starter kit of critical parts for each new model to assure the timely availability of such parts for repair. Today, many years later, our field reliability has reached levels below 10 parts per million (ppm) per year.

Companies have also intensified their training for repair shop personnel with seminars, videotapes, better instruction manuals, service bulletins, diagnostic instruments, repair tools and—above all—certification and recertification of such personnel. One large manufacturing company has even rated and ranked its service personnel in the field, not only for technical proficiency, but more important, for the image they portray of the company to the consumers serviced.

Service centers are also being rated for the timeliness and accuracy of service in consumer surveys. As warranties for product reliability keep getting extended—from three months in the 1960s to one year in the 1970s, two years in the 1980s, three years in the 1990s, and five to seven years in the 2000s—it is not a stretch to project lifetime warranties in the next decade. Similarly, service work guarantees are also being extended to lifetime levels.

Field Maintenance

The ideal for all products is 100 percent reliability for its specified life, measured in years. This is the objective of MEOST. As products become more and more complex, however, breakdowns do occur. In production, the old saw used to be: "If it ain't broke, don't fix it." As a result of this fallacious thinking, the ratio of breakdown maintenance to preventive maintenance is still 80:20 in the average company. Fortunately, with the development of total productive maintenance, the downtime due to process equipment failures has a goal of no more than 5 percent in progressive companies.

The same targets should apply in the field with an emphasis on preventive maintenance. This is done with scheduled servicing where a product is checked at periodic intervals to rework or remove critical parts before a specified age to substantially reduce the likelihood of failure. (The emphasis here is on critical parts. Research conducted on previous field maintenance practices found that 89 percent of the items that had been designated for removal had no decrease in reliability with age. There was no need, therefore, to impose a mandatory age limit on them.)

The automotive industry has used schedule servicing for many years. For airlines, scheduled servicing is mandatory. But for almost all household appliances it is, indeed, a rare practice. Manufacturers can enhance both customer loyalty and increase their own profitability with an intelligent scheduled servicing program offered to consumers at a nominal cost.

Maintenance Reliability

The reliability of the maintenance function is a concern in itself. An Air Force study found that maintenance-caused failures accounted for 2–48 percent of total failures. Balanced against the fact that maintaining military equipment is, on an average, eleven times the cost of procurement, this is an appalling figure. General Motors' Chevrolet division, as an example, identified twenty-two areas essential for its customers' total satisfaction. For each of these areas, specific evaluations were used to determine if a dealership was worthy of certification, which includes a review of the dealer's procedures and facilities as well as customer feedback on dealer performance.

Spare Parts

There are several issues that must be addressed on spare parts in the field:

- ❖ They should be adequate in quantity.
- ❖ Their availability to the user and servicer should be timely.
- ❖ Their quality should be assured (especially for a component that would be compatible with a revised product configuration or vice versa).
- ❖ Modularization in design is necessary for quick diagnosis and replacement by the user and/or the servicer.
- ❖ The requirement of parts availability for a minimum of seven to twenty years is mandated by governmental regulations.
- ❖ A balance between parts availability and minimizing parts inventory costs, mainly through reliability improvements, must be crafted.
- ❖ The minimizing (or even outlawing) of substitute, noncompany parts to maintain quality reputation of the company is good protective strategy.

Field Complaints

Many companies are smugly satisfied with their warranty costs. In my consultations with several of them, a one percent warranty cost, expressed as a percentage of sales dollars, is considered by them to be a mark of quality excellence. Their accounting departments factor in such costs as a matter of routine. But, when these warranty costs are translated into failures as a percentage of sales units, they can reach 5–15 percent. That still doesn't faze top management, who glibly proclaim that "such figures are the norm in our industry."

The Tip of the Iceberg

What these unthinking managers and their accounting minions do not estimate is the cost of customer defections (discussed in Chapter 6). In an oft-quoted study conducted by The Coca-Cola Company:

❖ Users who felt that their complaints were not satisfactorily resolved told a median of nine to ten people about their negative experience.

❖ Twelve percent of the complainers told more than twenty people about their extreme dissatisfaction and the negative responses they received from the company.

❖ Almost three out of four complainers who received positive responses from the company, along with meaningful corrective action (including financial compensation), were willing to go back to the company and become loyal customers.

❖ Consumers who were completely satisfied with the response from the company told a median of four to five people about their positive experience.

Using such figures as a baseline, it doesn't take rocket science for a company to estimate the tip of the iceberg representing warranty costs—in fact, you could estimate the whole iceberg if you take into account the multiplier effect of the lifetime lost revenue of a disgruntled customer in terms of:

❖ Years of adult life
❖ The percentage of friends and relatives (usually 10–15 percent) following the customer's advice to shun the company
❖ The loss of service sales
❖ The loss of parts sales
❖ The loss of finance charges

Unfortunately, with the dismal record of accountants who would rather be exactly wrong than approximately right, there is not much hope for companies to reform. They would rather collide with the iceberg and sink their Titanics.

Factors Affecting Field Complaints

Not all the field complaints are directly related to the product or nonproduct causes directly controlled by a company, such as billing, wrong destinations, or transport damage. There are several indirect influences as well, including:

- ❖ *Economic Climate.* Complaints tend to fall in a seller's market and increase in a buyer's market—both based on the laws of supply and demand.

- ❖ *Degree of Competition.* The greater the competition, the greater is the tendency to complain and then switch. There are fewer complaints for a company that is a quasimonopoly because customers look upon their complaints as futile. But look out if that monopoly is nearing its end!

- ❖ *Consumer Age.* Complaints tend to rise with the age of the consumer. Below the age of twenty-five, people seem to be more forgiving.

- ❖ *Consumer Income.* The higher the family income, the higher the complaint rate.

- ❖ *Unit Price of the Product.* The higher the unit price of a product, the higher the complaint rate. Furthermore—and this is another iceberg effect—the ratio of complaints to actual failures is surprisingly low. In products under $10 per unit, the ratio is less than 0.1. Even for a $1,000 product, it is 0.4.

- ❖ *No Trouble Found (NTF) Failures.* One of the perplexing frustrations for most companies is a category of field failures where 20–33 percent of the returned units seem okay, even after careful failure analysis. These are labeled "no trouble found" or "no apparent defect" (NAD), or "checks OK." In the past, companies used to dismiss such failures as their responsibility, blaming instead ignorant customers and quick-buck servicers and dealers with totally fictitious claims. In recent years, however, companies have taken a more enlightened position by declaring that customers must have had a good and sufficient reason for returning a failed unit. With the growth of electronic parts in more and more products, intermittent operations, where a failure can appear and disappear, are more prevalent. Progressive companies try "torture" tests on the returned product to convert an intermittent into a solid failure. One of the best torture tests is MEOST. Using it, I have been able to capture real failures on 50 percent of NTFs.

- ❖ *Subterfuge.* One cannot completely dismiss the fact that there is a small percentage of outright fraud perpetuated by customers, servicers, and dealers. While investigating the large number of returned units, one manufacturer of air compressors found that customers would buy the compressors, use them over the weekend, and return them to the dealer or retail shop as unsatisfactory. The phenomenon became known as "the weekend customer." One service station generated hundreds of service repairs under warranty with fictitious customer names and a terse analysis—"Don't work"—until he was caught.

CASE STUDY

Caterpillar, Inc.—A Benchmark Company in Field Service

Peoria-based Caterpillar, Inc., known all over the world for its tractors and earth-moving equipment, gained a reputation for meeting the Japanese challenge—number-two equipment maker Komatsu Ltd.—head-on and winning. What is less well known is its fantastic field service, the reliability of its products, and its partnership—no, marriage—to its dealers.

Reliability: Better Than the Best Automobile in the World

The much-vaunted reliability of the automobile ranges from eighty to 180 failures per 100 cars per year for an average of 400 hours of operation per year. In the lead is Toyota, with its Lexus and Camry models. This is roughly one failure per 400 hours.

Caterpillar, whose much more complex machinery operates continuously in a far more hostile off-the-road environment and is subjected to far more violent stresses, has achieved a failure rate of slightly more than two failures per 1,000 hours, beating the best automobile to the reliability punch.

Strategy to Serve Customers' Needs

Caterpillar's design strategy is to:

❖ Achieve a mean time between failure (MTBF) from 500 hours to 1,000 hours

❖ Develop microprocessor-based diagnostic capability in each tractor

❖ Redesign its products to permit faster modular exchange of electrical, hydraulic, and engine-driven train components

❖ Provide loaners to customers during repairs

❖ Improve fill rates for parts from 95 percent to almost 100 percent (Caterpillar guarantees a part to be sent to a customer anywhere in the world within forty-eight hours.)

Caterpillar's remanufacturing operations, which tears down and rebuilds field subsystems, prides itself in achieving even higher field reliability than its original manufacturing.

Caterpillar Dealerships—The Crown Jewels

Caterpillar's dealerships are, however, the pièce de résistance of its field operations. The reason for their success is that Caterpillar follows to textbook perfection its relationships with its dealer organizations:

❖ Dealers are exclusive to Caterpillar. They serve no competitor, nor do they deal with service parts from non-Caterpillar sources.

❖ Caterpillar limits its dealership numbers so that it can concentrate on truly serving them.

❖ Dealers perform vital final assembly, test, and turnkey operations.

❖ Dealers have close and intimate contact with Caterpillar customers. To a customer, a dealer is Caterpillar.

❖ The company is fantastically loyal to its dealers.

❖ The CEO has almost daily contact with dealers, knows them by name, and trusts them, so much so that he puts more faith in feedback from dealers than he does with internally generated reports.

As a result, Caterpillar has succeeded in building a legendary, loyal dealer network that, in turn, has generated a truly loyal customer culture all over the world.

SELF-ASSESSMENT/AUDIT ON FIELD OPERATIONS EXCELLENCE

Table 15-2 is a self-assessment/audit that a company can conduct to score its field operations. It has four key characteristics and fifteen success factors, each with five points, for a maximum score of seventy-five points.

Table 15-2. Field services: key characteristics and success factors (75 points).

Key Characteristic	Success Factors	Rating				
		1	2	3	4	5
10.1 Product Reliability	1. Multiple Environment Over Stress Testing (MEOST), derating, and failure analysis disciplines are fully utilized to achieve as close to the ideal of zero field failures as possible. 2. Mean time to diagnose (MTTD) and mean time to repair (MTTR) targets are met through built-in diagnostics, modularization, and ergonomics disciplines. 3. Product safety and product liability prevention disciplines have been factored into the design.					
10.2 Predelivery Services	1. Quality audits are conducted on the adequacy of: handling, packaging, packing, transport, storage, owner manuals, service manuals, and billing—well ahead of the consumer receipt of the product or service. 2. Installation at the distributor/dealer is made easy and simple with techniques similar to design for manufacturability (DFM). Similar instructions for assembly by consumers are made easy, simple, and mistake-proof. 3. Training, videotapes, warning labels, fail-safe mechanisms, and maintenance contracts prevent human errors on the part of consumers, where necessary.					
10.3 Services to Downstream Supply Chain	1. The principles of partnership to the company's downstream partners (e.g., commitment to partnership: ethics; trust; active, concrete help; mutual escalation of profits, etc.) are scrupulously followed. 2. Reductions of a large and unwieldy distributor/dealer/servicer base are in effect and this base is not shared with competition. 3. The company spends priority time with the downstream supply chain to build personal relationships to strengthen partnership. 4. The company forms joint councils with distributors, dealers, and servicers to improve and solidify the climate of partnership.					

| 10.4 Services to Users | 1. The vital importance of "frontline troops" is recognized and great attention is paid to how they are hired, trained, compensated, promoted, and empowered.
2. Management "walks the talk" with frontline troops and celebrates their contribution to customer loyalty.
3. Repair service is strengthened by:
❖ Companies establishing company-owned service stations
❖ Certification and rating of servicing personnel by the company
❖ Creation of exchange stations, floater units, and starter kits
4. Maintenance of product reliability is enhanced by:
❖ An emphasis on preventive maintenance rather than "breakdown maintenance"
❖ Ready availability of spare parts along with configuration control
5. The company estimates the approximate cost of customer defections and concentrates on winning customers back rather than using warranty costs alone as a field metric. | | | | | |

Sixteen

FROM THE BLACK HOLE OF LITTLE ACCOUNTABILITY TO SERVICE AS A PRODUCTIVITY CONTRIBUTOR

❖ ❖ ❖

The Fisher [representative] says: "Wait a minute. I did my job—fabricate a steel door and ship it to General Motors Assembly division (GMAD). It's GMAD's fault." So you go to GMAD and say: "Listen. One more lousy door and you're fired." He says: "Wait a minute. I took the Fisher

door and the car division's

specs. And I put them

together. So, it's not my

fault." So you go to the

Chevrolet guy and say:

"One more lousy door. . . ."

He says: "Wait a minute.

All I got was what GMAD

made." So, pretty soon

you're back to the Fisher

guy and all you're doing is

running around in great

big circles. . . .

—*FORBES* MAGAZINE

Storm Clouds in the Service Industry

The service industry now accounts for more than two-thirds of the gross domestic product of most industrial nations and for almost three-fourths of the workforce. Its value-added contribution is nearly ten times that of direct labor in manufacturing.

Yet, as Richard Quinn and Christopher Gagnon stated in a landmark article in the *Harvard Business Review*: "Daily we encounter the same inattention to quality, emphasis on scale economics, and short-term orientation in the service sector that earlier injured manufacturing." They add ominously: "If service industries are misunderstood, disdained, or mismanaged, the same forces that led to the decline of manufacturing stand ready to cut them to pieces."[1] The storm clouds are gathering. Consider the following:

❖ There are far fewer winners of the Malcolm Baldrige National Quality Award in the service sector than in manufacturing.

❖ A study by the Illinois Institute of Technology indicated that while blue-collar productivity in manufacturing has been consistently over 80 percent and rising, business (i.e., white-collar) productivity in office environments has been below 40 percent and falling.

❖ In quality, manufacturing has made steady progress. Yet, among the services that support manufacturing—from marketing to engineering, from accounting to personnel—quality is off their radar screen. Service quality is at least twenty to twenty-five years behind manufacturing quality.

❖ In manufacturing, cycle time and its synonyms—such as "just in time," zero inventories, stockless production, and lean manufacturing—have become household words. But cycle time in service industries is a foreign concept, relegated to another planet.

It is the objective of this chapter to elevate the importance of the internal customer—that is, the next operation as customer—in order to measure and greatly improve quality, cost, and cycle time in the service industry and in the business operations of a manufacturing industry. This chapter also briefly compares the relative merits of NOAC and business process reengineering (BPR) that has gained currency in recent years.

Objectives/Benefits of Next Operation as Customer (NOAC)

There are primary and secondary benefits to using NOAC.

Primary Benefits

❖ *Achievement of Customer (and Employee) Loyalty.* It is an article of faith that a company can only achieve customer loyalty if it succeeds in achieving employee loyalty. By moving the goal posts of the external customer into a close coupling with the internal customer, we achieve a synergy between the external customer, the internal customer, and internal suppliers—namely, the employees.

❖ *Workplace Satisfaction.* Chapter 9 indicated several ways to let the genie of employee empowerment out of the bottle and give employees the freedom and job excitement they so richly deserve. These ways to empowerment include open book management, self-directed work teams, and the minicompany. NOAC adds yet another dimension to that transformation—converting the boredom and frustration of white-collar employees to the joy of achievement in the workplace.

Secondary Benefits

❖ *Customer Knowledge.* There is a better knowledge of internal customers and their needs and of internal suppliers and their constraints.

❖ *Teamwork.* Departmental walls and functional silos are broken down. There is a refreshing change from win-lose contests to win-win teamwork, from firefighting to fire prevention.

❖ *Better Evaluations.* The company puts a much greater emphasis on internal customer evaluation than the obsolete concept of performance appraisal by a single supervisor.

The Discovering of NOAC by Serendipity

In the early 1980s, I visited Japan several times as a member of various study missions. We kept bumping into a phrase—the next operation as customer—an idea initiated by Dr. Kaoru Ishikawa, the father of the quality movement in Japan. But when I sought to explore how Japan's leading companies implemented this simple but elegant concept, I found that they had not made the transition from concept to practice.

Upon my return to the States, I started developing the principles and practices of NOAC in the business areas of my automotive and industrial electronics group at Motorola. NOAC had a successful launch. One reason for this was the inclusion of NOAC practices within the framework of Motorola's well-known participative management program (PMP)—an employee involvement initiative that's tied in with financial gain sharing. This financial incentive helped every PMP team put NOAC into effect.

Motorola University then took our NOAC development and packaged it to extend Motorola's famous Six Sigma process from a manufacturing and product focus to all white-collar operations. More than 50,000 Motorolans were trained in NOAC techniques. It is now mandatory that each administrative area in the company apply NOAC principles in its daily work. In the meantime, parallel developments have taken place in other leading companies. Terms such as *business process improvement, process management, mapping,* and *flowcharting* have entered our industrial vocabulary. It is to be hoped that just as DOE is becoming the centerpiece of quality, MEOST as the centerpiece of reliability, and cycle time as the centerpiece of manufacturing, so will NOAC become the centerpiece of all service operations, many of which are mired in antediluvian practices.

Basic Principles of Next Operation as Customer

Before developing a NOAC road map, its simple but powerful principles need to be enunciated:

1. *The Internal Customer as Prince.* It is universally recognized that the external customer is king. But the internal customer is given short shrift. In a logical customer-supplier flow, manufacturing is engineering's inter-

nal customer. In practice, however, engineering looks upon manufacturing as a second-class citizen. In NOAC, this attitude undergoes a metamorphosis. Manufacturing is now engineering's internal customer. If not a king, it is at least a prince. Manufacturing now measures and grades engineering performance on the basis of mutually agreed-on quality, cost, and cycle-time targets. The same applies to all administrative areas, each of which has a major internal customer.

2. *Process, Customer, and Supplier.* All work is a process. The process user receives input from an internal supplier—the previous operation—adds value to that input, and converts it into an output for the internal customer—the next operation. Figure 16-1 shows this relationship between the internal supplier, process user, and the internal customer.

3. *Internal Customers' Requirements, Measurement, and Feedback.* The supplier must assess the internal customer's requirements (or specifications) and reach a mutual agreement as to their validity and feasibility, consistent with the internal supplier's constraints and resources. (Lack of agreement can be kicked up to a higher "court"—the NOAC steering committee.) The method of measurement against the internal customer's requirements, the frequency of measurement, and the feedback mechanics from customer to supplier must be mutually determined.

4. *Consequences.* There should be a careful assessment of the consequences, both to the process user and to the organization, of meeting or not meeting the internal customer's requirements. This entails appropriate incentives and penalties. Incentives could include recognition, pay raises, and even promotion. Penalties could include poor performance appraisals, done by the internal customer (in addition to or in place of the boss), and/or no pay raises. The internal customer has the right to go

Figure 16-1. Next Operation as Customer: a schematic representation.

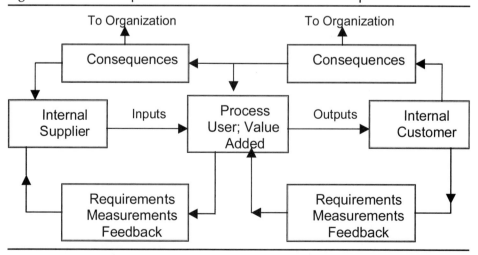

elsewhere within the company for the service or, in extreme cases, even outside the company for such service. (Sometimes, the mere threat of going outside is sufficient to shape up the internal supplier team.)

There are, however, pitfalls in assessing consequences. Often:

❖ Individual or team performance, either good or bad, is ignored.

❖ Poor performance is rewarded.

❖ Good performance is punished!

A company must be sensitive to this patent unfairness, because it results in a loss to the organization and depresses employee morale. Management must first research the system within which the supplier-customer link functions to make sure that the system is not the culprit rather than the individual or team. W. Edwards Deming asserts that most poor performance appraisals are the fault of the system, rather than the fault of the individual or the team.

5. *Continuous, Never-ending Breakthrough Improvement.* Measurements represent only a starting point, a baseline. The thrust of NOAC, however, is improvement—in quality, cost, cycle time, and effectiveness. There is no finish line to improvement. It is continuous and never-ending. However, a distinct feature of NOAC is out-of-the-box breakthrough creativity, where traditional and incremental approaches are cast aside and a revolutionary approach is put in place to achieve spectacular results.

6. *Employees as Partners.* In the final analysis, NOAC promotes partnership:

❖ First, between the internal customer and the supplier—who quickly find that interdependence and cooperation (rather than the dominance/subservience relationship of the past) is the key.

❖ Second, between management and the white-collar workforce. Having established overall goals and begun the monitoring of results, management gets out of the way, trusting employees and giving them freedom to innovate.

Typical White-Collar Services in Manufacturing Industries

It is easy to visualize relationships between external customers, internal customers, and internal suppliers in traditional service industries, such as banking, insurance, airlines, hotels, car rental companies, and a whole host of similar companies where there is no physical product generated. There are similar nonproduct services generated in manufacturing industries that are designated white-collar or business services. They are the focus of this chapter. Examples are:

Top management	Human resources	Finance/accounting
Marketing	Public relations	Information
Sales	Security	systems (IS)
Purchasing	Training	Product control
Inventory Control	Logistics	Maintenance

Internal Customer–Supplier Relationship Charts

Figure 16-1 was a schematic representation of NOAC, where the require-ments and measurements of the internal customer are fed back to the internal supplier, along with the consequences to the latter and to the organization. In actual practice, the relations between the internal cus-tomer and the internal supplier are more complex. For instance:

❖ There can be more than one customer, even among major cus-tomers.

❖ Often, there can be a role reversal, with the supplier becoming the customer and vice versa.

Figure 16-2 shows a typical internal-supplier relationship chart for the purchasing function. It has several customer-supplier links, some of them

Figure 16-2. Internal customer-supplier relationship chart for purchasing.

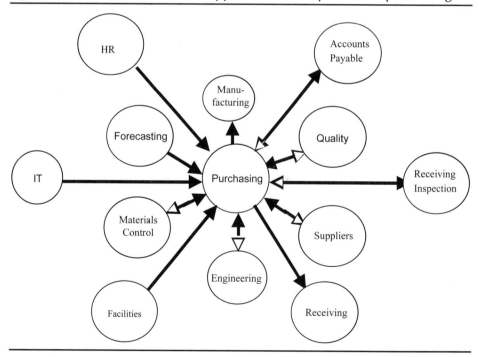

more important than others. An arrow represents each link, with the shorter arrows being more important than the longer ones. The black arrowheads indicate the purchasing department's customers, while the white arrowheads indicate purchasing as customer.

Several benefits accrue if each team or department draws such internal customer-supplier relationship charts. The team is able to:

❖ Develop a more comprehensive picture of a group's job.

❖ Identify key customers—and in many cases—reciprocal links.

❖ Clarify interrelationships that are not conveyed by traditional organization charts.

❖ Educate new employees.

❖ Clarify roles and responsibilities, with a focus on the external customer as king and the internal customer as prince.

❖ Move from local optimization to total optimization (as graphically portrayed in Figure 16-3). The first part of the figure, labeled A, optimizes each area's own interest. Skills are confined to a narrowly defined area of responsibility, while materials and information flow are disrupted because of people's narrow focus. The figure labeled B shows more understanding and coordination between teams/departments. Materials and information flow are smooth and linear because of people's customer-oriented approach.

An Organizational Framework for NOAC

NOAC is a process, not a program with limited goals or limited life. It needs organizational nourishment to sustain it through the vicissitudes of management changes, "program of the month" whims, and conflicting priorities. This organizational framework includes the following participants:

1. *Steering Committee.* Although NOAC can succeed without a top-level steering committee, initiatives taken solely at lower levels or by individu-

Figure 16-3. A comparison of local versus total optimization in NOAC.

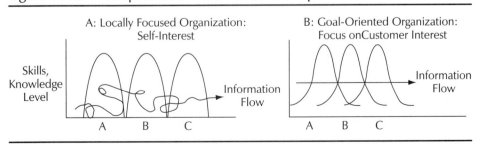

als, however, often fizzle out without the organizational lifeline of a steering committee. The steering committee should have the president of the company as chairperson and key members of the executive staff as members. A senior executive is selected as secretary. (At the division or plant levels, similar steering committees should be created and linked to the corporate committee.) The committee's tasks include:

❖ Drafting a mission statement applicable to all service (business) operations

❖ Defining critical business issues related to service (business) operations

❖ Establishing quantifiable goals for quality, cost, and especially cycle-time improvements, along with a timetable for their achievement

❖ Benchmarking other companies for best practices in service operations, determining their success factors, and developing plans to reduce the gap between the company and its benchmark

❖ Assigning a process owner to each key process

❖ Prioritizing and selecting a macroscopic cross-functional team for each key process

❖ Designing incentives/penalties for the process improvement teams

❖ Providing resources, training, and support to the teams

❖ Acting as a "supreme court" to settle lack of agreement and other controversies between internal customers and internal suppliers

2. *The Process Owner: Wiring the Organization Together.* To prevent problems and real external customer issues from falling through the organization cracks of vertical management, progressive companies have appointed a process owner for each cross-functional (or macroscopic) process. This person has the responsibility to assure that this "macro" process truly satisfies the end-customer and that suboptimization is avoided. An important task for the process owner is to select and lead a cross-functional team of people drawn from the appropriate functions.

3. *The Improvement Team.* Chapter 8 indicated that the team concept was the most effective form of organization, and it is the basic building block of NOAC. Whether the task is quality problem solving, value engineering, supply management, or white-collar operations, teams create a synergy where the whole is greater than the sum of the parts and where the bonds created between team members result in close, productive relationships. NOAC team responsibilities, under the guidance of the process owners, include:

❖ Determining major internal customers of a business process, their requirements, and measurements to gauge progress

❖ Flowcharting the current or "is" process, then moving to a value-added "should" process with a shorter cycle time, using improvement tools

❖ Using "out of the box" thinking for breakthrough improvements in quality, cost, cycle time, and effectiveness

An NOAC Road Map—The Improvement Cycle

With the above-mentioned organizational infrastructure in place, a ten-step road map for implementing NOAC in any business process is ready to be initiated. The ten steps are summarized in Table 16-1 and further elaborated on in the next sections.

Step 1: Selecting a Relevant Business Process/Service

The selection of a relevant business process or service is based on the priorities of attaining first maximum external customer loyalty, and second, the business priorities of the company. The selection is sometimes made by the steering committee or by the process owner–led team.

Table 16-1. The ten-step NOAC road map for business processes.

Step 1.	Select the business process/service with a compelling need for improvement and establish internal targets.
Step 2.	Identify the major internal customers of the business process/service.
Step 3.	Determine the requirements of the major customers. Reconcile the requirements of the major customers with the requirements and specifications of the external customer.
Step 4.	Reach agreement with the major customers on goals; the method of measurement to chart progress, feedback intervals; evaluation; and consequences.
Step 5.	Flowchart the entire "macro" process (i.e., an "is" chart).
Step 6.	Determine the typical cycle time for each step in the flowchart and the total cycle time.
Step 7.	Separate the value-added steps from the nonvalue-added ones.
Step 8.	Estimate the number of steps and the cycle time saved by eliminating the nonvalue-added steps.
Step 9.	Eliminate or reduce each nonvalue-added step using NOAC improvement tools and redraw the flowchart from "is" to "should."
Step 10.	Using "out of the box" thinking and creativity tools, formulate and implement a totally different and radical approach to providing the service, including even the elimination of such service.

Start by asking the same basic questions that a company as a whole asks itself regarding its business:

- ❖ What *is* our service?
- ❖ What *will be* our service?
- ❖ What *should be* our service?

Just as a company that fails to ask these questions of its business runs the risk of a narrow focus and loses its competitiveness, a NOAC team that fails to ask these questions of its service runs the risk of losing its relevance to customers.

The NOAC team should also formulate a mission statement as if it were an outside contractor wanting to renew its contract with its major customer within the company. Objectives, goals, strategies, tactics, and plans should follow.

Step 2: Identifying the Major Internal Customers of the Business Process/Service

As shown in Figure 16-2, there can be several internal customers for a service department or team, but only a few are the major customers on whom it must concentrate. Often, services pay scant attention to their real customers. As an example, I was consulting with a major university in the midwest about introducing Six Sigma into its operations, but the faculty could not accept the fact that their students were their customers, rather than being looked upon as "chattel."

Step 3: Matching Requirements of Major Internal Customers With Those of the External Customer

There are several considerations associated with customer requirements. They should be:

- ❖ Consistent and in harmony with the needs and expectations of the external customer
- ❖ Consistent with the corporation's objectives and goals
- ❖ Clear, firm, meaningful, and mutually acceptable to the internal customer and the internal supplier
- ❖ Comprehensive yet realistic and measurable

Consistency between internal customer's requirements and those of the external customer is very important, yet there are perils in getting there. As an example, a department generating service manuals was asked by its internal customer—a new product launch team—to reduce cost by reducing the number of charts, diagrams, and service hints. The chief customer officer of the company, touching bases with the company's major external customer, ascertained that completeness of servicing information was more important than cost reduction and vetoed the cost-cutting requirements of the internal customer.

Examples of internal customer requirements in service operations that can lead to customer "wow" are depicted in Figure 6-6 in Chapter 6.

Step 4: Reaching Agreement With Major Customers

It is possible that the internal supplier, because of resources, capital equipment, and time constraints, cannot meet some internal customer requirements. Usually, such "disconnects" can be mutually resolved by the customer and supplier. Sometimes, the steering committee is called in, acting as a supreme court. Agreement on goals is one objective. There must also be agreement on:

- ❖ Methods of measurement
- ❖ Feedback mechanisms
- ❖ Consequences

Measurement of Internal Supplier Performance

There are several measures by which an internal customer can assess the internal supplier's performance. These include:

- ❖ *Management by Objective (MBO)*. This is an old but worthy workhorse.
- ❖ *5-up and 10-up Charts*. This is Motorola terminology for charting progress in its manufacturing plants and its service operations worldwide by function/team. Table 10-3 in Chapter 10 on measurement is an example of a chart used to track quality, cost, and cycle-time effectiveness in several white-collar operations such as marketing, sales, engineering, purchasing (i.e., supply management), quality assurance, accounting, human resources, and information systems (MIS).
- ❖ *External Customer Complaints*. At best, this is a post-mortem.
- ❖ *Management Audits*: An actual management walkthrough of the process flow, from beginning to end, is an effective way to get firsthand experience of the difficulties employees face every day in performing their tasks.
- ❖ *Cost of Poor Quality*. Key elements of COPQ are listed in Figure 10-1.

❖ *Suggestions for Improvement Made by Support Service Personnel.* Companies measure the suggestion climate by counting the number of suggestions per employee per year. In manufacturing, the U.S. rate is a skimpy 0.1; in Japan it is over ten, as an average, with leading companies registering over fifty. The rate of implementation of these suggestions is less than 20 percent in the United States, but over 75 percent in Japan. Similar statistics are not even recorded in white-collar operations, both in the United States and in Japan.

❖ *General Statistics*: Several service operations use a variety of statistical measures to gauge NOAC progress. They include:

1. Ratio of number of employees in a given process to sales dollars generated by the process

2. Ratio of number of employees in the process to total number of company employees

3. Comparison of number of employees in the process to the number of employees in a similar process in a benchmark company

4. Impact of NOAC on business parameters (e.g., American Express, in sustaining process improvements in its customer service operations, increased its revenues by $17 million)

5. Industry reports such as those published by J. D. Powers and Consumers Union, as well as government regulatory agency reports to the media, among others [KB1]

Feedback From the Internal Customer

In the final analysis, the most important measure of NOAC progress is not what management thinks it is, but what the customer says it is. Suggested feedback techniques include:

1. *Internal Supplier Effectiveness Index (ISEI).* Chapter 10 developed a company effectiveness index (CEI) by which an external customer can evaluate a company's overall effectiveness vis-à-vis its competition. A similar internal supplier effectiveness index (ISEI) can be developed. Table 16-2 is an example of how the internal customer of a training project rated an internal supplier team that designed the training manual. The overall ISEI was a poor 53 percent. The team also received low scores for several individual requirements that the internal customer deemed important, especially meaningfulness, accuracy, and on-time delivery. The elegance of ISEI is the remarkable way in which it simultaneously analyzes:

❖ The relative importance internal customers attach to their priority requirements

Table 16-2. Internal supplier effectiveness index (ISEI) case study: training manual team rated by the development project manager.

Requirement	Customer Importance (I) Scale 1–10	Customer Rating (R) Scale: 1–5	Score S S = (I) × (R)
1. Quality			
❖ Comprehensiveness	7	4	28
❖ Accuracy	9	1	9
❖ Clarity	8	3	24
❖ Meaningfulness	10	2	20
2. Timelessness			
❖ (a) On-Time Delivery	8	1	8
❖ (b) Cycle Time	5	1	5
3. Cost (to Customer)	6	3	18
4. Dependability			
❖ Promises Kept	4	2	8
❖ Credibility	6	3	18
❖ Trustworthiness	7	2	14
5. Cooperativeness			
❖ Responsiveness	5	4	20
❖ Flexibility	4	3	12
❖ Approachability	7	5	30
❖ Courtesy	4	5	20
6. Communication			
❖ Listening	4	4	16
❖ Feed-forward Information	5	2	10
Total Score	Total (Y) = 99		Total (T) = 260

Internal Supplier Effectiveness Index $= \dfrac{T\%}{5y} = \dfrac{260\%}{99 \times 5} = 52.51$

- ❖ The resultant strengths and weaknesses of the internal supplier
- ❖ An agenda for improvement before the next feedback and evaluation

2. *Image Survey.* Another measurement instrument is the image survey, where the internal customer lists its major suppliers in priority order, along with the requirements for service from these internal suppliers in quantified terms. Table 16-3 is an example of a parts and service operation scoring its three internal suppliers—purchasing, engineering, and publications.

3. *Feedback From Several Major Customers.* Table 16-3 is an example of one internal customer evaluating three major suppliers. Table 16-4 is an example of a single supplier evaluated by its major customers. In Table 16-4,

Table 16-3. Image survey by an internal customer—a scorecard.

Internal Customer: Parts and Service

Internal Supplier	Customer Requirements	Supplier Performance: Actual 1st Qtr	2nd Qtr	3rd Qtr	4th Qtr	Image	Score*
1. Purchasing	Parts available (within 1 hour)	2 days	1.4 days	1.1 days	7 hrs	Rate of progress too slow	1
2. Engineering	Fix for repetitive problems (1 month)	2 mos	2 mos	1.3 mos	1.25 mos	Rate of progress fair, but not sufficient since engineering delays other Parts and Service suppliers	2
3. Publications	Field bulletins each month	1.5 mos	1.75 mos	1.25 mos	1 mo	Service stations pleased with progress	3

*Index for Score:
1 = Well below expectations; progress too slow.
2 = Below requirements; progress fair.
3 = Meets requirements; progress good.
4 = Above requirements; progress superior.
5 = Well above requirements; progress excellent.

the supplier is a team responsible for developing a participative management process (PMP) as part of the company's gain-sharing initiative. Its functions include information assistance, feedback surveys, and training to all PMP participants and to the PMP council task force. The requirements of these customer constituencies are timeliness, accuracy/completeness, cooperation, responsiveness, guidelines/options, and alternative and overall effectiveness. The internal customers use a rating scale of one to three. The scorecard shows the strengths and weaknesses of the PMP administration, the latter involving training in general and participative problem solving in particular.

Feedback Frequency

The frequency of feedback from customer to supplier varies with the complexity of the process, the urgency of the improvement, the difficulty of measurement, and the interpersonal relations between customer and supplier. The usual period is once a month. In the early stages of NOAC, some companies even have customer-supplier meetings each day for ten to fifteen minutes until they develop a more practical modus vivendi. In NOAC, as in most business dealings, integrity, mutual help, and trust form the solid foundations upon which the superstructure of partnership can be built.

Table 16-4. Measurement of an internal supplier—administration of participative management.

Rating Scale
3 Exceeds Requirements
2 Satisfactory
1 Needs Improvement
N/A Not Applicable

Service Provided by Priority Order by the Department Administering the Participative Management Program O/(PMP)

Definitions for Measurement Criteria

Timeliness. Ability to meet required service within agreed upon time

Accuracy/Completeness. Service rendered is within acceptable limits of user's needs and covers all requirements

Cooperation. Willingness and ability to work well with others

Responsiveness. Supportiveness, Approachability, Communicability Guidelines/Options

Alternatives. Service provider offers guidelines for user to follow, suggesting options or alternatives to complete a given task

Overall Effectiveness. A summary and integration of the above parameters

		Timeliness	Accuracy/Completeness	Cooperation	Responsiveness	Guidelines Options Alternatives	Overall Effectives
Information/Assistance	Group Staff Support	N/A	N/A	N/A	N/A	N/A	N/A
	Steering Committee Assistance	2.0	2.0	3.0	2.75	2.3	2.3
	Goal Committee Assistance	3.0	2.5	2.9	2.9	2.9	2.9
	Improvement Team Assistance	2.9	2.5	2.7	2.0	1.5	2.5
	Assistance During PMP Implementation	2.6	2.6	2.9	2.8	2.6	2.8
	Assistance in Culture Progression	2.5	2.5	2.5	2.5	2.3	2.5
Feedback Survey	Survey Results Preparation	2.9	2.6	2.8	2.9	2.4	2.7
	Survey Analysis	2.8	2.7	2.9	2.8	2.4	2.8
	Manager Briefing	2.8	2.6	2.8	2.8	2.7	2.9
	Results Management Support	2.6	2.76	2.9	2.9	2.6	2.7
Training	Training Classes	2.2	2.6	2.6	2.6	2.2	2.7
	Training Schedule	2.0	2.0	2.1	2.1	2.0	2.0
	Training Status Report	1.6	1.9	2.0	2.1	2.0	2.0
	Participative Problem Solving Training	N/A	N/A	N/A	N/A	N/A	N/A
Council Task Force Representation	Representation	2.6	2.0	2.5	3.0	2.5	2.8
	Knowledge of Subject	N/A	3.0	N/A	2.0	3.0	3.0
	Level of Participation	3.0	N/A	2.5	2.0	3.0	2.5
	Pilot Program Support Followup	3.0	2.5	3.0	3.0	2.5	3.0

Overall effectiveness of PMP department

Comments

Consequences

Consequences to the internal supplier and to the whole organization must be carefully crafted before the work of NOAC can begin, preferably by the steering committee. Table 16-5 is a list of consequences that can be considered for the supplier either performing or not performing to customer requirements and the impact of each on the organization.

Step 5: Flowcharting the "Macro" Process

The definitions for the flowchart include:

Term	Definition
Macroprocess	A process that cuts across departmental, business, or divisional boundaries.
Microprocess	A process confined to a department's or group's jurisdiction.
Block diagram	A diagram that traces the various paths that materials, paperwork, and information flow between suppliers and customers.

Table 16-5. Consequences of an NOAC supplier performing/not performing to customer requirements.

Consequences	Performing to Customer Requirements	Not Performing to Customer Requirements
To Supplier	❖ Recognition/commendation ❖ Positive performance appraisals ❖ Job enlargement ❖ Bonuses/merit raises, and gain sharing ❖ Promotion	❖ Counseling, coaching, and training ❖ Negative performance appraisals ❖ Job redesign ❖ No merit raises or bonuses ❖ Task transferred to another operation: 1. Within company 2. Outside the company*
Impact on Organization	❖ External customer satisfaction/loyalty enhanced ❖ Enthusiastic, productive employees ❖ Higher profits/ROI/market share ❖ Premium pricing possible	❖ External customer dissatisfaction ❖ Poor employee morale ❖ Employees nonparticipatory ❖ Lower profits/ROI/market share

*Note: Often, the threat of the customer going outside the company is enough to get the supplier in shape.

Flowchart   A pictorial representation of the detailed steps by
                      which a process works; it can uncover potential
                      sources of trouble sometimes called
                      "disconnects" or "white spaces."

Mapping   Another term for a block diagram or flowchart.
                      There are three types of maps or charts:[2]
                      1. The "is" chart (the current process flow)
                      2. The "should" chart (an improved process flow
                         implemented by the NOAC team but without
                         the necessity for additional resources)
                      3. The "could" chart (an improved process flow
                         requiring additional manpower or capital
                         resources)

Flowcharting is a key tool in NOAC. Workers who have been involved with an administrative/business process for years are amazed at how much flowcharting reveals that they did not know. It is important to verify the flowchart by actually walking through the process. The NOAC team can then determine whether:

❖ The flowchart can be changed from "is" to "should."

❖ The flowchart is acceptable, but its execution is poor.

❖ The process is poorly managed.

Step 6: Determining the Cycle Time for Each Flowchart Step and the Total Cycle Time

In Chapter 14, cycle time was described as the great integrator because it can be used as a single metric to measure quality, cost, delivery, and overall effectiveness. Because of this integrating power, it is used by several companies (starting with Motorola) as the only metric needed for any business process/white-collar operation.

The number of steps in flowcharting may vary from less than ten to several hundred. It is necessary to determine the typical cycle time for each step. (Some teams use three estimates—pessimistic time, optimistic time, and realistic time.) The total cycle time is the sum of the typical or realistic time of each step.

Step 7: Separating Value-Added Steps From the Nonvalue-Added Ones

It is surprising how many white-collar people are unaware of nonvalue-added factors in any business process. They include:

Waiting (queue) time

Setup/changeover time

Transport time

Downtime (equipment)

Inspection time

Priority shuffling

Shortages (material/manpower)

Rejection/rework time

Approval time

Poor job redesign

Many nonvalue-added steps in production, depicted in Figure 14-4, are also applicable in service operations.

Step 8: Estimating Cycle Time Saved by Eliminating Nonvalue-added Steps

As a rule of thumb, at least 60–75 percent of the total number of steps in a flowchart are nonvalue-added. Furthermore, they account for at least 50 percent of the total cycle time. These steps, therefore, can either be eliminated or substantially reduced. Many companies make the mistake of spending an inordinate amount of effort in mapping. However, as will be pointed out in step 10, a totally revolutionary approach to the business process may make mapping obsolete altogether. Hence, use mapping, but only as a fast preliminary exercise.

Step 9: Using NOAC Improvement Tools to Redraw the Flowchart From "Is" to "Should"

NOAC principles, organizational structure, and teams are important. Measurement is important. But as seen in Chapter 11, unless our good people are given tools—powerful tools—process improvement is likely to be modest and temporary. There are two types of process improvement tools—the more traditional tools, generally used in step 9, and the creative, breakthrough tools (see step 10). Each has been given a college-type grade from D minus to A plus.

❖ *The Seven Tools of Quality Control (QC)*. The Japanese have long used a collection of quality tools. In production, the following tools are helpful at a kindergarten level. In services they are slightly more useful:[3]

PDCA (plan, do, check, and act)	D minus grade
Check sheets	D grade
Bar graphs and charts	D grade
Histograms and frequency distributions	C grade
Pareto charts	B minus grade
Cause-and-effect (Ishikawa) diagrams	B minus grade
Control charts	C minus grade

Only two of these—Pareto charts and cause-and-effect diagrams—are worth mentioning:

1. *Pareto charts* separate the vital few causes (20 percent or less by numbers, but 80 percent or more by effect) of a problem. They are useful in service work to prioritize actions.

2. *Cause-and-effect diagrams* are useful in developing a laundry list of the causes of a problem, but if there are too many such causes, the team can be bewildered about where to start an investigation.

❖ *Statistical Process Control (SPC)*: C minus grade. In production, SPC—another name for control charts—is only useful for monitoring a product whose variation has been greatly reduced through DOE. In services it is even less useful.[4]

❖ *Brainstorming, Nominal Group Technique, Multi-Voting*: B grade. Brainstorming is an old technique to tap the creative thinking of the NOAC team to generate and then distill a large list of ideas and solutions. It is a useful starting point for improving a business process. One variant is a nominal group technique, where the team establishes a priority or rank order for each idea. Another variant of brainstorming is getting each team member to vote on each idea. The ones with the most votes are selected. The process generally yields two to five ideas.

❖ *Benchmarking*: B plus grade. The virtues of benchmarking as a problem-solving tool in manufacturing and in product design were extolled in Chapter 11. But benchmarking is equally useful in service operations. There is always out there, somewhere, a best-in-class company whose proficiency and success factors in business processes can be studied and emulated by a company. Benchmarking can offer quick results vis-à-vis the slower longitudinal progress an organization can make on its own steam.

❖ *Technology*: *Going Beyond Silicon and Software*: B grade. There is probably a more general understanding of improvement in support services through technology breakthroughs than of the other tools discussed in this chapter. Computers and the Internet has revolutionized white-collar operations even more than it has manufacturing. Yet, heavy investments in information technology have not resulted in the hoped-for bonanzas. Computerizing a business process to make it efficient, when it can be seriously redesigned or eliminated altogether, is worse than useless. Instead of embedding business processes in silicon and software, we should dismantle them, using creative approaches (discussed later in this chapter) to achieve the needed breakthroughs in quality, cost, and cycle time.

Step 10: Using Out-of-the-Box Thinking and Creativity Tools to Implement New Approaches to Providing Service

Most service improvement efforts confine themselves to mapping and cycle-time reduction. That approach can and does result in, say, a 50 percent reduction in cycle time. But it still represents thinking "within the box," where the NOAC practitioner draws a self-imposed boundary of conventional measures of improvement.

What is needed is a free flow of creativity, of radical ideas to achieve breakthrough improvements. Step 10, therefore, concentrates on these highly creative tools, as distinguished from those in step 9. These truly creative tools for implementing step 10 are:

❖ Total value engineering
❖ Design of experiments
❖ Process redesign

1. *Total Value Engineering (TVE)*: A plus grade. Chapter 11 discussed total value engineering as an alternative to traditional value engineering in product work. Traditional value engineering concentrates on cost reduction only, without lowering quality. Total value engineering, by contrast, concentrates on maximizing customer loyalty, even if it means incrementally added costs. Total value engineering has equal merit in services as a powerful creative tool. Its algorithm follows:

❖ What is the service?
❖ What does it do and what does it cost?
❖ Who is the customer?
❖ What else would do the job that would "delight" the customer?
❖ What will that cost?

Total value engineering (TVE) is disarmingly simple. Yet, applied to any business process or service, it packs a powerful punch. A cardinal rule is to challenge everything: Question every rule, every procedure, every system, and every assumption. It asks the first question: "Why is the process or step necessary?" It then questions the response with another "why," going five "whys" deep, probing, digging, and challenging. The creativity of TVE can raise questions such as:

- ❖ Can the process/procedure/system be eliminated altogether? (This is a zero-based approach.)
- ❖ Can it be substituted with a totally different and radical approach?
- ❖ Can the final customer's requirement be exchanged for a "wow"—an unexpected "delight" feature?
- ❖ If the process/procedure/system cannot be eliminated, can it be simplified, minified, or combined with another?
- ❖ Can it be executed in parallel with a previous process—instead of in series—to save time and costs and preventing hand-off discontinuities?
- ❖ Can it be performed better (with greater customer enthusiasm) by another department or team or even by another company?

2. *Design of Experiments (DOE) in Service Work*: A plus grade. The application of this powerful quality tool in production and in design has been amply detailed in Chapters 11 and 14. My research in the last ten years has been extended to the application of the Shainin/Bhote DOE to business processes, services, farms, schools, hospitals, and government. DOE methods include: multi-vari, components search, paired comparisons, B versus C, and scatter plots.

- ❖ *Multi-vari charts* break down a large and unmanageable list of problem causes into smaller and more manageable families of related causes. The most typical families are time-to-time, unit-to-unit, or within-unit causes. The multi-vari acts, therefore, as a problem filter, eliminating the unimportant causes. The technique finds numerous applications in administrative work and in all types of services. As an example, a large hotel was dissatisfied with pro forma guest responses to its customer satisfaction surveys. It decided to conduct in-depth personal interviews with its guests. It used a multi-vari stratification of ten categories of guest concerns:

Approaches/access to hotel parking
Front desk
Concierge
Room amenities (housekeeping)
Room service (food/beverages)

Restaurants

Hotel facilities

Business services

Climate of caring

Unexpected/unanticipated experiences

The sample size was 200 guests. In the free-flowing interviews, each category was given an importance of one to three by the customer, who then rated the hotel on a scale of one to five for each category. Two categories—the climate of caring and the unexpected/unanticipated experiences—resulted in the hotel's concentration on these "delights," with an accompanying substantial increase in occupancy rates and in repeat customers.

❖ *Components search* requires a good assembly and a bad assembly, with the ability to switch parts from the good to the bad and vice versa. The problem then follows either the switched part of the rest of the assembly. Components search can be used to differentiate between two dealers, two installers, two servicers, two clerks, two bank tellers, two branch managers, and so on, as long as there is a notable output (Green Y) difference between one who is good and the other who is poor. The process steps can be documented or switched sequentially from good to poor to determine the culprit cause (the Red X). As an example, a company scored its field servicers on the basis of customer ratings. It found a big difference between two of its service personnel in the same product in the same customer installation. The results came as a surprise because the servicer with the poor rating was considered faster and more accurate than the one with the higher rating. Further investigation revealed that the higher-rated servicer was friendlier, took the time to explain work to the client, and built personal relationships of a lasting nature.

❖ *Paired comparisons* is used when a product cannot be disassembled and reassembled without damaging or destroying it. The same technique can also be applied to two sets of human factors—say, in personnel work, in sales/marketing, in farms, in pharmaceuticals, and in many nonproduct applications. In fact, paired comparisons is one of the most versatile tools in NOAC, just as it is in product work. As an example, a personnel manager wanted to determine the reason her temporary employees had a high turnover rate, costing the company $800,000 per year. The temporaries had the same fringe benefits (as the smaller percentage that constituted their permanent employees). Further, no one had ever been laid off in the fast-growing company. A paired comparison, with eight temporary employees who had quit the fastest and eight temporary employees who had stayed the longest (three years and more), was conducted using the following parameters:

Commuting distance to work

Public transportation use

History of turnover with other companies

Level of education

Perceived quality of supervision

Disruptions with rotating shifts

Perception of pay relative to outside job

Perception of treatment compared to that of permanent employees

A Tukey test (see Chapter 14) of these eight parameters revealed that perception of treatment versus that for permanent employees was the Red X cause (end count of 7, with 95 percent confidence) and that commuting distance was the Pink X cause (number 2 cause; end count of 6, with 90 percent confidence). The company could not legally discriminate on its hiring because of commuting distance, but it found that perceived second-class treatment was based on the outlook of several managers who needed an "attitude correction."

❖ *B versus C,* as explained in Chapter 12, determines if an improvement in a "B" (for better) product over a "C" (for current) product was truly effective and permanent. B versus C is also an extremely versatile tool in administrative areas—to determine which of two advertising campaigns, two sales promotions, two personnel policies, two customer questionnaires, and so on, is better. The list of applications is endless. As an example, a company was considering two types of gain-sharing plans—B and C—for its employees. Management wanted to know which plan the employee would favor, without having to do lengthy surveys. Six employees were selected at random and the details of both plans explained. All six chose plan B over plan C. This gave management a 95 percent confidence that plan B would be the preferred choice of the rank-and-file employees. I have used B versus C as "my poor man's Gallup poll" on public issues with 95 percent confidence in the outcomes.

❖ *Scatter plots* can be used in service work to compare the relationship between two parameters, as each is varied over a range of levels, and to determine the correlation between the two.

Figure 6-2 in Chapter 6 depicted a scatter plot (or correlation study), where the correlation between customer satisfaction and customer loyalty was decidedly poor (a fat parallelogram), whereas Figure 6-2 showed a remarkably close correlation between customer loyalty and a company's profitability (a thin parallelogram).

3. *Process Redesign: Exploding Taylorism—Twice:* A plus grade. Process redesign is the equivalent of exploding Taylorism—twice. Let us look at how process redesign works in two cases: dismantling the assembly line and reducing support services.

❖ *Use in Dismantling the Assembly Line.* Chapter 8 was an eloquent testimonial to the evils of Taylorism on the production assembly line. What is not generally recognized, however, is that the "assembly line" syndrome is alive and well even in businesses processes. Each clerk, each white-collar operator, moves a limited task forward to the next operator, with only a tunnel view of the overall task and of the client or customer. This hand-off practice promotes errors, delays, discontinuities, and white spaces within the same department and between departments. In addition, the process gets cast in concrete with rules and regulations imposed by a stifling bureaucracy. Instead of each person performing a few steps in the process chain, assembly line fashion, one person can perform all the steps in the entire process. In fact, the same implementing concepts described in job redesign for manufacturing in Chapter 9 (Figure 9-3)— namely, combining tasks; forming natural work units; establishing client relationships and vertical job enrichment; and opening feedback channels—can, with equal validity, be applied to business processes.

As an example, a large life insurance company required thirty steps—involving nineteen people and five separate departments (e.g., credit checking, quoting, underwriting, and so on) to approve customer applications. The total cycle time ranged from five to twenty-five days. (The theoretical cycle time, that is, the actual working time, was only seventeen minutes!) The company completely revamped its approval process. It created case managers who could perform all the tasks for an insurance application. They were supported by powerful PC-based workstations, along with an expert system that acted as a coach. The results: Cycle time was reduced to two to five days. The company eliminated 100 field positions, despite a doubling of the applications volume.

❖ *Use in Reducing Support Services.* Process redesign can also be useful in reducing support services. It is well known that there are seven support service personnel for every production line worker. That is bureaucracy raised to the power of seven. Intelligent NOAC reverses the trend by:

❖ Giving inspection and test back to line workers, instead of an overhead quality function

❖ Giving cost accounting back to manufacturing to do its own cost accounting with its own computers and expert systems, instead of relying on a separate cost accounting department

❖ Disbanding production control and planning, with their cumbersome MRPII systems, and converting a cumbersome "push" system into an automatic "pull" system with Kanban

❖ Eliminating or drastically reducing the need for forecasts by sales departments by reducing cycle times from weeks and days into hours and minutes

❖ Disbanding the need for a separate MRO section within purchasing by giving departments authority to order inexpensive items directly with a simple credit card system

❖ Pulverizing accounts payable operations by cutting checks directly to partnership suppliers, based on quantities shipped by manufacturing

In short, support operations (such as quality assurance, accounting, purchasing, human resources, and others) should become one-or two-person operations—with highly-qualified people, who can command respect as coaches and helpers, not bloated control cops, information carriers and paper shufflers.

CASE STUDY

The Departmental Annual Budgeting Process

This is a time-worn ritual each year in thousands of companies: Each department is asked to prepare a budget for the next year. The submitted budget then travels in circles from each level in the hierarchy to the next higher level and back, until the budget is finally approved thousands of dollars and many months later. On a national scale, this antediluvian practice wastes billions of dollars.

The conventional solution would be to flowchart (i.e., map) this budgeting process and eliminate or shorten non-value-added steps, such as the ritualistic approval/disapproval cycles in the management chain. Automating the process using parallel steps and the concept of zero base budgeting could also shorten the cycle time. At best, these measures could cut the total cycle time from, say, six months to three months.

NOAC out-of-the-box thinking challenges the need for the whole kit and caboodle process. With a few of my clients, I introduced the following plan for redesigning the departmental budgeting process:

* The president or general manager determines the overall budget, based on sales forecast for the next year, tempered by a 5 or 10 percent factor of safety to allow for forecast inaccuracies.

* Each department is given a percentage increase/decrease over the previous year, again with a further factor of safety.

* Only a few exceptional departments, with unavoidable increases or decreases, are reviewed.

* There is no need to budget for subaccounts. The department manager is given total authority/responsibility to live within the total budget.

The result was that the budgeting process was reduced to one week! If forecasts change in the succeeding months, the percentages are automatically changed accordingly.

Business Process Reengineering (BPR): Fad or Breakthrough?

Business process reengineering was first introduced by Michael Hammer and James Champy in the early 1990s in their well-known book *Reengineering the Corporation*.[5] The advocates of BPR claim that the range of cost savings as a percentage of total business unit costs go from 13–22 percent. Its detractors claim business success, but at the expense of grave employee concerns over layoffs. Its objectives and methodology are similar to NOAC, but BPR:

* Places greater emphasis on dismantling the departmental organization structure and converting it to a totally process-oriented structure

* Totally transforms the way people are hired, trained, evaluated, and promoted

❖ Includes a greater commitment by the CEO to lead the change, especially in terms of actual time spent in BPR involvement

BPR Mistakes and Pitfalls

The history of BPR has not been encouraging. Three out of four programs have failed. Only 5 percent have succeeded. There are several reasons:

❖ The result, if not the emphasis, has been on downsizing, restructuring, TQM, and automation rather than helping all employees to move forward together.

❖ There has been a lack of senior executive sponsorship.

❖ Only 5 percent of employee time (as opposed to 50 percent) is spent on BPR.

❖ BPR looks for quick results: one year instead of three to five years.

❖ It targets nonstrategic businesses of little value to customers.

❖ It uses technology—that is, automation—as a substitute for process redesign.

Commonalties Among Successful BPR Companies

There are, however, notable successes where the dos and don'ts of BPR have been followed. These successes include:

❖ "Assembly lines" dismantled and total responsibility transferred to one person

❖ Vertical managerial hierarchy discarded, with workers making most of the decisions

❖ Parallel work instead of series, linear sequencing

❖ Triage concept of process work—one for simple, one for medium, and one for difficult processes—rather than one complex standard process

❖ Checks and controls reduced, with bureaucratic micromanagement replaced by trust

In conclusion, BPR appears to be too radical a structural metamorphosis from the departmental organization (prevalent for the past hundred years) to the enlightened process-oriented team organization that it rightly advocates. It also sets in motion a cultural metamorphosis that companies, mired in Taylorism, find difficult to implement. NOAC, on the other hand, is a simpler, easier, and less threatening approach to transforming services. Therein lies its appeal.

CASE STUDY

Solectron Corp.—A Benchmark Company in the Area of Service Operations

Solectron, a contract manufacturer for several Fortune 500 companies, is headquartered in Milpitas, California. It has the unique distinction of being the only company that has won the Malcolm Baldrige National Award twice in the twelve-year history of the prize. Its business performance has been stellar:

- ❖ Its stock rose 525 percent in three years.
- ❖ It has a compound annual growth rate of more than 50 percent.
- ❖ Its return on equity doubled—from 11.1 to 22.2 percent in five years.
- ❖ Its sales/employee increased 2.5:1 in four years.
- ❖ Its profit per employee rose from $2,700 to $8,000 in four years.
- ❖ Its sales per square foot went up from $180 to $390 in two years.
- ❖ Its profits per square foot increased from $10 to $24 in two years.

Solectron's devotion to the customer—both external and internal—is legendary. Every week it holds a meeting, presided over by its CEO, for about a hundred of its senior management, where each team/department presents the score it receives from its external and internal customers on alternate days. The meeting lasts only an hour, with each team allowed five minutes on the average to flash customer scores, called the customer satisfaction index (CSI). Quality and delivery performances are objective measurements; communication and service are subjective. Grading is on a severe scale. The only acceptable response is "A" (100) for maximum satisfaction. A grade of A minus (90) is lower than the corporate goal of 97. Grades of C and D are worth zero and minus 100, respectively. Only the poor scores are grilled by management at these meetings.

Solectron has a ten-step NOAC improvement process called quality improvement process, or QIP, for each of its

> business processes where several parameters are tracked. Table 16-6 gives examples from its sales, human resources, finance, information systems (IS), and safety/telecommunications departments.
>
> The stock market may go up. The stock market may go down. But the odds are that you will always win with Solectron!

SELF-ASSESSMENT/AUDIT ON SERVICE OPERATIONS EFFECTIVENESS

Table 16-7 is a company self-assessment/audit of its service operational effectiveness. It has three key characteristics and fifteen success factors, each worth five points for a maximum score of seventy-five points.

Table 16-6. Tracking business improvement parameters at Solectron.

Sales	Human Resources	Finance	MIS	Safety/ Telecommunications
❖ Internal customer satisfaction index (CSI)	❖ Internal CSI	❖ Internal CSI	❖ Internal CSI	❖ Accidents
	❖ Compensation competitiveness	❖ Receivables (days outstanding)	❖ Training hours	❖ Lost time
❖ Sales actual versus projection	❖ Benefits competitiveness	❖ Payables (days outstanding)	❖ Errors/line code	❖ Safety audits
❖ Quotes won/lost	❖ Benefit costs	❖ Number of reworked journal vouchers		❖ Safety training
❖ Quotation cycle time	❖ Employee opinion survey	❖ Month-end close cycle time		❖ Environmental compliance
	❖ On-time filling of positions	❖ Forecast accuracy		❖ Telephone response time
	❖ First-pass yield on interviews	❖ Order entry verification		
	❖ Employee turnover			
	❖ Late reviews of performance appraisals			

Table 16-7. Services: key characteristics and success factors (75 points).

Key Characteristics	Success Factors	Rating				
		1	2	3	4	5
16.1 NOAC Principles	1. Every business process has an internal customer and an internal supplier, with the former elevated to a status almost comparable with the external customer. 2. The internal customer formulates requirements in terms of quality, cost, cycle time, dependability, cooperativeness, etc. These requirements must be fair, meaningful, and mutually acceptable to both the internal and external customer in advance. 3. The internal customer, not the supervisor, becomes the main performance appraisal evaluator of the internal supplier. 4. The consequences of such appraisals both to the internal supplier and to the organization are clearly defined, including salary increases, bonuses, penalties, and task transference from supplier to another department/team or even to outside the company. 5. There is a continuous and never-ending improvement in all business processes. 6. NOAC knits all service operations together, generating team synergy and morale.					
16.2 NOAC Structure	1. There is an active steering committee of senior executives to lead, guide, and measure the effectiveness of all service operations. 2. Cross-functional teams are established to improve major business processes selected by the steering committee. 3. The teams are guided by process owners who monitor the preeminence of the external customer and smooth relationships between teams 4. Each team carefully constructs internal customer-supplier relationship charts (see Figure 16-2). 5. The assembly line concept has been dismantled and the concepts of process redesign are implemented.					

	6. The ways people are hired, evaluated, compensated, and promoted are radically changed to create a culture where people move toward "owning a piece of the action" (see Chapter 8).					
16.3 NOAC Implementation	1. The ten-step road map (Table 16-1) has been established and followed as a guide for all business improvement. Flowcharting is detailed and walked through to determine total cycle time—the principal metric in NOAC—to craft the "is" chart. 2. Nonvalue-added steps are eliminated or drastically reduced to craft the "should" chart. 3. Out-of-the-box thinking is employed to achieve breakthrough improvements using the creative tools of: ❖ Total value engineering ❖ Process redesign ❖ Job redesign ❖ Design of experiments ❖ Drastic reduction of service operation headcounts					

Seventeen

FROM MEDIOCRITY TO WORLD-CLASS RESULTS

❖ ❖ ❖

Never mind your golf swing
. . . let's see where your ball
lands. . . .

—CARL LINDHOLM,
MOTOROLA VICE
PRESIDENT

A Seesaw Battle Between Process and Results

Many management gurus decry senior executives paying attention only to formulating goals and measuring results but ignoring the all-important in-between process by which their people translate goals into results. As a consequence, the pendulum seems to be swinging the other way. It has become almost fashionable to ignore results. ISO-9000, QS-9000, and the Malcolm Baldrige National Quality Award guidelines pay little attention to results. Only the more recent European Quality Award has listed results as an important category in its guidelines. (The more recent Baldrige guide-line revisions, especially in 2001, do include results as the final category.)

Granted, results are effects, not causes; outputs, not inputs. Nevertheless, results do represent the final proof of achievement. They cannot be explained away. They draw a firm line in the sand. They constitute the bottom line.

The previous chapters concentrated on process initiatives. Chapter 10, in particular, concentrated on measurements—leading versus lagging indicators; weak versus robust indicators; generic versus specific measurements. It produced a harvest of metrics, but with the admonition that too many measurements mean no measurement. This last chapter focuses on results—in two parts. The first is on the primary factors: few in number, but vital in measuring the success of a business enterprise. The second is on secondary factors that embellish and circumscribe the primary factors. (These few primary factors and the larger secondary factors are intended as guidelines. A company can select its own primary and secondary factors from these lists.)

The Primary and Secondary Factors That Assess the Success of a Business

In the final analysis, there are four primary constituencies on which a business must concentrate in terms of results—customers, leadership, employees, and last, but not necessarily the least, financials.

1. *Customers.* In terms of results, the customer takes the top priority. Without customers, there can be no company, no results. Companies that worship only the god of profit wind up with neither profit nor customer loyalty. While there are several types of customers, as detailed in Chapter 6, only the external customer is highlighted in the primary parameters. Others are included in the secondary parameters.

Four primary factors are key to customer results:

❖ *Core Customer Loyalty.* Their retention rate as a percentage of the total customer base is crucial for the success of a company. Yet more than 90 percent of companies are either totally ignorant of their customer retention rate or its direct contribution to profit. Worse, they are incapable of measuring it.

❖ *Value-Added to the Customer.* A company's primary objective is to add value to the customer, as perceived by the customer. If that value added is insufficient, customer could and should change suppliers or take on the product/service himself. Yet very few companies attempt to define their value to the customer, and very few customers quantify the value they have received.

❖ *Customer Base Reduction.* Not all customers are worth keeping. In fact, the mark of smart companies would be to give their poor customers—the bronze and tin customers—to the competition. The concentration decision, as Peter Drucker calls it, enables a company to serve its core customers—the platinum and gold customers—much more effectively and profitably.

❖ *Public Perceptions of the Company.* When all is said and done, a company grows in stature or fades away based on the perceptions of the public-at-large about the company as a good employer and as a responsible citizen (the latter in terms of the environment, safety, freedom from lawsuits or unfavorable publicity, and social responsibility).

2. *Leadership.* Next only to customers, leadership is the most important criterion for results. In this first decade of the twenty-first century, more and more companies are earnestly trying to define the characteristics required for determining the leadership potential of their senior staff and for making the much-needed metamorphosis from management into leadership. A company without true leadership is rudderless. Four primary factors are vital to leadership results:

❖ *Releasing the Full Potential of All Employees.* The true measure of a successful leader is to release the creative genie of each employee currently trapped in a bottle of bureaucracy, and giving it full scope to flower and bloom for the betterment of the employee and of the company. This means breaking the chains of enervating Taylorism and instead encouraging, supporting, promoting, and celebrating the growth of each employee.

❖ *Transforming Managers Into Leaders.* Most employees estimate that the ratio of leaders to managers in companies is 1:99. The measure of a successful, results-oriented company is the extent to which this dismal ratio has been reversed to favor enlightened leaders over autocratic managers. This is, indeed, a relatively unexplored frontier for the vast majority of companies, and represents a major challenge in this twenty-first century.

❖ *Conveying Ethics, Trust, and Help: The Hallmarks of Leadership.* As stated in Chapter 7, the personal philosophies and values of true leadership are embedded in three key words—*ethics, trust,* and *help.* They are the hallmark of leadership. Without ethics, the company is not likely to survive in the long run. Without trust, employees will always feel alienation. Without active, concrete help, they will not rise to their maximum of their God-given potential. Leadership involves uncompromising integrity in all business dealings, even at the expense of the bottom line. It involves trust in people so that the latter can earn that trust. It involves helping people technically, managerially, administratively, or emotionally, as coach and guide, rather than as an overbearing boss.

❖ *Gain Sharing.* One of the distinctive differences between a truly successful company and a mediocre one is the differentials in gain sharing between senior managers and those employees who have earned it by their tangible contributions to profitability. The average company, if it considers gain sharing at all, grants its employees, who have made a direct contribution to the company's bottom line, a grudging 10 percent of base pay. By contrast, companies that have smiled all the way to the bank generously allow gain sharing to their deserving employees in the range of 25–80 percent of base pay. Their leaders recognize that if it is fair to give managers incentive pay, it is equally fair to give its line workers, who have earned it, a comparable incentive pay.

3. *Employees.* Behind customers and leadership, employees are next in the rank order of constituencies. How often have we heard the slogan: "Employees are our greatest asset."? Yet how often those words ring hollow and false in actual implementation. In this age, when the customer is elevated to prime importance, it is becoming axiomatic that you cannot have customer loyalty without a corresponding employee loyalty. Four primary factors are essential in generating employee loyalty:

❖ *"Create Joy in the Workplace"—Deming.* Most visitors to a company can sense the mood of the workers with a brief walkthrough. If employees have a hangdog, unsmiling expression, visitors know that the prevailing climate is one of boredom, even alienation. On the other hand, if they observe cheerful, enthusiastic faces, visitors can sense a happy, productive atmosphere. Deming has said that a company "must create joy in the workplace." (I use a rule-of-thumb to gauge a company's productivity and employee morale when I first visit its plant. I observe the first ten people I encounter. If they appear busy, happy, and "hustling," I gauge good productivity. If they are furtive, cowed, bored, or sullen, I sense a poor work climate and poor productivity. Progressive companies have productivity on the Bhote meter of over 70 percent.)

❖ *The Ten Stages of Empowerment.* Chapter 9 depicts the ten stages of empowerment of workers in a company—from stage one (passivity) to stage ten (industrial democracy). Most companies have advanced to stage four or five (problem solving) in the last fifteen years. But very few have conceded the reins of operational management to their employee teams. With the introduction of open book management, self-directed work teams, and minicompanies, an organization can make a leap of faith to the full potential of stages nine and ten.

❖ *Ratio of CEO Salaries to Line Worker Salaries.* No rational worker would ever begrudge the higher salaries of top managers who have the onerous responsibility of guiding companies in these turbulent times. But when the ratio of total CEO compensation to that of the lowly time worker far exceeds the Japanese figure of 10:1, or even the German figure of 30:1, to astronomical levels of 100:1 and 500:1 in the United States, the gap is the harbinger of a potential revolt. What is worse, those figures are in the public domain for the whole world to see. It contributes to worker dejection, frustration, and alienation. What loyalty can be squeezed out from these unfortunate workers? There is an even greater danger. Capitalism, for the last century, has been the engine of economic development. But we are beginning to sprout the seeds of its decline in the public mind. The outcry against corporate greed at the disastrous World Trade Organization conference in Seattle and at the World Economic Forum in Davos, Switzerland, cannot be airily dismissed. If free enterprise is to survive, this ratio must by design, be reduced, voluntarily by a company's board of directors; if not, by a disenchanted public outcry.

❖ *The Reduction of Organization Layers Between Top Management and the Worker.* One sure mark of a company's employee health is the number of organizational layers between the CEO and the lowly worker. For a large company (generally more than 3,000 employees), the typical number of management layers is fourteen to eighteen. This many a level leads to bureaucracy and micromanagement, with middle management acting mostly as carriers of information back and forth from top management to the line worker. Modern organizational development recommends no more than five layers, even for the largest of companies. When managers have to shepherd fifty to a hundred employees each instead of six to ten, they can no longer micromanage them. They must have faith in their employees, give them the freedom to attain corporate goals, and get out of their way. This transforms autocratic managers into coaches and teachers.

4. *Financials.* Most companies, financial analysts, and the stock market concentrate on a company's financials as the only tangible measure of its performance. Although financials are important, they are only lagging indicators, not leading ones. That is why they are listed last among the four primary categories in our group.

Three primary factors give a well-rounded perspective on financials:

❖ *Return on Investment (ROI).* The most common yardstick of financial performance is profit on sales. But since profit on sales is not tied to money invested, it is poor as compared to other parameters of financial performance, notably:

Return on investment (ROI):	Income after interest and taxes, divided by assets minus current liabilities
Return on assets (ROA):	Income divided by assets (i.e., inventory and receivables plus fixed assets)
Return on net assets (RONA):	Income before interest and taxes, divided by assets minus current liabilities
Return on equity (ROE):	Income after interest and taxes, divided by stockholders' equity

Of these, ROI and RONA are the most meaningful of yardsticks because of the multiplier effect of asset turns on profits, which is expressed as follows:

$$ROI = \text{Profit on Sales x Asset Turns}$$

Where:

$$\text{Asset Turns} = \text{Sales Assets (i.e., Inventory} + \text{Receivables} + \text{Fixed Assets)}$$

Even a mediocre profit on sales of one percent can be turned into an outstanding ROI of 20 percent with asset turns of twenty.

❖ *Market Position.* Several companies consider market share to be particularly attractive and strive mightily to increase it. However, as was pointed out in Chapter 10, market share is, at best, a cloudy crystal ball. It could give a company a false sense of well-being as its market share increases while customer defections may slowly be pumping air out of its balloon. By contrast, market position vis-à-vis other companies in the same business gives a clearer signal of performance. (General Electric's Jack Welch jettisons any of his businesses if it does not achieve the number-one or number-two position and maintain it in the marketplace.) A company that is fifth or sixth in market rank clearly has a tougher road to travel than its competitor whose market rank is one or two.

❖ *Value-Added Per Employee.* Frequently, companies use the metric of sales per employee per year as a measure of their financial/productivity performance. In the past two or three decades, that number has steadily climbed from below $50,000 to more than $800,000. However, this metric is only good to measure a company's longitudinal (i.e., year by year) progress against time. It cannot easily compare the productivity of different companies because their material dollar content, associated with outsourcing, varies widely from company to company and skews the numbers disproportionately. A truer productivity measure is value-added per employee—that is, sales dollars minus material dollars. Then comparisons across companies become meaningful. Their success factors. Later, in Table 17-4 we will see a summary of these primary results by the above four key characteristics.

Considerations in Choosing Secondary Parameters for Company Results

Secondary parameters are not as vital as the primary parameters for assessing a company's results, but they lend support to the primary factors and put results in a total perspective. They have been included as an extremely comprehensive guide to companies that may wish to consult them in assembling a results list of their own.

This secondary list of results is divided into two groups:

❖ By area (Table 17-5)—customers; leadership; organization; employees; tools; design; supply chain management; manufacturing; field; and services. These areas correspond to appropriate chapters in this book.

❖ By discipline (Table 17-6)—quality; cost/productivity; cycle time; innovation.

Guidelines

In choosing the specific milestones for the ratings of each parameter in the secondary list of results, the following guidelines have been chosen:

1. The ratings are based on personal experiences with more than 400 companies with whom I have consulted.
2. They are stretch and reach-out goals, especially for ratings of four and five (that is, for sigma levels five and six). There is no need for groveling in mediocrity.
3. The ratings apply more to larger companies with more than 10,000

employees. They can be modified and toned down for smaller companies.

4. Most parameters are measured in percentages per year or in ratios.

5. For some of the parameters, the evaluator (scorekeeper) is the customer. The usual scale is one to five, with one being the least effective and five the most effective.

6. Where the customer, as evaluator, can compare a parameter associated with a company's performance vis-à-vis its best competitor, a scale of minus five (very unfavorable) to plus five (very favorable) is used.

7. For other parameters, employees (at appropriate levels) are the evaluators. This is generally done in employee attitude surveys where training, in terms of background and meaningfulness and impact, is given by human resources departments prior to such surveys. Honesty, transparency, and freedom from spying or recriminations are emphasized at such training sessions.

8. For these secondary parameters, several characteristics have been combined to form a score based on multiplying the importance of each parameter and its rating by the evaluator. These are:

❖ *Leadership.* Table 17-1, section A (personal philosophies and values), section B (principles and corporate roles), and section C (corporate roles), as rated by employees.

❖ *Principles of Supply Chain Management.* Table 17-2 as rated by the customer-supplier council.

❖ *Principles of Next Operation as Customer (NOAC).* Table 17-3 as rated by the NOAC steering committee

9. Milestones in some sections are repeated in other sections for a better perspective. Sometimes the milestones so repeated are not exactly the same because of differences in the requirements of different departments.

In summary, there are the following secondary parameters from which a company can choose the most appropriate to fit its purpose and culture:

By Area	Number of Secondary Parameters
A. Customers	15
B. Leadership	14
C. Employees	14
D. Tools	28
E. Design	29

Table 17-1. Leadership rating: personal philosophies and values rating, leadership principles, and corporate roles.

Leadership	Importance 1–5 (I)	Rating = (R)					Score (S) = (I) × (R)
		1	2	3	4	5	
A. Personal Philosophies and Values* 1. Ethics and uncompromising integrity 2. Trust in employees 3. Active help in developing employee potential 4. Freedom to employees to pursue goals 5. Small set of superordinate values 6. Vision to explore uncharted paths 7. Inspiring employees 8. Relinquishing formal power 9. Warmth and genuine interest in people 10. Humility—by thought, word, deed 11. Superior listening skills 12. Coach, teacher (not boss)							
Total Score = $\dfrac{\text{Sum of (S)\%}}{\text{Sum of (I)} \times 5}$							
B. Leadership Principles* 1. Alignment of company as a good corporate citizen with the need for betterment of society 2. Maximizing stakeholder value 3. Focus on customer loyalty and retention 4. Maximizing customer loyalty by maximizing employee loyalty 5. Win-win partnerships with upstream and downstream suppliers 6. Goals—few in number, vital in importance							

Total Score $= \dfrac{\text{Sum of (S)\%}}{\text{Sum of (I)} \times 5}$								

C. Corporate roles
  1. Prioritizing employee initiative, creativity, and intrapreneurship
  2. Focus on corporate renewal
  3. Removing fear and bureaucracy
  4. Emphasis on training all employees
  5. Concentration on long-term profits
  6. Service to the community
  7. Converting social problems into business opportunities								

Total Score $= \dfrac{\text{Sum of (S)\%}}{\text{Sum of (I)} \times 5}$								

Note: The ratings and scores are determined by employee attitude surveys. Such surveys must be preceded by training sessions conducted by the human resources department on the full meaning and significance of each parameter.

By Area	Number of Secondary Parameters
F. Supply Chain Management	28
G. Manufacturing	24
H. Field	18
I. Services	13

By Discipline	
A. Quality	11
B. Cost/Productivity/Financials	14
C. Cycle Time	5
D. Innovation	12

SELF-ASSESSMENT/AUDIT OF PRIMARY RESULTS

Table 17-4 is a company's self-assessment/audit of its primary results. It has four key characteristics and fifteen success factors, each worth five points for a maximum score of seventy-five points.

In Table 17-4, the sigma levels associated with ratings of 1, 2, 3, 4, and 5 are 2, 3, 4, 5, and 6 sigma, respectively (similar to Table 5-3). Whenever ratings are nonquantitative, a subjective scale of 1–5 is used, with 1 being the least effective and 5 the most effective, as perceived by those surveyed.

(text continues on page 385)

Table 17-2. Principles governing supply chain partnerships: a rating.

Principles	Importance 1–5 (I)	Rating = (R)					Score (S) = (I) × (R)
		1	2	3	4	5	
1. Commitment to partnership by customer							
2. Ethics and uncompromising integrity							
3. Mutual trust, earned							
4. Agreements with a handshake (not hiding behind legal contracts)							
5. Active, concrete help on both sides							
6. Supplier cost (a ceiling, not a floor)							
7. Supplier profits (a floor, not a ceiling)							
8. "Open kimono" policy —sharing of strategies, costs, and technology							
9. Partnership—a long-term marriage							
10. Supplier—an extension of the company							
Total Score $= \dfrac{\text{Sum of (S)\%}}{\text{Sum of (I)} \times 5}$							

Note: The rating and scores are determined jointly by the customer-supplier council, which consists of members from the company, and from the partnership supplier members of the council.

Table 17-3. Next Operation as Customer (NOAC): a principle rating.

NOAC Principles	Importance 1–5 (I)	Rating = (R)					Score (S) = (I) × (R)
		1	2	3	4	5	
1. Internal customer as prince							
2. All work is a process, with the customer-supplier as key link							
3. Internal customer requirements, measurements, and feedback							
4. Consequences							
5. Continuous breakthrough improvements							
6. Employees as partners							
Total Score $= \dfrac{\text{Sum of (S)\%}}{\text{Sum of (I)} \times 5}$							

Note: The rating and scores are determined by the NOAC steering committee.

Table 17-4. Primary results: key characteristics and success factors (75 points).

Key Characteristic	Success Factors	Rating				
		1	2	3	4	5
17.1 Customer	1. Core customers retained per year: percentage (%) of total	?	60%	75%	90%	99%
	2. Value added to customers (as perceived by them)	1	2	3	4	5
	3. Customer base reduction: percentage (%) of total	0%	5%	10%	30%	60%
	4. Public perceptions of company as employer; on ethics, environment, social responsibility, etc.	1	2	3	4	5
17.2 Leadership	1. Percent (%) of full creativity of employees released	0%	5%	10%	25%	50%
	2. Percent (%) of managers transformed into leaders	0%	2%	5%	20%	50%
	3. Leadership: ethics, trust, help (as perceived by employees)	1	2	3	4	5
	4. Gain sharing: percent (%) of base pay for average worker	0	5%	10%	25%	>50%
17.3 Employees	1. Joy in the workplace (as perceived by employees)	1	2	3	4	5
	2. Stage of empowerment	1	5	7	8	9,10
	3. Ratio of CEO: lowest line worker total compensation (salary = bonuses, stock options)	500:1	200:1	100:1	75:1	50:1
	4. Number of layers between CEO and line worker (large companies)	>15	12	9	7	5
17.4 Financials	1. Return on investment (ROI)	<2%	5%	15%	30%	>50%
	2. Market position (vis-à-vis competitors)	>6th	5th	4th	3rd	1st, 2nd
	3. Value-added per employee/ Year: (dollars × 000)	30	70	150	300	>600

Table 17-5. Parameters for secondary results: by area.

Recommended Parameters	Rating				
	1	2	3	4	5
A. Customers					
A.1 Loyalty of core customers					
❖ Longevity: years	—	One Time	1	3	>5
❖ Repeat purchases: number of times	—	2	3	4	>5
❖ Referrals by customers: number	—	1	5	8	>15
❖ Company effectiveness index (Table 10-4): Score percentage (%)	—	40%	60%	80%	>95%
❖ Customer ideas for improvement: number	—	1	5	10	>20
❖ Frequency of senior management visits to customers: number per year	1	5	15	50	>100
❖ Amount of senior management time spent with customer: days per year	1	5	30	100	>200
A.2 Customer satisfaction					
❖ Differential rating vis-à-vis competition (−5 to +5)	−5	−1	+1	+3	+5
❖ Customer complaints/claims: percent (%) of units sold	20%	5%	1%	0.1%	0.01%
❖ Product recalls: number in 5 years	>5	3	2	1	0
❖ Lawsuits upheld against company: number in 5 years	>10	6	4	2	0
A.3 Stakeholder satisfaction					
❖ Stock appreciation per year: percentage (%)	0	5%	10%	20%	>50%
❖ Effectiveness of overall partnership with company	1	2	3	4	5
A.4 Internal customer					
❖ Internal supplier effectiveness index (Table 16-2)	—	40%	60%	80%	>95%
❖ Reduction in cycle time: percentage (%)	0%	25%	50%	80%	>95%
B. Leadership					
B.1 Personal philosophies and values: (Table 17-2 score)	—	25%	50%	75%	>90%
B.2 Leadership principles (Table 17-3 score)	—	25%	50%	75%	>90%
B.3 Corporate roles (Table 17-2 score)	—	25%	50%	75%	>90%
C. Organization					
C.1 Dismantling Taylorism: percent (%) dismantled	—	25%	50%	75%	>95%
C.2 Conversion of departments into cross-functional teams: percent (%) conversion	—	20%	40%	65%	>90%
C.3 Decentralization: percent (%) reduction in corporate staff	0%	20%	40%	60%	>80%

C.4 Employee hiring based on team player potential, customer sensitivity, entrepreneurship, growth potential: percent (%) departure from previous practices	0%	10%	20%	40%	>60%
C.5 Training: benefit-to-cost ratio	—	2:1	5:1	10:1	>30:1
C.6 Performance appraisal: percent (%) change from boss evaluation to internal customer evaluation	—	10%	25%	50%	>75%
C.7 Compensation: percent (%) change from pro forma minor merit increases to no merit increases; but generous incentives based on performance	—	5%	10%	25%	>50%
C.8 Promotion: percent (%) change from performance criteria to growth and leadership potential	—	5%	10%	25%	>50%
C.9 Safety/environment: number of accidents/injuries/EPA violations per year	>20	10	5	2	0
C.10 Time allocation: percent (%) of managers' time spent in meetings	>50%	30%	15%	10%	<5%
C.11 Percent (%) time spent on customer care (Table 6-2 score)	<5%	10%	25%	50%	>75%

D. Employees

D.1 Number of layoffs: percent (%) of employee population/year	>25%	10%	3%	1%	0%
D.2 Voluntary employee turnover: percent (%) of employee population/year	>30%	15%	7%	3%	<1%
D.3 Absenteeism: percent (%) of total work days/year	>10%	5%	3%	1%	<0.5%
D.4 Complaints/grievances: percent (%) of total number of employees/year	>20%	15%	7%	3%	<1%
D.5 Percent (%) yearly improvement in employee attitude survey scores	0%	3%	7%	15%	>30%
D.6 Employee readiness for empowerment (Table 9-3 score)	<9	9–18	19–27	28–30	>30
D.7 Percent of workers having near-full trust in company leadership	0%	5%	15%	40%	>80%
D.8 Number of jobs redesigned for vertical job enrichment: percent (%) of total jobs	0%	10%	20%	50%	>80%
D.9 Average number of hours/week of management by walking around (MBWA) for each senior manager	0	1	2	3	>4
D.10 Training: hours per employee per year	<5	10	20	50	>100
D.11 Training cost: percent (%) of payroll dollars	<0.5%	1%	3%	5%	>8%
D.12 Teaching employees financials: percent (%) of total number of employees	0%	5%	20%	50%	100%

(continues)

Table 17-5. (Continued).

Recommended Parameters	Rating				
	1	2	3	4	5
D.13 Self-directed work teams: percent (%) of total number of employees	0%	10%	20%	30%	>60%
D.14 Formation of number of minicompanies within company	0	2	5	10	>20
E. Tools					
E.1 Design of experiments (DOE)					
❖ Number of DOE projects completed in design as a percentage (%) of total projects	0%	1%	10%	40%	>80%
❖ Number of problems solved with DOE as a percentage (%) of total problems/year	0%	5%	15%	30%	>50%
❖ DOE: savings-to-cost ratio	0%	2:1	7:1	20:1	>100:1
❖ Number of line workers (direct labor) using DOE as a percentage (%) of total workforce	0%	5%	20%	40%	>75%
E.2 Multiple Environment Over Stress Testing (MEOST)					
❖ Reliability improvement over historic levels using MEOST: ratio	0	2:1	10:1	50:1	>100:1
❖ Warranty cost reduction through MEOST: percent (%)	0%	5%	30%	60%	>95%
E.3 Mass customization/quality function deployment					
❖ Number of options generated per product platform	0	5	25	100	>500
❖ Quality function deployment (QFD)					
❖❖ Cost savings compared to similar earlier products	0	1.1:1	1.3:1	1.7:1	>2:1
❖❖ Quality improvement over similar earlier products	0	1.3:1	1.5:1	1.8:1	>2:1
E.4 Total productive maintenance (TPM)					
❖ Overall equipment effectiveness (OEE)	—	<50%	675%	85%	>95%
E.5 Benchmarking					
❖ Number of benchmark projects/year	0	1	3	7	>10
❖ Benchmarking savings-to-cost ratio	0	1:1	3:1	8:1	>20:1
E.6 Poka-yoke					
❖ Number of poka-yoke projects completed/year	—	5	20	50	>100
❖ Quality improvement through poka-yoke: percentage (%)	—	10%	50%	100%	>300%
E.7 Next Operation as Customer (NOAC)/Business process reengineering (BPR)					

❖ Internal supplier satisfaction index (Table 16-2 score)	—	40%	60%	80%	>95%
❖ Cycle-time reduction: percentage (%)	0%	25%	50%	80%	>95%
❖ BPR cost reduction: percentage (%) of company sales	0%	3%	6%	12%	25%
E.8 Value engineering					
❖ Percent (%) value increase—as perceived by customer	0%	5%	20%	50%	>100%
❖ Savings by project: percentage (%)	0%	4%	7%	15%	>30%
❖ Value engineering savings, sharing: suppliers-to-company ratio	100:0	95:5	85:15	70:30	50:50
E.9 Supply chain management (SCM)					
❖ Overall material quality improvement: percentage (%)	0%	10%	50%	100%	>500%
❖ Overall cost reduction: percentage (%)	0%	5%	10%	15%	>25%
❖ Lead time (important parts): days	100	40	15	3	<1
❖ Supplier/distributor base reduction: percentage (%)	—	50%	65%	80%	>90%
E.10 Lean manufacturing/cycle time					
❖ Manufacturing cycle-time reduction: multiples of theoretical cycle time	100	50	20	4	1.5
❖ Business process cycle time: multiples of theoretical cycle time	100	40	15	3	1.5
❖ Design cycle time: multiples of theoretical cycle time	200	100	50	15	4
F. Design					
F.1 Organization effectiveness					
❖ Percent (%) of projects using concurrent engineering	—	40%	60%	80%	100%
F.2 Management guidelines effectiveness					
❖ Maximum percentage (%) of parts allowed in new design	100%	80%	675%	50%	25%
❖ Number of new products each year as a percentage (%) of total products	5%	10%	20%	35%	>50%
❖ Reverse engineering as a percentage (%) of total products	—	20%	50%	75%	100%
F.3 Design quality/reliability effectiveness					
❖ Defects at product launch: parts per million (ppm)	>100K	50K	10K	2K	<300
❖ First time overall yield at launch: percentage (%)	<40%	60%	80%	90%	>95%
❖ Months of engineering changes after job 1	>18	12	6	3	1
❖ Parts derating ratio	—	1.1:1	1.2:1	1.5:1	>2:1
❖ Reliability at product launch: percent (%) failure rate (i.e., infant mortality)	>30%	10%	2%	0.5%	0.05%
❖ Critical process parameters at launch: C_{PK}	<1.0	1.0	1.33	1.67	2.0

(continues)

Table 17-5. (Continued).

Recommended Parameters	Rating				
	1	2	3	4	5
❖ Ergonomics versus competition: customer perception (−5 to +5)	−5	−1	+1	+3	+5
❖ "Wow" features versus competition: customer perception (−5 to +5)	−5	−0	+1	+3	+5
❖ Product liability costs: percent (%) of sales	>500%	60%	20%	2%	0%
❖ Number of design flaws prevented per product through MEOST	—	0	2	4	>6
❖ Number of important parts with realistic specifications/tolerances: percentage (%) of total	—	5%	10%	50%	>80%
❖ Number of parts with classification of characteristics: percentage (%) of total	—	10%	50%	80%	100%
F.4 Design cost-effectiveness					
❖ Product cost versus competition: customer perception (−5 to +5)	−5	0	+1	+3	+5
❖ Design cost: actual versus target: percent (%) overrun/underrun	+50%	+10%	0	−10%	−25%
❖ Design for manufacturability (DFM) score (Boothroyd-Dewhurst scale: 1–100)	—	20	50	80	>90
❖ Early supplier involvement (ESI) savings: percent (%) of product costs	—	3%	6%	12%	>20%
❖ Value engineering savings: percent (%) of product cost	—	0	5%	10%	>20%
❖ Strategic business unit (SBU) reduction: percent (%) of total SBUs	—	1%	5%	10%	>30%
❖ Part number reduction: percent (%) of total parts	—	20%	40%	60%	>80%
❖ Number of parts cost targeted: percent (%) of total parts	—	5%	10%	25%	>50%
F.5 Design cycle-time effectiveness					
❖ Time to market: from concept to launch (months)	>25	18	12	9	<6
❖ Frequency of product introductions/year	<1	1	3	5	>10
❖ Time to recover (cost/time profile) costs of human inventory (months)	>15	12	10	8	6
❖ Design cycle time: actual versus target: percent (%)	+100%	+20%	0%	−20%	−50%
❖ Tooling time: percent (%) of total design time	80%	50%	30%	20%	10%
G. Supply chain management (SCM) G.1 Management effectiveness					
❖ Principles of partnership (Table 17-4)	—	25%	50%	75%	>90%

❖ Differential between price (market) erosion and supplier cost decreases	—	5%	0	+5%	>+10%
❖ Supplier/distributor evaluation of company (Table 13-6)	1	2	3	4	5
❖ Outsourcing: percent (%) of product costs	—	10%	50%	70%	90%
❖ Levels of subsuppliers influenced/ helped	0	1	2	3	4
❖ Supplier base: percent (%) reduction from historic levels	0%	25%	50%	75%	>90%
G.2 Active/concrete help to suppliers: effectiveness					
❖ Number of commodity teams	—	2	4	8	>12
❖ Man months of help per commodity team/year	—	4	10	25	>50
G.3 Quality/reliability effectiveness					
❖ Supplier defect levels: parts per million (ppm)	>10K	5K	1K	100	10
❖ Supplier reliability levels (failure rate/ in ppm/year)	>100K	30K	5K	500	<100
❖ Certified parts: percent (%) of total	—	30%	60%	80%	>95%
❖ Parts with c_{PK} 2.0: percent (%) of important parts	—	5%	30%	75%	>95%
❖ Number of partnership suppliers extending active help to their sub- and sub-subsuppliers: percent (%) of total supplier partnership base	—	<5%	10%	25%	>50%
G.4 Cost effectiveness					
❖ Material cost decreases/year: percentage (%)	0	5%	10%	15%	>25%
❖ Total ownership costs as a percentage (%) of purchase costs	>150%	140%	120%	110%	102%
❖ Material inventory turns/year	<3	10	40	80	>150
❖ Procurement administration costs as a percentage (%) of material costs	>20%	15%	10%	5%	2%
❖ Commodity team savings: percent (%) of material costs	—	5%	10%	30%	>50%
❖ Profit % increases for partnership suppliers	—	2%	5%	10%	15%
❖ Volume increases for partnership suppliers: percentage (%)	—	10%	75%	200%	>500%
❖ Financial incentives/penalties: percent (%) of total number of key parts	—	<1%	5%	20%	>50%
❖ Savings, sharing on value engineering, early supplier involvement ideas: company-to-supplier ratio	—	100:0	90:10	70:30	50:50
G.5 Cycle-time effectiveness					
❖ Lead time on important parts: days	100	40	15	3	<0.5
❖ Lead time from subsuppliers to partnership suppliers: days	100	40	15	3	<0.5

(continues)

Table 17-5. (Continued).

Recommended Parameters	Rating				
	1	2	3	4	5
❖ Cycle time from customer order to supplier order: days	>50	20	10	5	<1
❖ Number of partnership suppliers with lean manufacturing systems (Kanban)	—	10%	50%	75%	>95%
❖ Partial authorizations for partnership supplier raw materials: percent (%) of total partnership suppliers	—	5%	25%	50%	75%
H. Manufacturing					
H.1 General					
❖ Percent of manufacture in the country/region where the company's products are sold	—	5%	50%	75%	>90%
❖ Ratio of mass customization: mass production	1:20	1:10	1:1	10:1	20:1
❖ Ratio of Kanban to MRPII	0:100	1:20	1:1	20:1	100:0
H.2 Quality effectiveness					
❖ Cost of poor quality: percent (%) of sales dollars	>25%	15%	7%	3%	<1%
❖ Cost of poor quality per employee per day: dollars	>200	100	50	25	<10
❖ Total defects per unit (TDPU)	>8	1.5	0.5	0.1	<0.1
❖ Outgoing quality: parts per million (ppm)	>10%	3K	500	100	<10
❖ First-time overall yields: percentage (%)	<40%	70%	90%	95%	>99%
H.3 Cost effectiveness					
❖ TPM: overall equipment effectiveness (OEE)	—	<50%	65%	85%	>95%
❖ Savings/year of team projects: percent (%) of sales dollars	—	2%	5%	10%	15%
❖ Inventory turns/year	3	10	25	60	>100
❖ Number of suggestions/employee/year	0.1	0.5	5	50	>100
❖ Number of suggestions implemented/employee/year: percent (%) of total	—	10%	35%	60%	>80%
❖ Ratio of preventive maintenance to breakdown maintenance	0.100	20:80	50:50	80:20	95:5
H.4 Cycle-time effectiveness					
❖ Work-in-process (WIP) cycle time: days	>30	10	3	1	0.1
❖ Ratio of product flow stations: process flow stations	0:100	1:10	1:1	10:1	100:0
❖ Average setup times: minutes	>180	60	15	4	<2
❖ Average lot sizes: percent (%) of historic lot sizes	100%	80%	20%	5%	1%

❖ Reductions in master schedules: percent (%) of historic number	100%	80%	30%	10%	0%
❖ Necessity for forecasts	100%	80%	50%	20%	0%
❖ Number of multiple skills/employee	1	3	6	9	>12
❖ Customer lot size reductions: percent (%) of historic	100%	80%	60%	40%	20%
❖ Increases in frequency of customer orders: percent (%) of historic	0%	100%	200%	400%	800%
❖ Space reduction through cycle-time reduction: percentage (%)	0%	20%	40%	60%	80%

I. Field

I.1 Reliability effectiveness

❖ Number of critical parts derated: percent (%) of total critical parts	—	5%	40%	80%	100%
❖ Product safety versus competition: customer perception (-5 to $+5$)	-5	-1	$+1$	$+3$	$+5$
❖ Product liability prevention versus competition: customer perception (-5 to $+5$)	-5	-1	$+1$	$+3$	$+5$
❖ Failure analysis accuracy: percentage (%)	—	10%	50%	80%	100%
❖ Failure analysis speed: days	>30	20	14	7	3
❖ Reliability/year versus target: percentage (%)	10%	50%	100%	200%	1,000%
❖ Mean time to diagnose (MTTD) versus target: percentage (%)	>300%	150%	100%	75%	50%
❖ Mean time to repair (MTTR) versus target: percentage (%)	>400%	200%	100%	75%	50%

I.2 Predelivery services effectiveness

❖ Quality audit score: (packaging, storage, transport, etc.)	—	20%	40%	60%	>90%
❖ Installation/assembly effectiveness (dealer perception)	1	2	3	4	5

I.3 Services to downstream supply chain

❖ Reduction of distributor/dealer base: percent (%) of total	0%	5%	20%	40%	80%
❖ Priority time with distributors/dealers (distributor/dealer perspective)	1	2	3	4	5
❖ Effectiveness of joint councils with company (distributor perspective)	1	2	3	4	5
❖ Effectiveness of overall partnership with the company (Table 13-6)	1	2	3	4	5

I.4 Services to users

❖ Top management recognition of frontline troops (perceptions of customer contact employees)	1	2	3	4	5
❖ Ratio of company-owned service stations to independent service stations	0:100	2:95	10:90	30:70	50:50
❖ Ratio of preventive maintenance in the field to breakdown maintenance	0:100	20:80	50:50	80:20	95:5

(continues)

Table 17-5. (Continued).

Recommended Parameters	Rating				
	1	2	3	4	5
❖ Effectiveness of field repair—accuracy and speed (customer perception)	1	2	3	4	5
J. Services					
J.1 Next Operation as Customer (NOAC) principles (Table 17-5 score)	—	25%	50%	75%	>90%
J.2 Structure					
❖ Effectiveness of steering committee (perceived by NOAC teams)	1	2	3	4	5
❖ Effectiveness of process owner and teams (perceived by the steering committee)	1	2	3	4	5
❖ Customer-supplier relationship charts in effect: percent (%) of total teams	—	5%	20%	50%	>95%
❖ Processes redesigned: percent (%) of total processes	—	5%	20%	50%	>90%
J.3 Implementation/progress)					
a) Quality					
❖ Internal supplier effectiveness index (Table 16-2)	—	25%	50%	75%	>90%
❖ Image survey (Table 16-3)	1	2	3	4	5
❖ Major customer feedback	—	1	1.5	2.5	3
❖ Industry reports (e.g., J. D. Powers, Consumers Union, etc.) versus competition (-5 to $+5$)	-5	-1	$+1$	$+3$	$+5$
b) Cost					
❖ Service team generated sales increases: percent (%) of total sales dollars	—	2%	5%	8%	>15%
❖ Savings generated by service teams: percent (%) of total sales dollars	—	3%	7%	10%	>20%
c) Cycle time					
❖ Number of process steps reduced: percent (%) of total steps	—	5%	25%	50%	>80%
❖ Cycle time reduced: percent (%) of total cycle time	—	20%	50%	75%	>90%

Table 17-6. Parameters for secondary results: by discipline.

Recommended Parameters	Rating				
	1	*2*	*3*	*4*	*5*
A. Quality (company as a whole)					
❖ Cost of poor quality: percent (%) of sales dollars	>25%	15%	7%	3%	<1%
❖ Cost of poor quality (COPQ) per employee per day: dollars	>200	100	50	25	<10
❖ Field reliability: parts per million (ppm)/year	>15%	5K	1K	100	<10
❖ Outgoing quality: parts per million (ppm)	>10K	3K	500	100	<10
❖ Total defects per unit (TDPU)	>8	1.5	0.5	0.1	<0.1
❖ First-time overall yields: percentage (%)	<40%	70%	90%	95%	>99%
❖ c_{PK}s on critical/important parameters	—	<1.0	1.33	2.0	>3.0
❖ Product liability costs: percentage (%) of sales	>500%	60%	20%	2%	0%
❖ Ergonomics versus competition: customer perceptions (−5 to +5)	−5	−1	+1	+3	+5
❖ "Wow" features versus competition: customer perceptions (−5 to +5)	−5	−1	+1	+3	+5
❖ Safety features versus competition: customer perceptions (−5 to +5)	−5	−1	+1	+3	+5
B. Cost/productivity/financials					
❖ Profit: percentage (%) of sales after tax	—	2%	6%	12%	>20%
❖ Profit growth/year: percentage (%)	—	1%	5%	10%	20%
❖ Asset turns	<1	2	4	8	>15
❖ People turnover/year: percentage (%)	>30%	15%	5%	2%	<0.5%
❖ Sales/employee/year: (dollars × 000)	50	100	300	700	>1000
❖ Satisfaction of all stakeholders (1 to 5)	1	2	3	4	5
❖ Product cost: percentage (%) of sales	>80%	75%	65%	55%	50%
❖ TPM: overall equipment effectiveness (OEE)	—	<50%	75%	85%	>95%
❖ Cross-functional team savings: percent (%) of sales	—	2%	5%	10%	15%
❖ Inventory turns per year	3	10	25	60	>100
❖ Supplier cost reductions/year: percentage (%)	0	5%	10%	15%	>25%
❖ Value engineering savings: percentage (%) of product costs	—	1%	5%	10%	25%
❖ Early supplier involvement (ESI): savings of product cost	—	2%	5%	10%	>20%
❖ Sales wins-to-: sales losses ratio	50:50	60:40	70:30	80:20	>90:10
C. Cycle time					
❖ Manufacturing cycle time (raw materials, WIP, finished goods): days	>60	20	6	3	<1

(continues)

Table 17-6. (Continued).

	Rating				
Recommended Parameters	*1*	*2*	*3*	*4*	*5*
❖ Business process cycle time reduced: percent (%) of historic	—	20%	50%	75%	>90%
❖ Design cycle-time reductions: percent (%) of historic	—	20%	50%	75%	90%
❖ Lead time from partnership suppliers (A items): days	100	40	15	3	<0.5
❖ Lead times from subsuppliers to partnership suppliers (A items): days	100	40	15	3	<0.5
D. Innovation					
❖ Sales/year of new products: percentage (%) of total sales	0%	5%	10%	25%	50%
❖ Frequency of stream of new products (with less than 25 percent change from previous models) per year	—	1	2	3	4
❖ Number of "skunk works" projects launched per year	—	1	4	8	>15
❖ Number of ideas lifted from competition through reverse engineering per year	—	5	10	20	>50
❖ Number of patents generated per year: percent (%) of technical population	—	1%	3%	5%	>10%
❖ Number of patents commercialized per year: percent (%) of technical population	—	—	1%	2%	>3%
❖ Number of benchmark projects/year	—	2	5	10	>20
❖ Number of suggestions/employee/ year	—	0.1	1	15	>50
❖ Number of suggestions/employee/ year: implemented	—	—	0.3	10	>40
❖ Number of customer "wow" features added per major product	—	—	1	2	>3
❖ Number of customized features (differentiators) added to a standard product	—	2	4	8	>20
❖ Number of incentive payments to partnership suppliers/year for ideas accepted.	—	2	10	20	>50

CASE STUDY

General Electric—A Benchmark Company in the Area of Results

General Electric, as a whole, and Jack Welch, its retiring chairman in particular, have been held up as model phenomena of the corporate world. Every management textbook pays due homage to their achievements. It is true that GE's success has sometimes been on the backs of people who've been used and discarded. (In January 2001 GE announced the layoff of 75,000 employees—one of the largest layoffs in corporate history.) Jack Welch has been derisively labeled "Neutron Jack" for saving buildings while destroying people. It is also true that many of GE's businesses were jettisoned for not climbing to number one or number two in market position.

Nevertheless, GE's results have been spectacular. The prestigious magazine *The Economist* calls "General Electric more country than company and its boss, Jack Welch, a legend. . . . If there is a corporate equivalent to the American presidency it is the post of chairman and chief executive of General Electric."[1]

GE's results are nothing short of the unbelievable. According to the company's 2000 annual report:[2]

❖ Revenues rose to more than $129 billion, up 16 percent.

❖ Earnings increased to $12.7 billion, up 13 percent.

❖ Net earnings per share grew to $1.27, up 19 percent.

❖ Dividends per share were 0.57, up 17 percent, and its stock split 3 for 1.

❖ Operating margin rose to over 18.9 percent, up two full points.

❖ Fifteen of GE's top businesses posted double-digit earnings (the company had struggled for eleven years to reach 10 percent).

❖ GE shareowners who held stock for five, ten, and twenty years were rewarded with an average annual ROI of 34 percent, 29 percent, and 23 percent respectively.

❖ The returns on a share of GE stock in three previous years were 51 percent, 40 percent, and 45 percent, respectively.

❖ GE has averaged a 24 percent per year total return to shareowners for the last eighteen years.

❖ The generation of $15.4 billion in free cash flow was up by $3.6 billion from 1999—a 31 percent increase.

❖ GE grew fourfold in Europe from a small presence in 1992 to $24 billion in 2000.

❖ Earnings in Japan were more than $300 million, with a projection of double that figure within three years.

❖ Twenty-six percent of GE's top executives are women and minorities.

❖ The center of gravity of GE has shifted from a provider of products to a company that is overwhelmingly one of services. Two-thirds of its revenues come from financial, information, and product services. GE is using its expertise and resources in making its customer assets more productive; and by so doing, reducing their capital outlays.

❖ GE was *Fortune*'s "Most Admired Company in America" for the fourth straight year. It was also *The Financial Times*' "World's Most Respected Company" for the third time. What a stellar record!

Conclusion

These seventeen chapters have been a long, dynamic, and exciting journey toward the Ultimate Six Sigma—toward a company's "true north." On that journey, it leaves behind the early and inconsequential milestones of ISO 9000, the Malcolm Baldrige National Quality Award, and the hyped Six Sigma of Upstart Consulting Companies.

It goes much, much further. In its pursuit of total business excellence, it offers a sturdy lifeline to the many companies thrashing to be rescued from the rough seas of shrinking profits, shrinking customer pools, and shrinking investor pools. Downsizing is dumb. Across-the-board cost reductions are dumb. My Ultimate Six Sigma shows how to resuscitate a company with concentration on: (1) customer retention, (2) drastic reduction in the cost of poor quality, (3) total productive maintenance, (4) the gold mine of supply chain management, (5) and a stream on new products, with small changes, but fast to leave competition in the dust. Collectively, these techniques can double, triple, and quadruple the anemic profits of most companies.

The Ultimate Six Sigma also targets a cultural revolution in our often stated, but little utilized asset: our people. There is a perennial question put to our management teams: "How much more can be squeezed out of the lemon?" The answer is that there is unlimited juice in this "lemon" called "our employees," and none of it has anything to do with "squeezing" at all. Enlightened leadership must firmly believe that there is an ocean of creativity, passion, and energy in our people, and that that ocean has no bottom. Effective leadership, to use another metaphor, frees the "creative genie" of all employees, imprisoned today by bureaucracy and neglect, with no asymptotic barrier to their growth potential.

The goal of the Ultimate Six Sigma can be realized when "TGIF" (Thank God It's Friday) can be transformed in peoples' spirits to "TGIM" (Thank God It's Monday); when employees are so enthused that they cannot wait for the weekend to be over so they can get back to the joy and productive excitement of the workplace.

In the final analysis, the Ultimate Six Sigma offers a way in which the grinding scourges of poverty, disease, illiteracy, and pollution affecting almost three-quarters of mankind can be solved—not by sterile governmental actions or by pious U.N. resolutions, but by applying the magic of free enterprise and business skills to transform these pressing social challenges into profitable businesses for the betterment of society.

REFERENCE NOTES

❖ ❖ ❖

Chapter 2

1. Schaffer and Thomson, "Successful Change Programs Begin With Results," *Harvard Business Review* (May/June 1992).

2. Jay Mathews and Peter Katel, "The Cost of Quality," *Newsweek* (September 7, 1992).

3. Gilbert Fuchsberg, "Quality Programs Show Shoddy Results," *The Wall Street Journal* (May 14,1992).

4. A. Blanton Godfrey, "Strategic Quality Management, Part 1," *Quality* (March 1990).

5. Ibid.

Chapter 5

1. Keki R. Bhote, "Plan for Maximum Profit: The 12 Critical Success Factors That Guarantee Increased Profits From Total Quality," in Volume 5, *The Total Quality Portfolio* (Zurich, Switzerland: Strategic Direction Publishers, 1996).

2. Keki R. Bhote, "The Quality Project Alert: The Early Warning Signals for Any Quality Initiative in Danger of Costing More Than It Earns," in Volume. 6, *The Total Quality Portfolio* (Zurich, Switzerland: Strategic Direction Publishers, 1996).

Chapter 6

1. Keki R. Bhote, *Beyond Customer Satisfaction to Customer Loyalty—The Key to Greater Profitability* (New York: AMACOM, 1996), p. 31.

2. Ibid, p. 32.

3. Frederick F. Reicheld and W. Earl Sasser, "Zero Defects: Quality Comes to Services," *Harvard Business Review* (September–October 1990).

4. A. T. Kearney, *The Customer Satisfaction Audit* (Zurich, Switzerland: Strategic Direction Publishers, 1994).

5. Ward Hanson, "All Yours," *The Economist* (April 1, 2000).

6. Ibid.

7. REL Consultancy Group, "Study on Customer Retentions," *Quality Digest* (June 1995).

8. H. James Harrington, "Seeking Out Errors," *Quality Digest* (May 1999).

Chapter 7

1. Bennet Davis, *TWA Ambassador* (December 1995).

2. Robert W. Galvin, *Idea of Ideas* (Schaumburg, Ill.: Motorola University Press, 1991).

3. Harry Mark Petrakis, *The Founder's Touch: The Life of Paul Galvin of Motorola* (New York: McGraw Hill, 1965).

4. Keki R. Bhote, *Strategic Supply Chain Management—A Gold Mine for Profitability* (New York: AMACOM, forthcoming).

5. Christopher A. Bartlett and Sumantra Goshal, "Changing the Role of Top Management Beyond Systems to People," *Harvard Business Review* (May–June 1995).

6. Peter F. Drucker, "Change Leaders," *Inc.* (June 1999).

7. Mary Walton, *The Deming Management Method* (New York: Putnam Publishing, 1986).

8. James O'Toole: *Leading Change* (Boston: Harvard Business School Press, 1996).

9. Bill Ginnodo, "Leading Change: An Interview With Motorola's Bob Galvin," *Commitment Plus,* Pride Publications (May-June 1996).

Chapter 8

1. *The Washington Monthly* (June 1986).

2. Michael Hammer and James Champy, *Reengineering the Corporation* (New York: Harper Business, 1993).

3. Tom Peters, *Thriving on Chaos* (New York: Alfred A. Knopf, Inc., 1987).

4. Hammer and Champy, *Reengineering the Corporation.*

5. A. T. Kearney, *Seeking and Destroying the Wealth Dissipators* (1985).

6. Gary D. Zeune, *Outside the Box—How to Beat Your Competitor's Brains Out* (1996).

7. Ibid.

8. Jan Carlzon, *Moments of Truth* (Cambridge, Mass.: Ballinger Publishing Co., 1987).

9. Masaki Imai, *Kaizen* (New York: Random House, 1986).

10. Motorola Benchmarking Team internal report.

11. Peters, *Thriving on Chaos*.

12. Mary Walton, *The Deming Management Method* (New York: Putnam Publishing, 1986).

13. Yankelovich Poll: Christopher.

14. Peters, *Thriving on Chaos*.

Chapter 9

1. Scott Myers, *Every Employee a Manager* (New York: McGraw Hill, 1970).

2. Roy Walters, Internal publication (Roy Walters and Associates).

3. James F. Lincoln, *Incentive Management* (Cleveland: Lincoln Electric Co., 1951).

4. Jack Stack, *The Great Game of Business* (New York: Currency Doubleday, 1992).

5. John Case, *Open Book Management: The Coming Business Revolution* (New York: Harper Business, 1995).

6. Chris Lee, "Open Book Management," *Training Magazine* (March 1995).

7. Stratford P. Sherman, "The Mind of Jack Welch," *Fortune* (March 27, 1989).

8. Jack Osborne, et al, *Self-Directed Work Teams* (Homewood, Ill.: Business One Irwin, 1990).

9. Kiyoshi Suzaki, *The New Shop Floor Management* (New York: Free Press, 1993).

10. Ricardo Semler, *Maverick: The Success Story Behind the World's Most Unusual Workplace* (New York: Warner Books, 1993).

11. Ibid.

12. Jack Stack, *The Great Game of Business* (New York: Currency Doubleday, 1992).

Chapter 10

1. Scott Myers, *Every Employee a Manager* (New York: McGraw Hill, 1970).

2. Special Committee on Financial Reporting, *Improving Business Reporting—A Customer Focus, Meeting the Information Needs of Users* (AICPA, 1994).

3. Nancy L. Hyer and Urban Wemmerlov, "Group Technology and Productivity," *Harvard Business Review* (July-August 1984).

4. *Blueprints for Service Quality*, American Management Association briefing, 1991.

Chapter 11

1. Keki R. Bhote, *World-Class Quality—Using Design of Experiments to Make It Happen* (New York: AMACOM, 1991).

2. Keki R. Bhote, *World-Class Quality—Using Design of Experiments to Make it Happen*, second ed. (New York: AMACOM, 2000).

3. Ibid, Chapter 22.

4. Ibid., Chapter 10.

5. Ibid., Chapter 11.

6. Ibid., Chapter 12.

7. Ibid., Chapter 13.

8. Ibid., Chapter 14.

9. Ibid., Chapter 15.

10. Ibid., Chapter 16.

11. Ibid., Chapter 17.

12. Ibid., Chapter 18.

13. Ibid., Chapter 19.

14. Ibid., Chapter 21.

15. Ibid., Chapter 22.

16. James H. Gilmore and Joseph B. Pine II, "The Four Faces of Mass Customization," *Harvard Business Review* (January-February 1997); and Edward Feitsinger and Hau L. Lee, "Mass Customization at Hewlett-Packard: The Power of Postponement," *Harvard Business Review* (January/February 1997).

17. Yoji Akao, *Quality Function Deployment* (Cambridge, Mass.: Productivity Press, 1990); and Bob King, *Better Designs in Half the Time* (Methuen, Mass.: Goal/QPC, 1987).

18. Seichi Nakajima, *TPM Development Program* (Cambridge, Mass.: Productivity Press, 1989); and Nachi Fuji Koshi, *Training for TPM* (Cambridge, Mass.: Productivity Press, 1990).

19. Bhote, *World-Class Quality*, second ed.

20. Ibid.

21. Ibid.

22. Shigeo Shingo, *A Revolution in Manufacturing: The SMED System* (Cambridge, Mass.: Productivity Press, 1985).

23. Bhote, *World-Class Quality*, second ed.

24. Michael J. Spendolini, *The Benchmarking Book* (New York: AMACOM, 1992).

25. Keki R. Bhote, "The Critical Success Factors in Benchmarking," in Vol. 4, *The Benchmarking Portfolio* (Zurich, Switzerland: Strategic Direction Publishers, 1995).

26. Shigeo Shingo, *Zero Quality Control, Source Inspection and Poka-Yoke* (Cambridge, Mass.: Productivity Press, 1986); and Nikkan Kogyo Shimbun, *Poka-Yoke* (Cambridge, Mass.: Productivity Press, 1988).

27. Lawrence D. Miles, *Techniques of Value Analysis and Engineering,* second ed. (New York: McGraw Hill, 1972)

28. Keki R. Bhote, *Total Value Engineering* (Singapore: Singapore National Productivity Board Conference, September 1990).

Chapter 12

1. Keki R. Bhote, *World-Class Quality—Using Design of Experiments to Make It Happen,* second ed. (New York: AMACOM, 2000)

2. Ibid, Chapter 15.

3. Ibid, Chapters 16 and 17.

4. Ibid, Chapter 18.

5. Ibid, Chapter 19.

6. Jeffrey Boothroyd and Peter Dewhurst, *Product Design for Manufacturing and Assembly* (Boothroyd and Dewhurst, 1987).

7. For a more detailed treatment on the methodology of total value engineering, please see Keiki Bhote, *Strategic Supply Chain Management—A Gold Mine for Profitability* (New York: AMACOM, forthcoming), Chapter 19.

8. Nancy L. Hyer and Urban Wemmerlov, "Group Technology and Productivity," *Harvard Business Review* (July-August 1984).

9. Raymond J. Levulis, *Group Technology* (K.W. Tunnell Consulting, Co.).

10. Hyer and Wemmerlov, "Group Technology and Productivity."

Chapter 13

1. Peter Hines, "Toyota Supplier System in Japan and the U.K.," *Lean Enterprise Research Centre Research Paper* (U.K.: Cardiff Academic Press, 1994).

2. James P. Womack and David T. Jones, *Lean Thinking* (New York: Simon and Schuster, 1996).

Chapter 14

1. Richard J. Schonberger, *World-Class Manufacturing Case Book* (New York: Free Press, 1987).

2. Tom Peters, *Thriving on Chaos* (New York: Alfred A. Knopf, Inc., 1987).

3. James P. Womack and David T. Jones, *Lean Thinking* (New York: Simon and Schuster, 1996).

4. For a detailed treatment of control charts and their limitations, please see Chapter 20 in Keki R. Bhote, *World-Class Quality—Using Design of Experiments to Make It Happen,* second ed. (New York: AMACOM, 2000).

5. For a detailed treatment of precontrol, please see Chapter 21 in Bhote, *World-Class Quality,* second ed.

6. For a detailed treatment of clue generation tools with case studies and workshop examples, please see the chapters devoted to it in Bhote, *World-Class Quality,* second ed.

7. Shigeo Shingo, *A Revolution in Manufacturing: The SMED System* (Cambridge, Mass.: Productivity Press, 1985).

Chapter 15

1. For a detailed treatment of "B" vs. "C," see Keki R. Bhote, *World-Class Quality—Using Design of Experiments to Make It Happen,* second ed. (New York: AMACOM, 2000).

Chapter 16

1. Richard Quinn and Christopher Gagnon, "Will Services Follow Manufacturing Into Decline?" *Harvard Business Review* (November-December 1986).

2. For a comprehensive treatment of organizations as systems, with flowcharting/mapping as an important technique, an excellent reference is Geary Rummler and Alan Brache, *Improving Performance—How to Manage the White Space on the Organization Chart* (San Francisco: Jossey-Bass, 1990).

3. Keki R. Bhote, *World-Class Quality—Using Design of Experiments to Make It Happen,* second ed. (New York: AMACOM, 2000), Chapter 3.

4. Ibid, Chapter 20.

5. Michael Hammer and James Champy, *Reengineering the Corporation* (New York: Harper Collins, 1993); and Michael Hammer, *Beyond Reengineering* (New York: Harper Collins, 1996).

Chapter 17

1. "The Man Who Would Be Jack," *The Economist* (December 2, 2000).

2. General Electric Annual Report 2000.

Index